Encyclopedia of Label Technology

Other Labels & Labeling books:

ENCYCLOPEDIA OF LABEL TECHNOLOGY
Michael Fairley

THE HISTORY OF LABELS
Michael Fairley and Tony White

DIGITAL LABEL AND PACKAGE PRINTING
Michael Fairley

ENVIRONMENTAL PERFORMANCE AND SUSTAINABLE LABELING
Michael Fairley and Danielle Jerschefske

CONVENTIONAL LABEL PRINTING PROCESSES
John Morton and Robert Shimmin

LABEL DESIGN AND ORIGINATION
John Morton and Robert Shimmin

LABEL DISPENSING AND APPLICATION TECHNOLOGY
Michael Fairley

CODES AND CODING TECHNOLOGY
Michael Fairley

LABEL EMBELLISHMENTS AND SPECIAL APPLICATIONS
John Morton and Robert Shimmin

BRAND PROTECTION, SECURITY LABELING AND PACKAGING
Jeremy Plimmer

DIE-CUTTING AND TOOLING
Michael Fairley

MANAGEMENT INFORMATION SYSTEMS AND WORKFLOW AUTOMATION
Michael Fairley

For the latest list please visit: **www.labelsandlabeling.com**

Encyclopedia of Label Technology

Michael Fairley LCG, FIP3 and FIOM3

Technical editing and additional contributions by
Andy Thomas and **Barry Hunt**

Encyclopedia of Label Technology

Second edition published 2014 by:
Tarsus Exhibitions & Publishing Ltd

Previously published as Encyclopedia of Labels and Label Technology

Printed by CreateSpace, an Amazon.com company.

ISBN 978-1-910507-00-1

Preface

It is now some ten years since the first ever Encylcopedia of Labels and Label Technology was published, bringing label terminology from around the world together to produce one standard international reference source. Such has been the book's success that it has been regularly used by thousands of label converting companies, industry supplies and label users over the intervening years.

However, much has changed in the world of labels since the book was first published: digital printing has grown from a niche technology to become an everyday commercial process; conventional analogue printing has also undergone considerable change, with servo drive revolutionizing press set-up and running; pre-press and origination is now significantly different; laser die-cutting is beginning to become widely used; and a whole new world of interactive and smart-phone enabled labels has emerged.

With so many developments and changes it was inevitable that a thoroughly updated and revised edition of the Encyclopedia would become necessary. Over one hundred new entries and references have now been added, many of the original entries updated, with a number of superseded terms removed altogether. Additional diagrams and illustrations have also been added where necessary to further aid knowledge and understanding.

Recognizing that many label converters are also now moving beyond labels into other areas of narrow-to-mid-web converting it was also decided to include references to sachets, pouches, flexible packaging, small folding cartons and lidding.

It is therefore hoped that this latest edition of the Encyclopedia will continue to play a significant role in helping the label producer and user industries to be better educated in understanding the terminology and technology of the fast-changing world of labels.

Michael Fairley
Director, Labels & Labelling Consultancy
Director, Strategic Development, Labelexpo Global Series

About the Author

Michael Fairley has been writing and speaking about label and packaging materials, technology and applications since the 1970s, both as the founder of Labels & Labeling and other print industry magazine titles and as an international consultant writing label industry market and technology research reports for the likes of Frost & Sullivan, Economist Intelligence Unit, Pira, InfoTrends and Labels & Labelling Consultancy.

He now works as a consultant to Tarsus Exhibitions & Publishing - which organizes the Labelexpo shows, Label Summits and publishes Labels & Labeling magazine - as well as regularly speaking at industry conferences and seminars.

He is a Fellow of the Institute of Package / Packaging Society, Fellow of IP3, a Freeman of the Worshipful Company of Stationers, an Honorary Life Member of FINAT and a Licentiate of the City & Guilds of London Institute. He was awarded the R. Stanton Avery Lifetime Achievement Award in 2009.

Acknowledgements

This Encyclopedia owes much of its existence to all those companies and individuals that have worked with both Labels & Labeling and Packprint World, providing information for feature and technical articles as well as educational and reference material. Their collective information and expertise has been distilled into this major work of reference.

Additionally, acknowledgement is made to the many speakers that have contributed papers to international label conferences and Label Summits run by Labelexpo. All have been studied as part of the research for this publication, as well as the content of many varied market and technical research reports.

Apart from the research and writing of the initial Encyclopedia publication in 2004, the latest edition of the Encyclopedia has been substantially updated and re-written, with the drafts being subjected to further technical editing by Andy Thomas, the Group Managing Editor of Labels & Labeling, Barry Hunt, Technical Editor of Labels & Labeling and, for the first edition, John Davies, formerly Business Development Manager at Samuel Jones.

Other companies and individuals have assisted with information and/or diagrams that have been incorporated or adapted to help illustrate the Encyclopedia – including Avery Dennison, HP Indigo, Gallus, Mark Andy, BST Pro Mark, InfoTrends, Phazer Technologies, Converting Technology International, Product & Image Security Foundation, UPM Raflatac, Vandagraf International, and Ron Spring - while Greg Smye-Rumsby of Concept Design has comprehensively drawn multiple diagrams from often rather vague or scribbled notes and sketches.

Extensive thanks are due to all those companies and individuals that have helped to get the Encyclopedia of Label Technology to its latest finished format.

A

A4 – One of the ISO standard international paper sizes, A4 (210 mm x 297 mm or 8.25 x 11.75 inches) is the commonest size used in cut sheet office and stationery printing and for sheeted pressure-sensitive paper (sometimes film) containing blank die-cut labels for processing through copier, laser and ink jet printing systems. The U.S. equivalent standard size of 8.5 x 11 inches (letter) is slightly wider than A4, but shorter in depth.

ABIEA – The Brazilian Pressure Sensitive Label Converters Association. Founded in 1986, the association represents its members' interests with government, unions and other parties.

ABLATION LAYER – A special thin film layer, called an infrared-ray ablation layer, on which images can be formed directly from computer when irradiated by infrared laser. Used in computer-to-plate technology. The ablation layer is provided on top of the conventional photosensitive resin layer. Only the laser irradiated layer is ablated (combustion, sublimation and decomposition, melting and evaporation), so that this layer plays the role of the conventional negative film.

ABRASION – The process of scuffing, scratching, wearing down, marring, or rubbing away

ABRASION RESISTANCE – The ability of a printed ink or coating to inhibit deterioration of the surface by friction or rubbing. May also be called 'rub resistance' or 'scuff resistance'.

Special laboratory or quality control equipment has been developed to test for abrasion resistance of inks and coatings. The term may also refer to the mechanical resistance of the surface of a label, the dependent variables being the surface structure of the label, the nature and pressure of the abrasive medium and the duration of the abrasive action.

The principle of abrasion or rub resistance testers is that a sample of the print or coating is placed face down onto to an unprinted sample of the same substrate and the two rubbed together for an agreed number of cycles. The degree to which a transfer of color to the unprinted sample takes place is inversely related to the rub resistance. This method of testing is more relevant for paper substrates rather than flexible films.

ABRASIVENESS – The tendency of some papers, surface coatings, treatments or inks to abrade or wear away die edges, slitter wheels, printing surfaces, etc, by friction.

ABRASIVE WEAR – Wear that occurs when a hard rough surface slides across a softer surface.

ABSORBENCY/ABSORPTION – The property of a porous material, such as most papers, which causes it to take up liquids or vapors in contact with it. This property enables the vehicle or liquid portion of, say, letterpress inks to be absorbed into the paper, leaving the pigment (or coloring matter) on the surface, together with sufficient vehicle to key the pigment to the paper surface.

In optics, absorption is the partial suppression of light through a transparent or translucent material.

ACCELERATED AGING – The subjecting of a label face material, label stock, ink or adhesive to far more severe conditions than are likely to be found normally, in order to predict long-term ageing behavior under normal conditions.

ACCELERATED DRYING – The use of electromagnetic radiation to speed up the drying

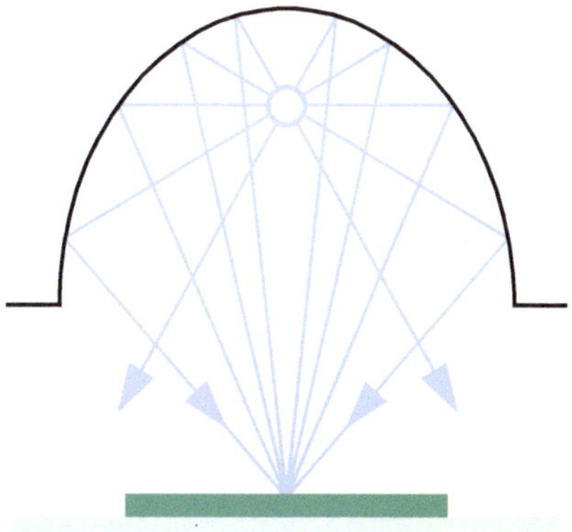

Accelerated ink drying using UV radiation.

(curing) of an ink, varnish, coating or adhesive. The different forms of electromagnetic radiation are UV, EB, IR and RF energy, which all fall outside the normal visible light spectrum.

UV (ultra violet) curing of inks is extensively used in the label industry. The inks contain special formulations that remain in their liquid form almost indefinitely until they are exposed to UV radiation of the correct wavelength. Then a very rapid chain reaction causes them to cure (dry) almost instantly.

Electron beam (EB) curing is similar to that of UV, except that a photoinitiator is not required in the ink, which means the inks are less expensive. However, the EB curing units have a very high capital cost, are much more bulky and operate more effectively in web-fed applications. They are used in some adhesive coating applications.

Infra red (IR) drying speeds up the setting and drying of inks through the use of heat, rather than through a curing reaction. It is not an instantaneous drying system, but simply introduces heat to speed conventional drying.

Radio frequency (RF) wavelengths generate heat within an ink or coating that contains molecules of water or alcohol. The RF energy output causes the molecules to vibrate and create internal heat through internal friction.

The use of electromagnetic radiation for the accelerated curing of inks, varnishes, coatings and adhesives is also referred to as radiation curing or 'Radcure'. *See also UV curing.*

ACCELERATOR/ACCELERANT – A material (catalyst) added to a liquid composition to change the whole mass into a solid *(see crosslinking)* or to speed up its cure rate. Used in pressure-sensitive adhesives and label inks.

ACETATE FILM – A transparent, gloss or matt label face material used in applications where clarity or fragile characteristics may be required in a pressure-sensitive label. The high transparency and purity of cellulose acetate films makes them a low-cost alternative for 'no-label' look applications. Acetate face materials do not suit harsh environments because of their sensitivity to solvents and poor long-term UV stability. Used with a high-strength solvent acrylic adhesive, acetate doubles as an over-laminate, which is virtually impossible to remove without damage. Good tear resistance and brittleness are positive characteristics in the growing market for clear or opaque tamper-evidence seals. Labels converted with die-cut security slits using a 50 micron (2.00 mil) clear acetate film effectively destruct when tampered with.

Acetate is printable by most conventional printing processes without corona pre-treatment and takes hot foiling with good results. The material also accepts variable-data imprinting, ballpen or pencil writing. Its high rigidity means it is readily applied by automatic machinery, but the film is not conformable and is unsuitable for squeezable containers.

Advantages and benefits of acetate films

Printing and application performance
- ■ Die-cutting
- ■ Resistance to stretch
- ■ Dispensability
- ■ Resistence to heat
- ■ Dimensional stability
- ■ Tear resistance
- ■ Solvent resistance (poor or caution)

End-usage criteria
- ■ 'No-label' look
- ■ Moderate cost
- ■ Tear resistance
- ■ Solvent resistence (poor or caution)
- ■ Conformability (poor or caution)
- ■ UV stable (poor or caution)

Key: ■ Good ■ Poor or caution

ACID-FREE PAPER – A paper of high purity, free from acid producing chemicals and which is used in applications where acid may damage materials with which the paper or label is in contact.

ACOUSTIC MAGNETIC TAG – A hybrid electromagnetic electronic article surveillance (EAS)

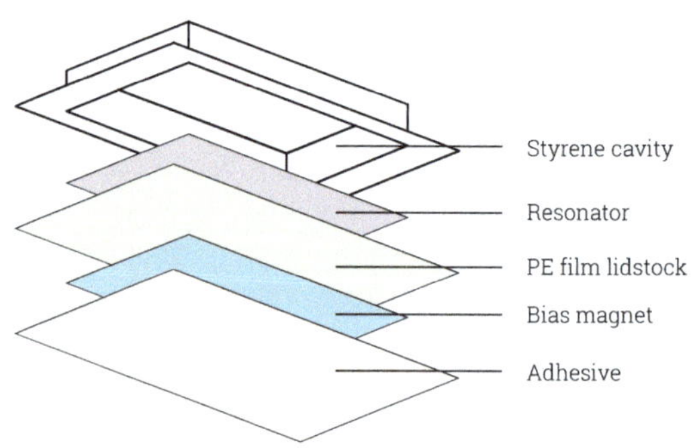

Acoustic magnetic tag construction

system in which metal tags can be magnetised and will vibrate precisely in response to a radio signal. May also be called acoustic harmonic or acousto-magnetic (AM).

Central to AM technology is a low power, narrow-band pulsed emission of electromagnetic energy. Between each energy pulse, a narrow-band receiving antenna 'listens' for the unique tag resonance, or 'signature' produced when an AM tag enters the detection field. The system analyses this signature to verify the presence of a tag and trigger an alarm to alert security staff of a possible theft. *See also EAS systems.*

ACRYLATES – Acrylate copolymers are esters of acrylic acid characterized by their high clarity and resistance to heat, light, ultra violet and oxidation. A very wide range of acrylic monomers and co-monomers enables formulators to provide pressure-sensitive adhesives for different applications. Acrylate adhesives are available for solution, emulsion and hot-melt systems.

ACRYLIC ADHESIVE – A transparent adhesive, which is adequately resistant to UV radiation and to relatively high temperatures. Whilst acrylic adhesives have a relatively low initial tack compared with rubber-based adhesives, their final adhesion is generally much stronger. Acrylic adhesives may be either dispersion-based or solvent-based. *See also Adhesives, pressure sensitive.*

ACRYLICS – Polymers based on acrylic monomers which are used in the label and packaging industries as a base for lithographic paste inks, solvent and water-based flexographic printing inks, overprint varnishes, pressure-sensitive tape and label adhesives, hot-melt adhesives and as a processing aid in the manufacture of clear rigid PVC films and bottles.

ACTIVATION – The application of heat to a heatseal coating to make it fluid and tacky.

ACTIVATION/DE-ACTIVATION – The process of changing the state of an electronic article surveillance (EAS) tag from inactive/active to active/inactive.

Active labels – Labels with their own power source able to indicate and communicate changes in the internal or external environment of a package.

ACTIVE TAG – An active tag (or live tag) is a radio frequency electronic article surveillance (EAS) tag used in retail stores that is presently active or in a state whereby it can be sensed by detection equipment. The tag returns a signal of its own and, unless removed or de-activated when a product is legitimately sold to a customer, can provide evidence of attempted store theft when the still active tag enters a field produced by an antenna system located at the store exit or checkout, thereby triggering an alarm. EAS tags may also be found in acoustic magnetic or radio frequency types.

The term active tag is also used to describe an RFID tag that comes with a battery that is used to power microchip circuitry and transmit a signal to a reading device, which may be up to 25 (82 ft), 30 (98 ft) or more meters away. Such tags are expensive and are therefore mainly used for tracking expensive items over a wide range.

ADDITIONAL CONTENT LABELS – Smart phones equipped with QR code or NFC readers can access online additional content from a label. *See also QR codes, Mobile action codes.*

Additional content is now provided on labels through the use of smartphones *Source: DigiMarc*

ADDITIVE COLORS – In color reproduction the light colors which, when combined, make up white light. The additive primary colors are red, blue and green (RBG). When lights of these colors are added together, they produce the sensation of white light to the viewer.

Computer monitors use additive color, while the printing process uses subtractive color. This causes inconsistency between what a designer sees on his monitor and what comes off the printing press.

ADHERE – To stick or hold fast. Occurs when two surfaces are stuck together by adhesion or an adhesive.

ADHEREND – The surface or substrate to which an adhesive is applied; the surfaces which are bonded together.

ADHESION – Relates to the bond that holds two surfaces together. In label terms it is a measure of the strength in which a label sticks to another surface or material; the interaction between adhesive and substrate.

ADHESION, PEEL – A measure of the force required to remove an adhesive coated material from another surface (usually a standard test plate) under specified angle and speed conditions, such as FINAT Test Method FTM 1.

ADHESION, SHEAR – A measure of the time which is needed to slide a specific area of a pressure-sensitive label from a standard flat surface in a direction parallel to that surface. Weight and heat are sometimes used to speed up the test.

ADHESIVE – A substance capable of holding or bonding materials together by surface attachment or adhesion. The principal uses of adhesives include the labeling of bottles, jars, drums, cases, shrink-wrapped packages, pallet and box ends, components or products, and in all kinds of manufacturing assemblies and products, as well as the forming and sealing of corrugated cases and folding cartons, the lamination of paper to paperboard and foil, and the lamination of plastic films for flexible packaging.

The vast majority of labels require an adhesive to bond the label to the container, or to some other surface. Being between the label material and the container or product surface, the adhesive must be compatible with both. If all the variations of label substrate (paper, paper-backed foils, metallized materials, plastics, synthetic papers, fabrics, metal foils, etc) and container/surface type (metal cans, glass bottles, plastic bottles, wrapping films, corrugated cases, etc) are then incorporated into the adhesive requirements, the challenges for adhesive manufacturers are quite formidable.

To meet the demands of different label types, different surfaces to be bonded, and different user performance characteristics, such as surface or environment conditions, the label industry uses four main types of adhesive: hot-melt, water-based, solvent-based and curable adhesive systems. These are explained below.

Hot-melt adhesives are thermoplastic materials with 100% solids that are heated to temperatures above their melting point and applied to substrates in the molten state. Unlike water-based or solvent-based adhesives, hot-melt adhesives do not require drying; they set as quickly as they can cool down to their solidification point, which makes them ideal for picking up labels in high-speed labeling lines. They are popular because of their high setting speed and because they form a virtually invisible line on glass, metal, PET and other plastic containers. In the self-adhesive sector, hot-melt adhesives can have very high tack, which makes them ideal for some very difficult applications.

Key criteria in the use of hot-melt adhesives are temperature – which controls viscosity – and adhesive film thickness, which affects speed of setting, tack and open time. The thicker the adhesive is applied, the longer it will take to set.

Water-based adhesives have had a dominant place in the label industry for many years and are used in wet-glue, self-adhesive and gummed paper formulations.

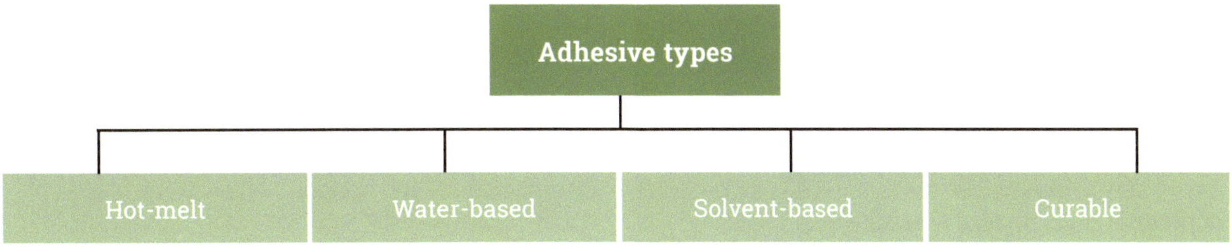

Four main types of adhesives used by the label industry

They are made up of materials or compounds that can be dissolved or dispersed in water to become tacky, then form a bond and dry by losing water through evaporation or by penetration into the label substrate. At least one surface must be absorbent or porous to form a strong bond. Water-based adhesives are available in a variety of chemistries and compositions and are categorized as either natural or synthetic polymers, as follows:

WATER BASED ADHESIVES

Synthetic polymers	Natural polymers
Polyvinyl acetate (PVA)	Casein
Acrylics	Dextrine/Starch
Polychloroprene	Natural Rubber Latex
Polyurethane dispersions	

Water-based adhesives may be either synthetic or natural polymers

Solvent-based adhesives are noted for their fast bond strength, good heat resistance, and versatile adhesion to a wide range of substrates and tolerance to a wide variety of production conditions – including low temperature or high humidity. However, solvent-based pressure-sensitive formulations have been widely displaced by water-based and hot-melt systems for economic as well as ecological reasons. Solvent rubber self-adhesives do not have good heat resistance, which limits their shelf life. With solvent-based adhesives solvent recovery and/or incineration is now essential to meet clean air legislation requirements. This is expensive and can only be justified for large output operations.

Curable adhesives that use UV or electron-beam systems to 'cure' or set the adhesive are a relatively new development and are used in some specialty tape applications and have also found some (limited) application in the self-adhesive label sector. Curable adhesives are cross-linked during the curing process. *See also Adhesive, pressure sensitive and Adhesive performance.*

ADHESIVE BLEED – The exudation of an adhesive from the edge of label stock or labels resulting from adhesive flow, which may be aggravated by high temperature or pressure and can cause converting problems due to adhesive pick-up on machine rollers. This is often seen as a discoloration of paper face stock (also called adhesive migration or penetration) and can cause converting problems. *See also Adhesive strike through.*

ADHESIVE, CLEAR – An adhesive, which forms a clear, colorless and transparent coating. An adhesive used for a 'no-label' look when used with a clear plastic face material.

ADHESIVE, COLD TEMPERATURE – An adhesive that will enable a pressure-sensitive label to adhere or stick well to a cold surface in cold or 'chill' temperature conditions, i.e. between + 5°C to – 5°C (41°F to 23°F) where there is often condensation.

ADHESIVE, DELAYED ACTION HEATSEAL – A heatseal adhesive, which remains tacky for a period of time after removal of the heat source. Traditionally used for labels that were applied to products which themselves may be heat sensitive, i.e. the end seal of a pleated packet of biscuits.

ADHESIVE FILM – A plastic film, which adheres to absolutely smooth surfaces (e.g. glass and painted metal) without the use of an adhesive. May also be referred to a 'self cling' film. Used for applications such as window stickers where no adhesive residue is required on the glass after the sticker is removed.

ADHESIVE-FREE AREA/ADHESIVE-FREE STRIPES – Pressure-sensitive labels which are provided with partially adhesive-free areas for ease of application and removal. Areas are strip coated

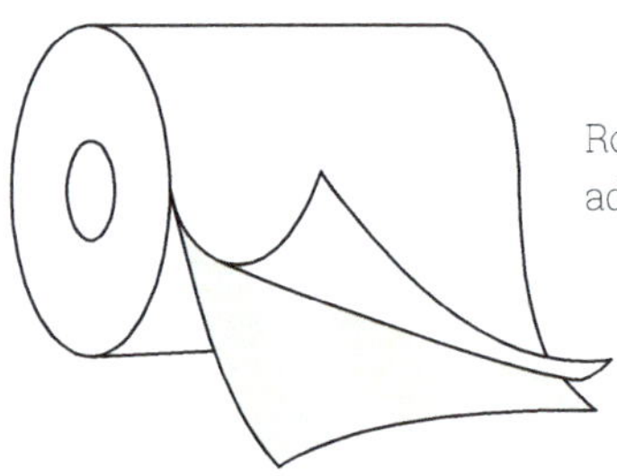

Roll labelstock containing adhesive-free stripes

alternately with and without adhesive. Such constructions provide non-adhesive areas on the label, typically a free edge, to allow for easier lifting of the label from the backing. Examples include usage in a very cold environment where the user wears gloves when applying the labels, and for some specialist price labels and receipts, etc. *See also Pattern coated.*

ADHESIVE, HEATSEAL TYPES – Adhesives that dry after coating onto label materials and which are rendered tacky or 'active' by the application of heat, i.e. delayed action heatseal. Prior to the development of self-adhesives, heatseal adhesives were widely used for manual and semi-automatic labeling, but are now almost extinct. Price-weigh labeling was the principal application. Other uses included the end seals on pleated packs and for labels around rolls of photographic films or skeins of wool.

ADHESIVE, INSTANTANEOUS HEATSEAL – Both aqueous and hot-melt types were developed for automatic heatseal labeling machines, where a heated platen, label and labeling surface can all be pressed together. The adhesive bonds as soon as the heat is removed.

These labels were used for biscuit end seals, toilet roll end seals and filmic pleat wrapping of cakes. Several variations of adhesive recipe were developed for different film surfaces, including a pure gum arabic adhesive that was both heatsealable and remoistenable for rolls of photographic film.

Instantaneous heatseal labels have been largely superseded by self-adhesive labels but are still used while heatseal labeling equipment survives in pharmaceutical and specialist applications, such as coffee pack seals.

ADHESIVE, HIGH TEMPERATURE – An adhesive designed to enable a pressure-sensitive label to retain its properties at high temperatures up to 232°C (450°F).

ADHESIVE, PEELABLE – A pressure-sensitive adhesive with a low ultimate adhesion to a wide range of surfaces. It can be peeled-off or removed from most surfaces easily and should preferably not leave a residue. *See also Adhesive, removable.*

ADHESIVE PERFORMANCE – Label adhesives are a key element in the label application process and have to perform as required on the labelling line and throughout all handling, shipping and end-use stages. Important features of a label adhesive are that:

- It must 'wet' the label surface, i.e. flow out onto the label substrate to which it is applied. It must not reticulate, but should give an even coating without faults.
- It must have a good 'initial tack' to enable the adhesive to grip the container or product surface during automatic application.
- The initial bond must be maintained, as applied, until the adhesive has time to bed down. There must be no 'edge lifting' or blistering.
- It must meet any required end-usage performance criteria – food contact, chemical or product resistance, water resistance, heat or cold storage, etc.

With the wide range and variety of end-use and performance requirements of label adhesives there are of necessity also a wide range of special adhesive performance needs. Pressure-sensitive adhesives for example, may be categorized by several criteria, initially according to their permanence or removability. The choice of adhesive will then be dictated by the nature of the substrate and any influences to which the label is exposed. Some examples of pressure-sensitive adhesives include:

Permanent adhesives usually stick quickly to the surface that they are being applied to and, after the normal bedding down period, their performance increases to form a fiber-tearing bond. They are designed to adhere at ambient temperatures. Permanent adhesives are also used with fragile substrates to provide tamper-evident label properties.

Removable adhesives have low tack and peel values to avoid damage to labels or goods when the labels are removed. The removal may be after many months and should be accomplished without leaving adhesive residue, e.g. retail gift labeling. Adhesives may be natural rubbers or acrylic polymers.

Filmic adhesive grades can be permanent, semi-permanent or removable and are used specifically with synthetic film labelstocks.

Deep freeze adhesives are designed to stick onto frozen meat packs with a labeling temperature down to -30ºC (-22ºF) and ultimate adhesion tolerances down to -50ºC (-58ºF).

Repulpable adhesives are specifically designed so as not to interfere with the paper repulping process used during the recycling of paper materials. A major requirement of such adhesives is that they must not produce 'stickies' in the special pulp filtration process designed to tolerate them. This is not the majority of papermaking equipment.

Semi-permanent adhesives are adhesives that remain removable and re-positionable for some time after application before becoming fully permanent. Often used where the pack to be labeled is not a constant shape and an operator needs to re-position labels that have been applied off-centre, e.g. cheese packs.

UV-curable adhesives have been developed to stand up to the toughest challenges and provide outstanding heat, plasticizer and chemical resistance. Originally offering lower cost but now more expensive, they do offer energy savings and improved coating efficiencies and can be used where hot air drying systems would damage the substrate, as with synthetic backing liners. Their use with hot-melt type adhesives improves the higher temperature performance of some label grades and also gives good peel and shear properties.

The range of standard pressure-sensitive labeling adhesives is wider than that of any other label technology, enabling the pressure-sensitive sector to offer more labeling solutions, with a wider range of face materials, and application onto a more diverse range of products, packs and containers. This can be seen in the diagram.

As far as wet-glue label adhesives are concerned the three main adhesive performance groups are:

Glass container label adhesives. A full range of high performance water-based casein and hot-melt adhesives are used for the glass bottle and jar labeling industry, as well as synthetic or resin-based adhesives, between them offering properties that include high performance, superior water resistance, moisture and ice resistance throughout the label application, bottle line handling and conveying, palletizing, shipping and end-use stages. If labels need to be removed from returnable bottles after use then an all purpose casein adhesive can be used.

Plastic container label adhesives. Resin-based adhesives for labeling plastic containers such as PET, treated polyolefins, etc, are designed to provide good wet tack and adhesion, excellent machinability and cleanability when labelling at high speeds on rotary machines. Hot-melt and water-based adhesives are used for labeling difficult to adhere plastic bottles with paper or film labels where they may be wide variations in temperature, as well as stresses on the label itself.

Metal container label adhesives. A wide variety of hot-melt and water-based labeling adhesives are available to the can market, including hot-applied pick-ups for affixing the leading edge of the label to the can on roll-through labeling equipment, as well as hot-melt and water-based adhesives for the label overlap seam bond. Overlap adhesives are formulated to give excellent adhesion.

ADHESIVE, PERMANENT – An adhesive which has a high ultimate adhesion which usually cannot be removed intact or only with enough force to overcome the adhesive bond – in which case the label may be damaged.

ADHESIVE, PRESSURE-SENSITIVE – An adhesive which in its dry state at room temperature is always tacky and will adhere or stick to a wide range of diverse surfaces by simple contact under light pressure. Used for pressure-sensitive labels, tapes, graphic arts, automotive, healthcare and other applications where an instant bonding to a variety of dissimilar surfaces upon contact is required.

The simplest method of categorizing pressure-

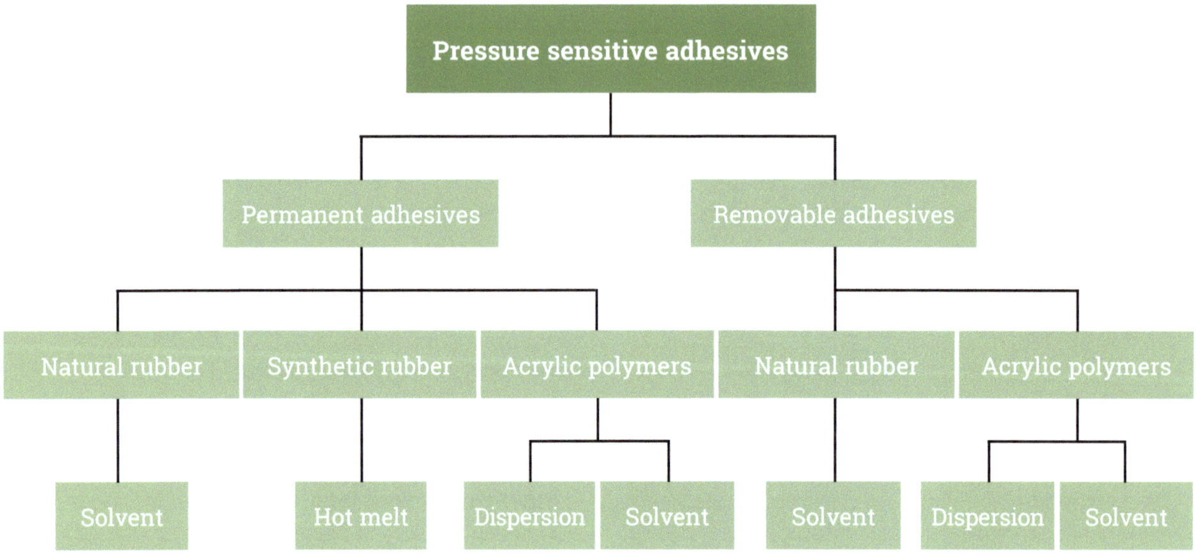

Pressure-sensitive adhesives can be classified according to their permanence or removability, with individual adhesive types for specific applications.

sensitive adhesives is to class them according to their permanence or removability. Within these two categories there are a range of different adhesive types – natural rubber, synthetic rubber and acrylic polymers – which in turn may be solvent-based adhesives, hot-melts or dispersion (emulsion) adhesives.

The choice of pressure-sensitive adhesive will be dictated, for example, by the nature of the substrate and by the influences to which the label is exposed. There are also regulations and standards that must be observed. Factors to be considered include;

- The required adhesion (permanent, removable, semi-permanent)
- Whether there is heat exposure
- The nature of the surface to be labeled and any possible pre-treatment of the surface to be labeled prior to application (paper, plastic, glass, metal)
- The conditions under which the surface is to be labeled (wet, dry, hot, cold)
- The prevailing conditions during the service life of the labeled product
- The method of labeling or dispensing used.

Many other factors may need to be considered when choosing an adhesive type. All available information regarding usage and application requirements should be made available to the laminate supplier when ordering. *See also Adhesives and Pressure-sensitive adhesives*

ADHESIVE, REMOVABLE – An adhesive with a low ultimate adhesion to a range of surfaces which can generally be removed or peeled-off intact.

ADHESIVE, REPOSITIONABLE – An adhesive which enables a pressure-sensitive label to be removed and repositioned shortly after application, prior to the development of the final or ultimate adhesion bond. Wrinkled or crookedly-applied labels coated with a repositionable adhesive for example, can be removed from the labeled surface for a short period of time following application and repositioned. Stronger label materials, being more tear-resistant, aid removability, as can temperature.

Some acrylic deep freeze adhesives are repositionable at ambient or cool temperatures and most general purpose permanent adhesives can be repositioned in cooler environments. The period during which removal of the label is possible depends on the type of print carrier (face material) used.

ADHESIVE RESIDUE – The pressure-sensitive adhesive remaining behind on a surface due to cohesive or priming failure when a pressure-sensitive label is removed from that surface. May also refer to unwanted adhesive residue on the backing due to a poor release coated liner or backing.

ADHESIVE SEPARATION – Partial retention of the adhesive by the print carrier (face material) and the substrate when the label is removed or placed under load.

ADHESIVE SPLITTING – A failure within the adhesive mass when labels are under stress or removed. If splitting occurs, part of the adhesive will remain on the labelled surface and part on the face material. The appearance is like stringing of adhesive or like stringing of chewing gum.

ADHESIVE SQUEEZE – Is seen when surplus adhesive spreads into the gap created by the cutter action. Can also be a problem with a tightly wound reel of laminate when the adhesive collects on the edges of the reel.

ADHESIVE STRIKE THROUGH – Term used when an adhesive penetrates through the face material of a pressure-sensitive lamination.

ADJUSTABLE ANVIL ROLLER – A special anvil roller that enables operators to make adjustments to the depth of a cut without changing the anvil roller and/or cutting tool. The diameter of the supporting ring can be radial expansion or contraction. These systems change the effective undercut in order to compensate for variation in thickness/calliper of the liner and of course the wear of the cutting lines.

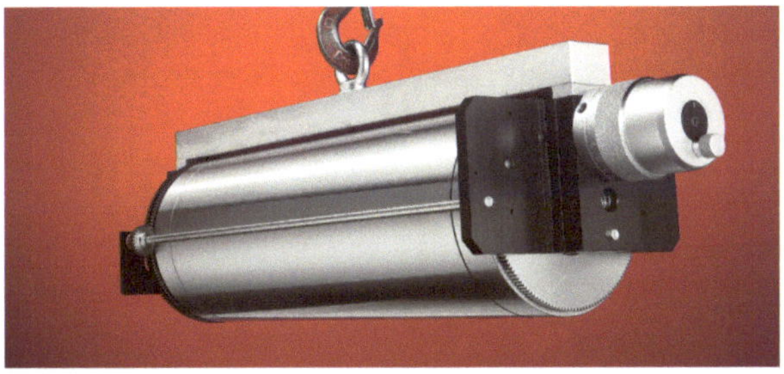

Illustration shows an adjustable anvil roller.
Source: Kocher & Beck

AEROSOL LABELING – Regulations covering the composition, labeling, marketing, classification and description of aerosol dispensers are set out by national governments or other organizations, such as the EEC. The labeling requirements of the various regulations may include information that aerosol dispensers be clearly and indelibly marked with their content by weight and volume, with warnings against misuse.

Additional information to be shown includes a marketer's identification, batch coding, safety phrases and a net contents declaration. If aerosol containers are used for chemical products reference should also be made to the appropriate chemical warning symbols. *See also Chemical warning labels.*

Protect from sunlight and do not expose to temperatures exceeding 50°C. Do not pierce or burn after use

Example of labeling information to be placed on aerosol containers

Examples of safety information that may be required on a typical aerosol dispenser label depending on the content or application are:

- Safety information
- Caution (Statutory and other recommended wording)
- Use only as directed information
- Pressurized container
- Do not spray onto food preparation surfaces
- Protect from sunlight and temperatures above 50°C. Do not pierce or burn
- Do not spray onto a naked flame, or any incandescent material
- Do not use on floors, inside baths or shower trays
- Flammable – together with the flame symbol
- Keep out of the reach of children

Exact requirements for labeling aerosols should be checked against specific country or EEC regulations, e.g. The Aerosol Dispensers (EEC Requirements) (Amendment) Regulations 1996.

AFTA – Australian Flexographic Technical Association. *See also FTA.*

AGING – Properties of materials can change with the passage of time due to their own inherent instability or to external effects such as light, UV radiation, temperature, humidity or oxygen. The progressive change in materials' properties as time elapses (either naturally or accelerated by changing normal conditions) is called aging.

AGING, ACCELERATED – The process of simulating natural aging through the use of exaggerated ambient conditions. Keeping samples in an elevated temperature, suited to their adhesive base, for set time periods can simulate longevity in normal conditions. Other methods include pressure tests and UV exposure.

AGE TIGHTENING – Also known as Age Up, this is the phenomenon in which release values of a pressure-sensitive adhesive increase with time. It can result from increasing association of the adhesive with the release composition due to adhesive flow and/or chemical reaction between the adhesive and excess crosslinker, and/or silicone post-cure around the adhesive.

AIAG – Automotive Industry Action Group. The Group was founded in 1982 by a group of visionary mangers from Daimler Chrysler, Ford Motor Company and General Motors to provide a forum where members co-operate in developing and promoting solutions that enhance the prosperity of the automotive industry. Its focus is to continuously improve business processes and practices involving trading partners in the supply chain.

AIAG members play a key role in the development of new technologies, as well as the standards, guidelines and processes that govern their usage. Members work collaboratively to develop solutions to new challenges. And to develop the insights, training, conferences and tools needed to operate responsibly – and profitably.

A full list of relevant automotive/label industry publications can be found on the AIAG web site.

AIDC – Automatic Identification and Data Capture (AIDC). An industry term used to describe the identification and/or direct collection of data into devices such as computer systems or programmable logic controllers, without using a keyboard.

Key goals of AIDC are to eliminate errors associated with identification and/or data collection; to accelerate the through-put process.

Technologies included within the AIDC family are card technologies, data communications technologies, barcode technologies, RFID technologies, Smart Networks, Cloud computing, Internet of Things, RTLS, emerging technologies and the support and supplies needed to support the industry. *See also AIM.*

AIM – Association for Automatic Identification and Data Capture Technologies. An international trade association made up of the leading providers of Automatic Identification and Mobility hardware, software and solutions – including resellers, system integrators, solutions providers, distributors, manufacturers, ORMs and consultants. It was established in 1972.

AIM is the leading international authority on:

- Barcodes
- RFID
- Mobile computing

AIM members are also leading solutions providers of voice systems, magnetic striping, mobile computing, OCR and contact memory.

An AIM North America chapter was established in 2003 to focus on the day-to-day business needs of the sales channel in the AutoID markets in the United States, Canada and Mexico.

AIM Europe was also formally chartered in 2013. It serves as a single voice for the region in developing relationships with the European Commission and other regional and national bodies. *See also AIDC.*

AIM – The European Brands Association. It represents brand manufacturers in Europe on issues affecting their ability to design, distribute and market their brands. Membership is some 1,800 companies of all sizes in 21 countries. AIM worked with CIES, the GMA and the Food Marketing Institute to develop GPP's A Global Language for Packaging and Sustainability: A framework and a measurement system for our industry

AIR KNIFE – A constant stream of compressed air applied to a paper web during roll coating to level and remove any excess of coating material. Used in the manufacture of some types of coated papers. Other methods of paper coating include cast, blade, roll or brush coating.

AISLE WIDTH – The spacing between antennas (the detection gates or pedestals) in a store using electronic article surveillance (EAS) technology.

ALPHANUMERIC – A set of characters or bars in a barcode which represents both alphabetic and numeric characters as well as symbols, such as punctuation marks.

ALUMINUM BLISTER FOIL – Provides a good barrier to moisture, vapour and gases and is suitable for high speed packing applications on rotary and flat bed blister packing machines.

The blister packs are compact in size and economical when compared to strip packs. Tablets can be removed easily from blister packs, as the hard aluminum foil breaks quickly when the tablet/capsule is pressed from the PVC side. Barrier properties of blister packs are comparatively lower than strip pack due to usage of PVC film on one side.

Aluminum blister foils may be unprinted or printed as strip foil as required. Many are now printed using inkjet technology.

ALUMINUM FOIL – A thin aluminum foil, either with or without paper backing. Used as a print carrier for special self-adhesive products where a metallic image is required and for bottle decoration.

ALUMINUM NECK FOILS – Gold, silver or colored non-laminated or laminated foils used for decorating the neck of beer bottles and some wines and soft drinks.

Neck foiling creates a metalic finish to the bottle neck in order to give a 'de-luxe' prestige labeling appearance to bottles containing premium brands, such as champagne. Additionally, the neck foil also acts as a tamper-evident seal. The most common neck foils are embossed with a 'grain' pattern, which assists application and improves the end result. *See also Neck foils.*

Neck foiling creates a metallic finish to the bottle neck.

AMBIENT TEMPERATURE – A term that actually means the temperature of the surrounding air but is often used to describe a comfortable working temperature range.

AMERIPEN – The American Institution for Packaging and the Environment. A North America-based action group established in 2011 to promote positive, science-based guidance, to ensure profitable and sustainable packaging decisions for the region. It includes Coca-Cola, DuPont, P&G and others.

AMETIQ – Associaci n Mexicana de Etiqueteros, the Mexican label association formed in 2008 on the initiative of six label manufacturing companies.

The Association's main objectives are to strengthen the flexographic label and tag conversion industry in Mexico and to integrate suppliers and converters interested in achieving the same goals.

AMPLITUDE MODULATION (AM) SCREENING – Conventional halftone screening in which the image is broken up into halftone dots of differing size.

ANALOG COLOR PROOF – An off-press color proof made from separation films.

ANCHORAGE – The adhesion of a material to a web substrate such as that of an adhesive to the face stock, or release coating to the liner. Anchorage can often be improved through the use of surface treatments.

ANFEC –Associatión Nacional De Fabricants Etiquetas En Continuo is the National Association of Spanish Label Converters. It was founded in 1998 with headquarters in Barcelona. The initial number of members was 49 printers and 17 suppliers.

ANILINE PRINTING – The early name given to a relief printing process using rubber plates and fast-drying liquid inks. The forerunner of flexography.

ANILOX INKING – That part of a flexographic inking system that consists of an etched metal or ceramic coated metal roller with cells of a fixed size and geometry which transfers a measured amount of ink to the printing plate.

ANILOX ROLLS – An engraved metal or ceramic-coated roll used to meter ink to the raised (image) areas on the relief printing plate used in the flexographic inking system. Each type of flexographic press uses an anilox roll, the surface of which is engraved at one of three angles with a pattern of tiny cells of fixed size and depth that transfer the ink to the plate. The cells are so small that they can only be seen under magnification. The size and number of these cells determine how much ink will be delivered to the image areas of the plate, and ultimately to the substrate. An anilox roll today is either copper engraved and then chrome-plated, or ceramic-coated steel with a laser-engraved cell surface.

It is usual to specify anilox rollers by a line screen count, which is the number of cells in a given linear distance; the higher the line screen count the larger the number of cells, and the smaller the cells. The smaller the cells, the smaller the volume of ink carried in the cells. The depth and shape of the cells is also a factor in the efficient delivery of a uniform ink film to the printing plate. Cell depth ratios are critical in choosing the right anilox roller for a particular application. To avoid anilox moiré, film or plate screen angles should be at least 7.5º away from the anilox cell angles.

The line screen count of anilox rollers – expressed as lines per cm or lines per inch – determines the resolution of the printing. Cell depth is generally measured in microns and, when related to the cell's opening, produces a depth-to-opening ratio; the most efficient openings producing a depth-to-opening ratiobetween 20 and 35 per cent.

Anilox rolls are carefully selected for specific types of printing, substrates, and customer requirements. The printer may well perform test runs to determine the ideal anilox for producing the desired ink distribution for halftones, spot color and solids.

ANILOX SLEEVES – Anilox sleeves are not new. They have been under development, testing, trials and use for a number of years, with many manufacturers of 'gearless' presses particularly looking at the benefits of sleeve technology, such as ease of register, overall quality, low weight, maintenance, etc, as well as lower shipping costs and storage capabilities. Although early attempts at anilox sleeve manufacture were somewhat hit-and-miss due to a variety of reasons there have subsequently been dramatic improvements in the material, construction and stability of anilox sleeves have been made.

An anilox sleeve. *Source: Harper*

In use, sleeves need to be expanded to fit securely on the press mandrels. This may be undertaken with a mechanical mandrel that expands by hydraulic action, or through the use of a press mandrel with an inner compressible layer, or bladder, that is activated by air. Mandrels used with anilox sleeves are critical to the success of the process, with diameter, circularity and concentricity all important.

The relative newness of current sleeve technology provides little documentation on long-term success or sleeve longevity. However, it is anticipated that as anilox sleeve technology becomes more mature and prices decrease, then sleeves will become much more acceptable.

ANSI – American National Standards Institute: The Institute oversees the creation, promulgation and use of thousands of norms and guidelines that directly impact businesses in every sector. They are also actively engaged in accrediting programs that assesses conformance to standards – including globally recognized cross-sector programs such as ISO 9000 and ISO 14000 management systems that are used in the label and packaging industries.

ANTENNA – A device that has the capability to radiate and/or receive electronic or radio frequency signals. To be effective all electronic article surveillance (EAS) systems require detectors containing radio antennas to be placed at or near store exits or checkouts – whether for customers or staff. These create or receive radio signals at a required frequency, producing a measurable response in an electronic tag or label (which has not been de-activated). This then emits an audible and/or visible alarm, which alerts staff or security personnel to the theft, or can be additionally linked to security cameras. Detectors are available in different shapes and sizes, which can be customized or colored in line with the decor of the store.

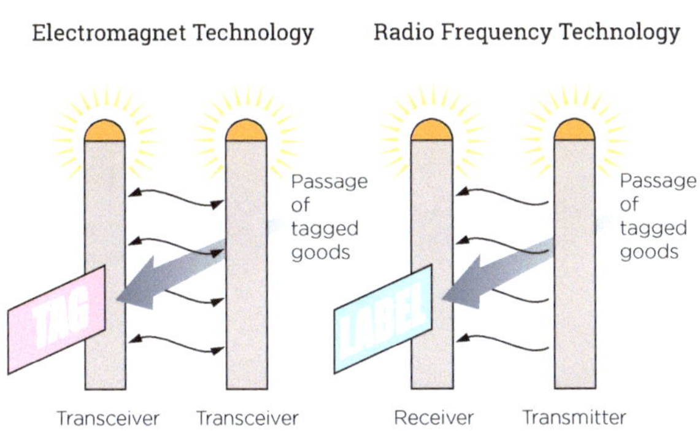

Electromagnetic detection gates have two transceivers which produce a magnetic field which is disturbed by alarm label.

Radio frequency detection gates have two antennae, one to transmit and one to receive. The EAS label disturbs the signal.

When using RFID tags the antenna is the conductive element that enables tags to send and receive data.

ANTI-COPY PAPERS – Papers that cannot be forged or counterfeited using color photocopiers.

ANTI-ERASABLE INKS – Inks that are damaged if any attempt is made to alter variably printed information (e.g., by means of mechanical abrasion or erasers). They can also be formulated so that they react to solvents. Most commonly used to print backgrounds on documents.

ANTI-MICROBIAL COATING – A coating which inhibits the growth of harmful bacteria including MRSA, C-diff, Listeria, E-coli and Salmonella. The active ingredients are silver ions which disrupt bacteria reproduction. FDA compliant for direct or indirect contact with food, medical or pharmaceutical products.

ANTI-MICROBIAL TAGS – Labels or tags containing a coating of chlorine dioxide which are designed to increase the shelf-life of soft fruits by significantly reducing levels of mold growth inside packs.

Moisture from the fruit activates the powder coating which cascades through the pack on a time-release basis. Used with products such as strawberries, blackberries or raspberries, the tags are also being developed for reducing mold or odor in applications such as shoes and clothing.

The chlorine dioxide coatings used with anti-microbial tags are permitted by the FDA and EPA.

ANTIOXIDANT – A chemical substance that is added in small quantities to materials (such as plastics) to inhibit or retard breakdown due to oxidation, particularly during processing. Antioxidants are found in very many polyolefin film stocks and rubber-based adhesives. Some of these will inhibit or 'poison' platinum catalysed, vinyl addition curable silicones.

ANTI-STATIC CONTROL – Static electricity is the result of an imbalance between negative and positive charges on the surface of a substrate. The static charge builds up on the surface and needs to be discharged by a combination of an electrical charge and ionized air. Static can be particularly troublesome with filmic substrates, with static charges generated whenever two surfaces are in contact and then separated, so causing a dragging or holding effect. Films containing amine-based antistats can also inhibit cationic cure silicones.

Static can additionally be a problem with sheet-fed applications. Here, static causes the sheets to be electrically attracted because of the opposing charges on each sheet.

Anti-static units on a web- or sheet-fed press, coating line, slitter, etc., are used to remove static charges.

ANTI-STICK COATINGS, TOOLING – A coating engineered to reduce the ability of other materials to stick to it. For certain applications such as hot-melt adhesives, multi-layered labels and labels with high application of adhesive (> 18 – 20 gr/m^2), tooling suppliers will offer the possibility to coat cutting dies. These coatings should prevent adhesive and ink adhering to the cutting edge and adhesive bridging to take place between the label and the waste matrix.

ANTI-THEFT MEASURES – In addition to electronic article surveillance (EAS) technology a variety of other measures may be used to combat retail theft. These range from structural packaging design (exaggerated size and tamper proofing) through CCTV surveillance systems to loss prevention programs for employees. *See EAS entries.*

ANVIL CUT – A die-cut in a pressure-sensitive label which is made through all of the components of the label stock, including the liner.

ANVIL POSITION – A cutting cylinder, whether rotary die, magnetic cylinder, sheeter, etc., which is placed in the bottom position of the cutting unit, whereas the anvil roller is placed in the top position of the cutting unit.

ANVIL ROLLER – Hardened steel roller upon which the bearers of a rotary die cutter ride, which also provides the hardened, smooth, surface for die-cutting.

APP – App is a shortened version of the term 'application' and is made available through an application distribution platform. A mobile app is a computer program that is designed to run on smartphones, tablet computers or other mobile devices.

APPLICATION – Refers to the process of applying or adhering a pressure-sensitive label to a surface.

APPLICATION TEMPERATURE – The temperature of the surface to be labeled and the self-adhesive labels at the time of application. All adhesives have recommended minimum application temperatures, which are ascertained by means of tests, carried out in ideal conditions (i.e. smooth, clean, dry surfaces, etc.).

Temperature of the environment or container when a self-adhesive label is applied.

APPLICATOR ROLLER – A coating, print, tint, lacquer or varnish roller, or a laminating roller that actually applies any of these to a substrate.

APPLICATOR, SELF-ADHESIVE – A table-top, stand-alone, or in-line device/machine which is used to automatically apply pressure-sensitive labels from a roll of labels onto a container or product. There are many different types of pressure-sensitive applicators, but they all have one thing in common: a means of peeling the release liner or backing away from the labels. This is usually accomplished by unwinding a reel of die-cut labels and then pulling the web under tension around a stripper plate or beak. As the backing is taken around the sharp angle of the plate the front edge of the label is peeled away. Stripper plates may be found in a variety of configurations – fixed plate, floating plate, pneumatic, etc.

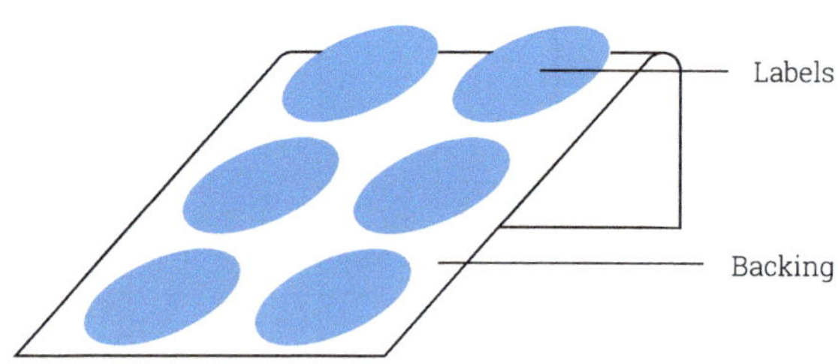

Rigid stripper plates for example, can be used for very uniform products, while spring biased plates may be recommended for certain side labeling applications. Floating stripper plates, either solenoid or pneumatically-operated, may be used to place labels accurately into recessed areas on a pack at a required point. Powered stripper plates need to be accurately controlled by, say, electronic timers, so that the label is pre-dispensed or maintained in contact with the pack for the correct period.

Once the labels have been detached from the backing there are various ways of feeding them forward and pressing (applying) them on to the pack, container or product to be labeled in the correct position. The effectiveness of the labeling operation is determined by the uniform application of pressure to the label in order to ensure a positive contact of the adhesive with the surface of the product being labeled.

In their simplest form various rubber or foam rollers and/or drum application devices are used to fix the label to the product, perhaps also supplemented by brushes. Other methods of uniformly applying

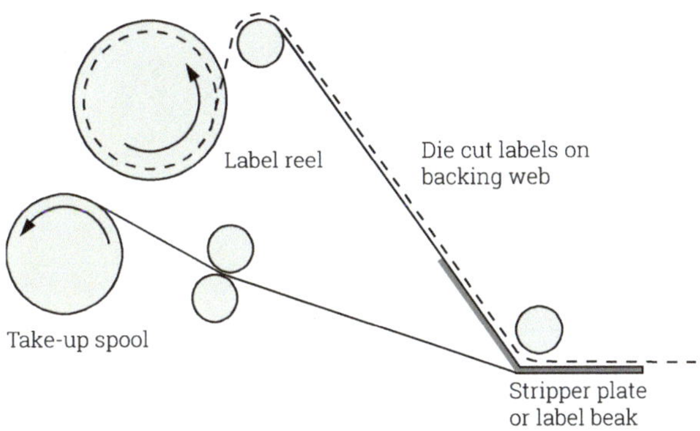

Principle of applying self-adhesive labels using a stripper plate or beak. Neck foiling creates a metallic finish to the bottle neck.

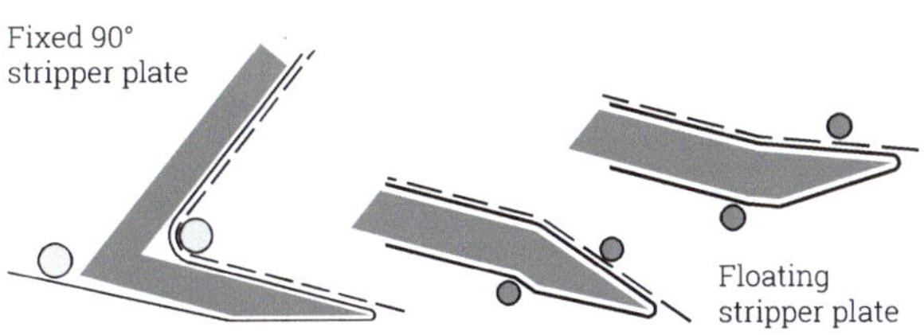

Different configurations for stripper plates/beaks

pressure to the label to ensure that it sticks with a positive contact to the product or container include pressure pads, pneumatic plungers, air jets, knurled rings or powered rotating brushes. Alternatively, the drum and released onto the container when it is in the correct position. Once the label has label may be held in position on a vacuum box or been applied

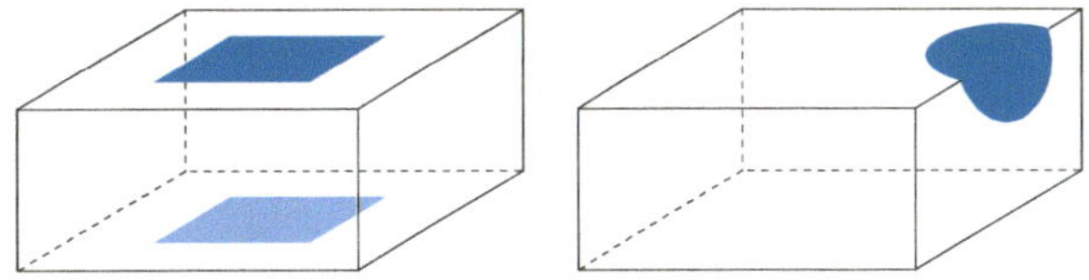

Application of labels to the top and bottom and around corners of objects

the backing paper is rewound on to a take-up spool.

Roller or drum application. Roller and/or drum application devices are one of the most common methods of affixing pressure-sensitive labels to products or containers. Ideal for flat or cylindrical packages, they may also be used to label concave and convex surfaces. By employing two application heads it is possible to apply labels to the top and bottom of objects simultaneously with only one pass through the machine. By setting the applicator heads horizontally it then becomes possible to apply labels to the sides of the objects.

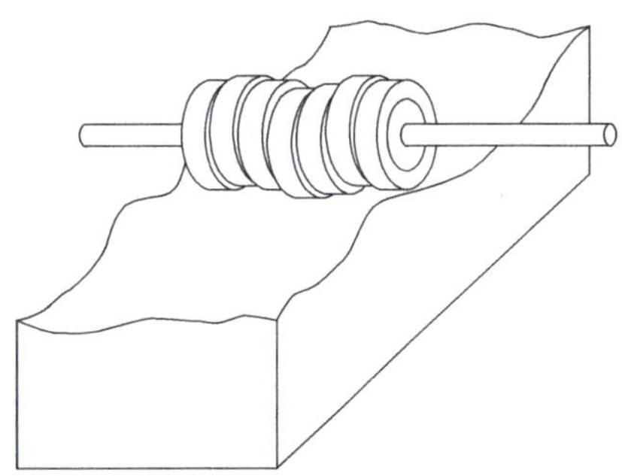

Applying labels to uneven surfaces.

With two units facing each other it is possible to label both sides of the object simultaneously. With some systems it is feasible to apply labels on the corners of boxes, such as medicines or confectionary box sealing. A variation of conventional roller application used on some machines is the use of rollers made up of knurled aluminum rings, so allowing packs with uneven or flexible surfaces to be handled. Powered rotating brushes, sometimes profiled to the corner of the product, may also be used to press the label into difficult areas.

Pressure-pad or plunger application. In some methods of label application a pad or plunger is used to fix the labels in position. Here the dispensed labels are sucked onto a vacuum pad, perhaps on the end of a plunger, and then applied to the product when it is in the correct position by releasing the suction. The result is a highly accurate labeling system. With such systems it is possible to select any surface, direction or angle, up or down, simultaneously to both sides, for label application, e.g. to place labels at an inclination of 45 degrees.

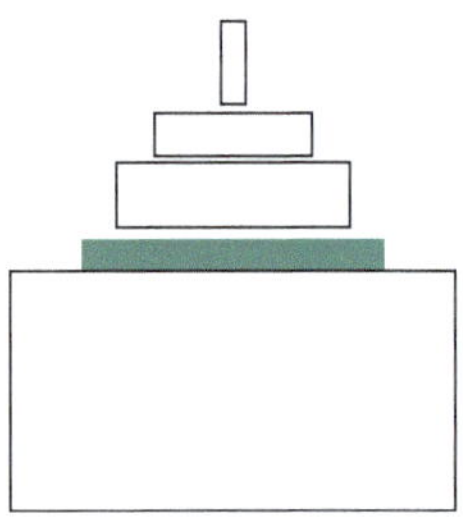

Use of a pressure pad to accurately apply labels.

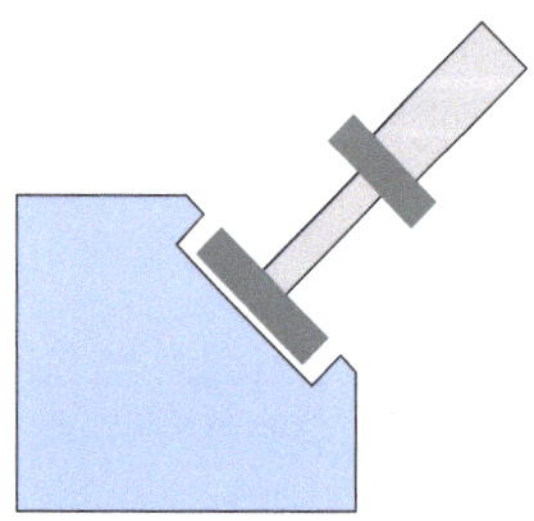

Application of label at an angle or into recesses.

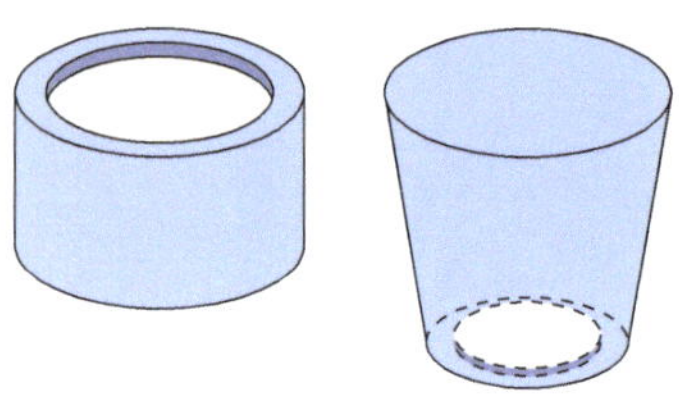

Labels applied in a recess.

Applying labels to uneven surfaces.

By employing a longer plunger arm it is possible to apply labels into deep recesses or even into the bottom of something like a cup or glass. Depths up to 150mm (5.1 inches) or so may be reached.

Air jet/air blow. This is another frequently used method of label application in which the labels are stripped as usual and then, after release, are again held on a low vacuum, but this time are blown onto the container by applying air pressure. Using this method of label application it is possible to apply labels in positions that would be virtually impossible by other methods. Absolutely no damage will be done to the surface or product to be labeled.

A variation of air jet application is where the application head is built into a mechanical arm which can then be positioned to label by air jet, products such as television tubes.

Label and product control. To ensure that labels are applied accurately and consistently to products the self-adhesive label applicator may be fitted with various label or product tracing/control devices.

Label tracing devices monitor the gap between the individual die-cut labels and interrupt the web feed at the end of each label. This resets the system to receive the next start signal from the product tracing

Use of end of roll and missing label detectors.

devices, which include micro-switches, photocells or spot color readers. Other controls on the web may be used to detect missing labels, the end of the reel or web breaks.

With many applicator systems the actual product or container triggers the application of the label through the use of product tracing/ control devices. This means that even with erratic or irregular product flow the system will automatically compensate and only dispense a label as required. Many different scanning/sensing technologies are available on self-adhesive applicators, from photocells with reflector mirrors, transmitter/receivers, proximity switches, spot or color readers or, for certain applications, micro-switches. For semi-automatic applications footswitches may be used. *See also Labeling machines.*

AQUEOUS – Water based or water containing. In the label industry the term is usually related to either aqueous adhesives or aqueous inks (flexographic or inkjet) in which water is used as the carrier or vehicle.

ARTWORK – The original design, drawings,

pictures and text produced by the designer or artist. All elements of design from which the black-and-white origination and printing plates are made. The production of finished material suitable for reproduction by any printing method or media. This may be presented as a black and white art sheet, with color overlays, or in disk or CD format, or even transmitted electronically for computer printout.

ASCII – Initial letters for 'American Standard Code for Information Interchange'. This is a standard 7-bit coding system used within the computer industry to convert keyboard input into digital information and to communicate between devices. An ASCII file is one that contains text only, as opposed to an electronic file that contains text and graphics. It covers all of the printable characters in normal use and control characters such as carriage return and line feed.

The full table contains 127 elements. Variations and extensions of the basic code are to be found in special applications. Extended ASCII has an 8-bit code.

ASPECT RATIO – The term given to the ratio of the height to that of the width in a barcode symbol.

ASSET PROTECTION – The generic name given to electronic article surveillance (EAS) systems used to protect office equipment and other company assets. Asset protection labels and tags generally use different frequencies from normal retail EAS.

ASSOCIATION OF POSTCONSUMER PLASTIC RECYCLERS (APR) – APR represents companies that acquire, reprocess and sell the by-product of more than 90 percent of the post-consumer plastic processing capacity in the United States, Canada and Mexico. It is working with the APR to establish RCA criteria for PET thermoform containers

AUGMENTED REALITY – Augmented reality, or AR, uses a smart phone, tablet computer or other camera-equipped device to super-impose an interactive virtual representation over an existing printed image, such as on a label or pack, and then adding explanatory data. End users must download a software application (app) or browser plug-in in order to have the augmented reality experience. *See App.*

AUTHENTICATION LABELS – Labels used to confirm, through the use of one or more overt or covert authentication features, that products reaching the consumer are genuine. Overt authentication features are those that can be checked and verifies by eye or touch – such as a color-shift film or hologram – while covert authentication features require some kind of authentication tool and a trained inspector. Authentication features may also be at the forensic level for verification in a laboratory environment as part of a covert investigation. *See also Security labels, Security design, Security inks, Security materials.*

AUTOCLAVE – A pressurized, steam-heated vessel used for sterilization of health-care, hospital and other products which need to be supplied or used in sterile conditions. Labels used in sterilization applications must endure a cooking process under superheated steam at 250°F (121°C) for about 30 minutes or to 270°F (132°C) for about 3 minutes to reduce the bacterial population to almost zero. There must be no package or label failure due to the high temperature steam exposure or due to a reaction at the high moisture levels. Other sterilization methods are gas and radiation.

AUTO DISCRIMINATION – The ability of a barcode reader to distinguish automatically between two or more different types of barcodes or symbologies.

AUTO-ID CENTER – A non-profit collaboration between private companies and the academic world that is developing an infrastructure for the global tracking of goods using RFID tags.

AUTO-ID LAB – A special RFID interest group established by the Auto-ID Lab at the Massachusetts Institute of Technology (MIT) which is developing tools and processes to help companies find the best way to flag items and design antennas for objects and packaging that are tough to tag. They have a long history in the field of networked RFID and pioneered the tern Internet of Things.

The Lab's primary research partner is GSI, a not for profit organization that is renowned for establishing standards for global commerce, such as barcodes to the retail industry some 40 years ago.

Automated ink management – Specifically engineered for small ink batches, an automated, lean ink management dispensing system provides label converters with a fast, precise and waste-free means of achieving spot colors on demand. Suitable for any ink volume requirements, such systems offer optimal dosage accuracy for a waste-free process that can typically improve ink yields by up to 30 per cent.

AUTOMATIC DE-ACTIVATION – Integration of the label or tag deactivation operation into the check-out operation at point-of-sale (POS). The barcode is read and the electronic article surveillance (EAS) tag deactivated simultaneously. There is no additional deactivation procedure required. Such systems streamline the check-out operation and also prevent certain types of retail theft.

AUTOMATIC IDENTIFICATION – A means of automatically collecting data for direct entry into computer systems without any human involvement. Relevant technologies of automatic data capture include barcodes, optical character recognition (OCR), magnetic ink character recognition (MICR), RFID, biometrics and voice recognition.

AUTOMATIC INSPECTION SYSTEMS – Systems designed to provide complete quality assurance and process control workflow in a label plant. Automatic inspection solutions enable printers to have more and better tools for meeting end-user customer demands for delivering process control and quality assurance at:

- 100% inspection for complete quality assurance on press and post press
- On-press solutions for process control
- Off-line pre-press verification.

Automatic inspections systems need to provide reliable quality assurance and process control. They should be cost-effective to install and operate; need to be operator friendly and operate without constant attention; cut waste; and build profitability by reducing the cost of quality.

Systems should be able to deal with labels of any size, on any label stock – including paper, reflective materials, transparent films, foils and flexible materials – and with all types of jobs, particularly critical jobs such as pharmaceutical, cosmetics, food and drinks labels. *See also Statistical process control, Six Sigma, TQM and Production control.*

AUTOMOTIVE LABELS – There are over 40 million passenger cars a year produced worldwide each year – the main output of the global automotive industry – as well as some 17 million or so commercial vehicles, with the main automotive manufacturing countries being Japan, the USA, Germany and Taiwan. General Motors, Ford, Daimler Chrysler and Toyota are the dominant manufacturers. Add to the list of passenger cars all the motorcycles, tractors, outboard engines, lawnmowers and so on, as well as the spare part and after-market requirements, and it can be seen that the automotive and related combustion engine markets are significant markets for the label industry.

Almost all of these passenger cars and other units require one, two or multiple high performance durable labels, increasingly with security features. Vehicle Identification Number (VIN) labels have also becoming a standard requirement across most segments of the automotive industry, while new regulations on end-of-life recycling instruction labels are expected to add further to the ever-growing label

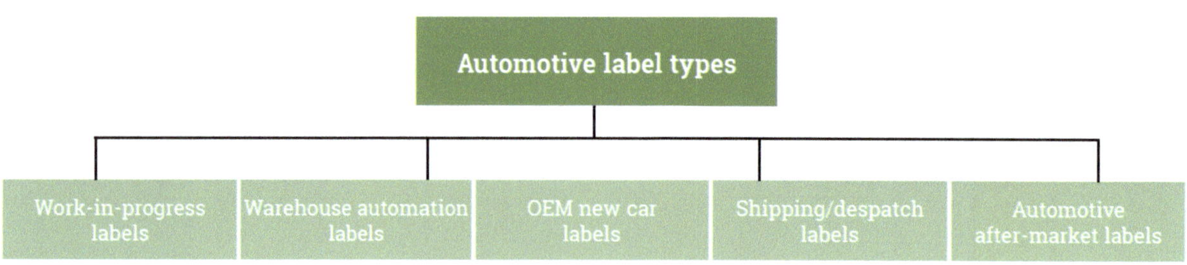

The need for automotive and related labels can be split into distinct categories.

needs of the sector. The annual value of label sales into the automotive sector is estimated to be in excess of US $120 million.

The need for automotive and related labels can generally be split into several distinct categories:

Work-in-Process (WIP) labels. Used during the manufacturing and assembly process from components supply through to the finished vehicles. Process tracking labels. Includes barcoded and sequentially coded tracking labels used throughout the supply chain and assembly. Very often these labels remain in place throughout the working life of the vehicle, but have no further functionality.

Warehouse automation labels. Shelf, rack, bin, location, etc labels, used in a variety of applications and may incorporate over-lamination onto a polyester or other synthetic substrate.

OEM, new car labels. Includes VIN labels – up to 12 per car fixed to body parts. Under the hood (bonnet) labels in the engine compartment (fuse chart, lubricant, cooling, nameplates and rating plates, emission control, etc). Component part authentication labels (exhaust pipes, body panels, windscreens, transmission housings, complete engines, seat belts, electronics assemblies, etc). Specifiers must take into consideration the needs of climatic and engine temperature changes, abrasion, oil and grease contamination, as well as chemical degradation.

Shipping/despatch labels. Includes AIAG labeling requirements, handling, address and contents labels.

Automotive after-market labels. Includes genuine service parts, replacement component/ spare part labels and accessory labeling, authenticity and warranty labels. May need to incorporate security features for authentication, holograms, void seals, safety sleeves, packaging seals and other security features.

The automotive sector is a key end-user of high-performance labels requiring durability, harsh environmental resistance, long life and permanence of manufacturing and /or authentication data. In addition, the trend to track original equipment parts during and after manufacture can be met effectively with high performance filmic WIP labels using electronically-generated information. In the spare parts and after-sales market, labels are also being increasingly used to label and 'authenticate' genuine approved replacement parts as distinct from counterfeits. *See also VIN labels.*

AXIAL – When viewing a cylinder the outer side is seen as a circle. A radius is a line drawn from the centre of the circle to the border or circumference of the circle moving parallel to a radius is called 'radial'

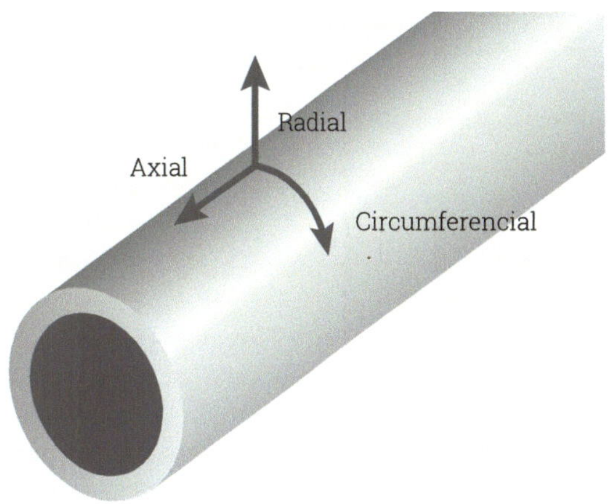

Picture shows axial, radial and circumferential aspects. *Source: Inline Devices*

movement. The axle of a metal cylinder with shafts on each end moving parallel to the axle, that is, from one shaft to the other, is called 'axial' movement.

AXIS – Axis of revolution of a gear; center line of a shaft.

B

BACKING – Refers to the carrier sheet or web of a material used in pressure-sensitive lamination that carries the permanently tacky adhesive coated face material. Usually has a release (silicone) coating applied to it so that the adhesive will not stick too tightly to it.

BACKINGLESS LABEL (LINERLESS LABEL) – A pressure-sensitive material without a backing or release liner to protect the adhesive before use. In such cases the adhesive is normally protected by a silicone coating on the face of the label which, when wound on itself, stops the face and back of the label sticking together.

Backingless pressure-sensitive labels first came to the fore in the early 1980s when Waddingtons in the United Kingdom developed a technology and coating system (Monoweb) to produce backingless labels, which found application in label user companies such as Heinz.

Today, backingless labels are most commonly found in the form of pressure-sensitive thermal labels used in print and apply price-weigh labelling and small, portable, thermal label dispensers. However, there are now a number of companies that now offer linerless solutions suitable for product decoration. *See Linerless.*

BACKING PAPER – Also referred to as the release liner, this is the component of the label laminate that protects the pressure-sensitive adhesive during manufacture, converting, printing and up to the final point of application. The backing is peeled away from the adhesive immediately prior to its application to a container, product or surface.

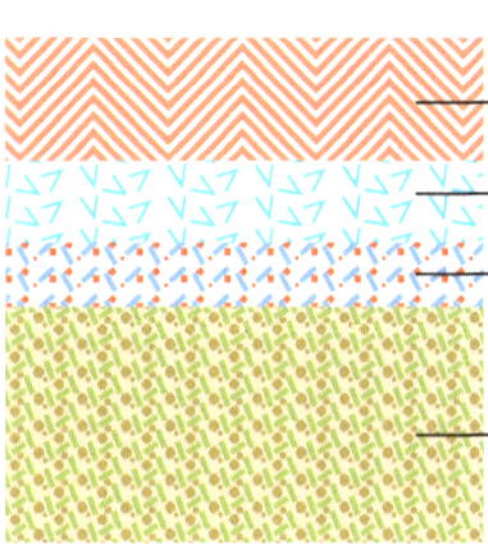

Construction of pressure-sensitive laminate showing backing paper and silicone release coating

BACKLASH – Refers to the striking back of connected wheels in a piece of mechanism- such as a label press – when pressure is applied. Another source defines it as the maximum distance through which one part of something can be moved without moving a connected part. In the context of machinery gears backlash, sometimes called lash or play, is clearance between mating components, or the amount of lost motion due to clearance or slackness when movement is reversed and contact is re-established. For example, in a pair of gears backlash is the amount of clearance between mated gear teeth.

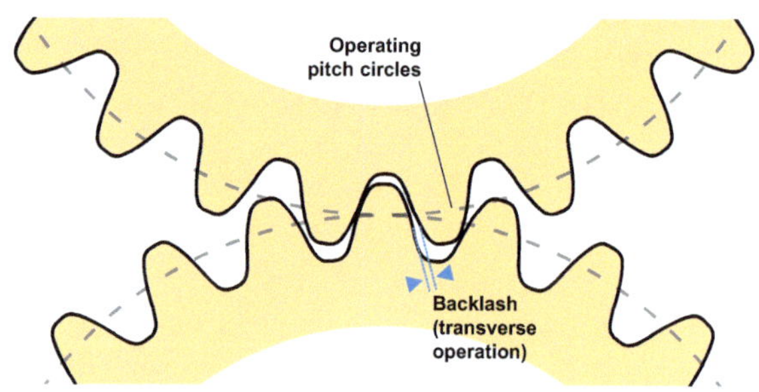

Theoretically, the backlash should be zero, but in actual practice some backlash must be allowed to prevent jamming. It is unavoidable for nearly all reversing mechanical couplings, although its effects can be negated. Depending on the application it may or may not be desirable. Reasons for requiring backlash include allowing for lubrication, manufacturing errors, deflection under load and thermal expansion.

Factors affecting the amount of backlash required in a gear train include errors in profile, pitch, tooth thickness, helix angle and center distance, and run-out. The greater the accuracy the smaller the backlash needed. Backlash is most commonly created by cutting the teeth deeper into the gears than the ideal depth. Another way of introducing backlash is by increasing the center distances between the gears.

BACK PRINT – printing a brand mark or adhesive reference onto the back of a self-adhesive label laminate for product recognition by end users.

BACK PRINTING – This is achieved by delaminating a self-adhesive laminate on a suitable label press, printing onto the adhesive and then re-laminating the roll before further printing and die-cutting. This technique is often used for labels applied to clear bottles or plastic film so that the printing on both side of the label (face and adhesive) saves using another label.

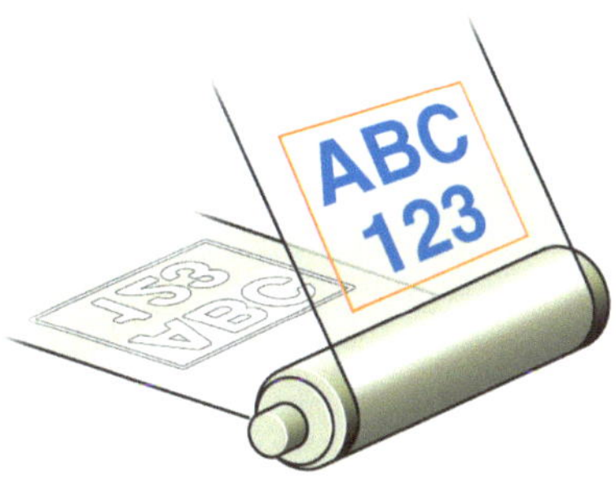

Printing on the back of a clear label substrate.

If this technique utilizes a clear label material, reverse printing offers a completely protected print image that is scratchproof, with the finish of the label film providing a glossy or matt surface to match the container. May also be called reverse-side printing.

For back or reverse-side printing the image is laterally reversed, and colours are printed from the darkest to the lightest; the label or web is viewed from the non-ink side of the substrate.

BACKSIDE GRINDING – A process to obtain a consistent and accurate die height for flexible cutting dies. The back side of the dies are ground using a large grinding wheel

BACK SLIT – A cut in the release liner or backing, usually along the web, but can be on the back of sheeted pressure-sensitive laminate to allow the face stock to be easily peeled away by hand when die-cutting has not be used.

BANDING – Effect created on some low-end computer printers which show gradation of colors or shades in distinct contrasting bands rather than one continuous flow.

BAND LABEL – A paper wrap-around label for use in banding together items such as envelopes, bulk bank notes, wool skeins, etc.

BAR – The dark element or lines of a printed barcode symbol. *See Barcode symbol.*

BARCODE – A standard method of identifying the manufacturer and product category of a particular

Barcodes consist of alternating thick and thin vertical dark and light bars

item. Invented in the early 1950s' barcodes have had a significant impact on accelerating the flow of products and information in the global business community. There are over 220 different known barcodes, although only a few of these are currently used; even fewer are widely used.

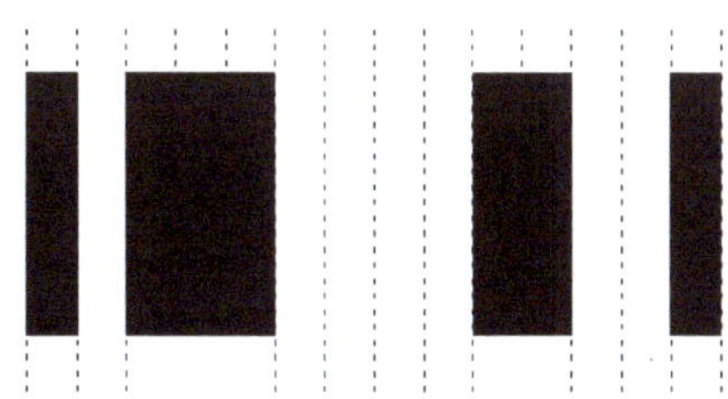

Barcodes are made up of modules of light or dark bars - a module being the smallest increment used by a barcode to represent data.

Barcodes consist of a series of thick and thin alternating vertical dark and light bars which are printed on to labels used for retail, business and commercial applications and which can be automatically read to provide a fast, accurate and

Code 39

2 of 5 straight

2 of 5 interleaved

2 of 5 non-interleaved

Universal Product Code (UPC)

Codabar

* C O D E 3 9 *

Code 39. Density 5.4

DLA90083M7978

Logmars 13 digit US Mil-Std-1189

Sick Code (Pharmaceutical)

5 012345 678900

EAN Version L reversed

EAN Version L 13

Barcodes in common use

efficient method of collecting and recording data.

Current uses of barcodes include the whole field of high street retailing, outer case coding, inventory control, asset management, job costing and pricing, production control, scheduling, materials control and management, in the parcel distribution services and in many other time consuming operations that would benefit from automating the process.

Barcodes have both a height and width, although there is normally no interpretive information in the height of the code. However, the height should be sufficient to allow the code to be read efficiently. The dark and light spaces in a barcode represent digitally encoded information, which can be 'read' by devices (barcode readers; barcode scanners), which scan a beam of light across the bars and so pick up signals from the various pulses of reflected light. The dark bars in the code absorb light while the white spaces reflect light, so providing a cost-effective method for automating data collection.

Most barcodes are made up of modules of light or dark bars – a module being the smallest increment that can be used in a barcode to represent data. Generally, codes are made up of exact module widths. As they are magnified or reduced in size, the module relationship between bars remains constant. The widths of individual bars and spaces however, are increased or reduced.

Special number and code systems for use in retailing have been developed and these are based on either the North American UPC (Universal Product Code) or the EAN (European Article Number) code. The choice of code for industrial applications normally comes down to a short list of two or three codes; Interleaved 2 of 5, Code 39 and the EAN code.

The main drawbacks of barcodes are that they do not identify unique items and that scanners need a line of sight to be able to read them.

BARCODE DENSITY – The number of characters that can be represented by a barcode per linear inch (25.4mm).

BARCODE READER; BARCODE SCANNER – A device used to identify and read a barcode. May also read, assemble, check and collate information of greater complexity than ever before. Equipment may be hand-held (where it is generally more convenient to take the reader to the product), desk-top fixed work station where the barcoded product is passed by hand under or over the reader, or conveyor-mounted where the coded product can be taken to the reader.

Information from the readers or scanners may be part of a total set of data that can be collated, analysed and then put to use in the running of a business or factory.

BARCODE SYMBOL – Special number and code systems for use in retailing have been developed and these are based on either the GS1 North American UPC (Universal Product Code) or the GS! EAN (European Article Number) code. In these systems, each retail product is assigned a unique manufacturers number which can be used instead of the full product description in all transactions that take place between manufacturer, distributor and retailer.

The manufacturer then allocates a second part of the number to various products himself. At this stage the code is purely number based to aid communication and does not yet depend on a symbol, barcode or scanner.

0 12345 67890

UPC symbol is made up of 10 characters that appear in figures below the symbol.

Once numbers have been allocated it is possible to convert this to a special barcode known as The Symbol, which is made up of a combination of thick and thin vertical bars. The structure and layout of both UPC and EAN symbols is very similar, shown here using a UPC symbol.

The various elements that make up a UPC symbol.

The UPC symbol is made up of ten characters that appear in actual figures below the symbol. A number system character appears halfway up the left hand margin. The various elements that make up the UPC symbol are: (a) a number system character which defines the application, e.g. retail groceries, drugs, etc. (b) start and stop bars to signal to the scanner the beginning and end of the code (c) number system character code bars (d) check digit code bars (e) actual number (f) corner marks.

The EAN symbol is also made up of ten characters. These again designate the manufacturer and the product – as with the UPC symbol – and appear in actual figures beneath the code.

5 012345 678900

The EAN symbol is also made up of 10 characters, plus country of origin and number system.

This time however, there are two additional characters at the start of the code. The first appears to the left of the guard bars; the second is incorporated into the characters beneath the code. These additional characters represent the country of origin and the number system. A check digit character also appears at the end of the line. This means that the EAN Symbol has thirteen numeric characters; the UPC Symbol has ten (plus one halfway up the margin. *See also EAS, GS1 and UPC.*

BARCODE VERIFICATION – A barcode only becomes effective if it can be read or scanned first time every time. A badly printed barcode results in low first reading rates, imprecise formation of symbols and to the user being dissatisfied. Acceptable barcodes are those that result in: dimensional accuracy and consistency; high print quality; a high print contrast ratio.

To ensure that barcodes will scan first time and meet their acceptance requirements it is normally necessary to have some kind of quality control device. When used in conjunction with the printing of barcodes these devices are called verifiers. Verifiers check that the barcode matches the specifications within the permitted tolerance levels, as well as confirming that the bars and spaces correctly correspond to the human readable information.

Barcode verification equipment can be categorized under five headings: optical measurement equipment, wand verifiers, opto-electrical equipment, laser verifiers and on-line verifiers.

BAR LENGTH – Pertains to a barcode. The bar dimension perpendicular to the bar width.

BAR WIDTH – The width or thickness of an individual bar in a barcode symbol. The thickness of a bar measured from the edge closest to the symbol start character to the trailing edge of the same bar. The number of possible bar width variations within a particular symbol depends on the symbology used.

BAR WIDTH REDUCTION – Reduction of the nominal bar width dimension (of a barcode) on film, masters or printing plates to compensate for printing gain.

BARRIER COAT – A coating applied to a material to prevent absorption or interaction between the material and any subsequent coating. In the label industry, a barrier coating may be applied to the back of the face material to provide improved opacity to the face material and/or to prevent migration between the adhesive and the face material or to improve the keying of the adhesive to the face material. May also be known as a sealer coat, primer or tie coat.

BASE – The major constituent, other than pigments and filler, that makes up the non-volatile portion of an adhesive, coating or sealer compound.

BASE – The supporting material for a flexible cutting die. Thickness varies depending on the application and total die height. Total cutting height – height cutting line = base.

BASIS WEIGHT – The weight in grammes per square meter of a label face paper in the metric system. *See also Grammage or GSM.*

BATCH COUNTER – A device used on sheeting or sheeter/stacker machines to count and batch sheeted or fan-folded labels.

BATCH PROCESSING – Processing carried out automatically by a computer on a number of files or a series of tasks without operator intervention.

BEARER BARS – Continuous strips of plate material usually placed on the outside of printing areas to minimize plate bounce and over-impression by helping to carry part of the impression pressure or load during printing. Also the part of a rotary cutter designed to support the cutter against the anvil and provide sufficient space between the peak of the cutting edge and the surface of the anvil to ensure adequate penetration of the cutter.

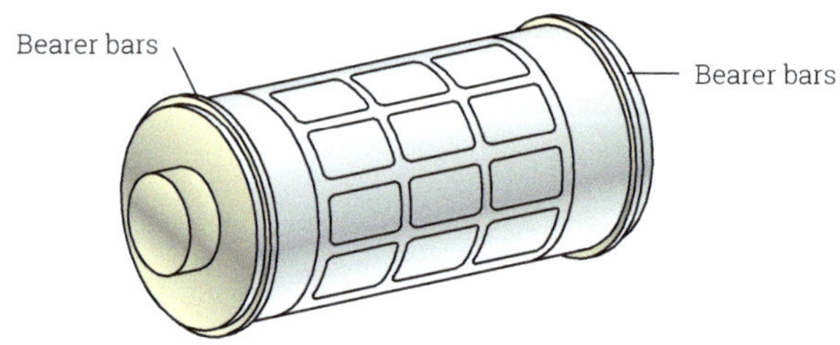

Bearer bars on the outside of printing and die cutting cylinders.

BEARINGS – The hard rings, often including balls or rollers, which provide a smooth rotary movement to cutters and printing cylinders that come into contact with each other during printing and die-cutting.

BEARING BLOCK – A block with internal bearings that holds the cylinders – magnetic cylinder, rotary die, anvil roller, etc.- in position in a die cutting station.

BEND – for envelope dies. Envelope holders require a clamping system to ensure that flexible cutting dies stay in place. The bend varies depending on the type of holder used and is normally on the leading edge of the die.

BIAXIALLY ORIENTATED – Label face and liner (polypropylene) film materials, which have been mechanically stretched (orientated) in both the machine and transverse directions during the manufacturing process. By doing this, the film's molecules are aligned in the cross (transverse) machine direction as well as the machine direction. Once biaxial orientation has taken place, the film exhibits a marked improvement in optics, strength, moisture barrier and low temperature durability. The increased tensile strength of the film improves the on-press stability, with less stretch and subsequently improved print performance.

There are two distinct methods of orientating (OPP) film; the 'blown tubular' and 'tenter frame' processes.

BIOPP/BOPP – Biaxially orientated polypropylene. *See above.*

BI-DIRECTIONAL READING – The reading of data successfully whether the reading or scanning motion is from left to right or from right to left, (as in supermarket barcode readers).

BI-DIRECTIONAL SYMBOL – A barcode symbol that can be read equally successfully from either direction.

BINARY – A numbering system in which numbers are expressed as combinations of the digits 0 and 1 based on powers of 2. In computer usage these are represented electrically by 'on' and 'off' – or in barcode terms by narrow and wide bars and spaces.

BIOCODE – A patented system for marking products and/or individual batches of products. One of the latest and most sophisticated product identification technologies to help manufacturers distinguish genuine articles from counterfeit articles.

Marker chemicals are added directly, in tiny concentrations (parts per billion or parts per million) to both liquid and solid products. The marked product can then be easily identified in the field using antibody-based test kits. In addition to incorporating the marker into products, the method lends itself to security labeling and a range of proprietary security labels incorporating the technology are now available from suppliers.

BIODEGRADABLE – The ability of a material, such as paper or plastic film, pressure-sensitive label stock or printed labels, to break down or decompose under natural conditions or processes.

BIOMETRICS – Technologies by which people can be identified by their individual physical or behavioral characteristics. These characteristics can be based on an individual's signature, fingerprints, retinal eye pattern, hand geometry, voice or keyboard rhythm. Biometrics has applications in passports, identity cards, access systems and for some high security label solutions.

BIOMONITORING – Biomonitoring systems include food quality indicators, temperature fluctuation indicators, breathable film labels, antimicrobial labels, gas scavenging labels and absorbing labels. Growth of on-pack biomonitoring systems is forecast to be 20 per cent per annum over the next five years. *See also Antimicrobial labels, Food freshness indicators, Smart active labels and Temperature indicators.*

BIT – A single binary element, with the word bit derived from an abbreviation of the words Binary digIT. A binary digit, which is the fundamental building block of digital computer systems, can only have one of two values, 1 or 0, depending on whether it is switched on or off. When combined into greater units – bytes, kilobytes, megabytes, and gigabytes – bits enable computers to process, display and store large amounts of information. Combinations of bits and bytes are made up as follows:

- Byte – 8 Bits
- Kilobyte (KB) – 1,024 Bytes (not 1,000 as often thought)
- Megabyte (MB) – 1,048,576 Bytes
- Gigabyte (GB) – 1,073,741,824 Bytes.
- Terabyte (TB) – The prefix 'tera' represents the fourth power of 1,000 and means 1,012 in the International System of Units (SI), and therefore one terabyte is one trillion bytes.

BIT DEPTH – In digital printing the bit-depth determines the number of output levels that can be printed. While a typical laser exposure unit is either 'on' or 'off', the light intensity of LEDs can be modulated to create different output levels. A laser for example has, by definition, a bit depth of 1, which means it can produce 2 output levels. The laser is either 'on' or 'off', i.e. the device dot is black (laser 'on' and therefore toner is deposited) or white (laser 'off' and therefore no toner is deposited). In an LED imaging system the light intensity can be modulated. If the imaging system has for example a 3 bit dot density, then the number of available output levels equals 23 = 8.

BIT-MAPPED FONTS – A font on which the characters are formed by patterns of tiny dots. They are always in a specific point size, unlike outline fonts, which are scalable.

BIT-MAPPED IMAGE – Type or graphics from computer paint programmes formed by a series of dots and jagged in appearance. An image with 1 bit of color information per pixel.

BLADE COATED PAPER – A printing or label face paper which has been coated either on or off the paper machine through the use of a flexible metal trailing blade to level and smooth the wet coating so as to give a good smooth coated surface.

BLANKET – Used in offset printing. A smooth rubber-faced fabric which is clamped around a cylinder and to which the image is transferred from the printing plate, and subsequently from which it is transferred to the paper being printed.

Trelleborg offset blankets

BLEACHED KRAFT – The standard backing liner material for sheeted self-adhesive label materials due to its dimensional stability and flatness properties. Mostly clay coated to hold out expensive silicone coatings and to provide an even coating with a controlled release level.

BLEED (ADHESIVE) – Also known as Edge Ooze. This is more likely on softer adhesive coatings, but can also indicate higher adhesive coat weight. It can be made worse by tightly wound reels stored in direct sunlight or in very warm conditions, or by poor reel slitting.

BLEED (ADHESIVE) – Penetration of an adhesive through the base paper due to either an inadequate paper structure or poor barrier coating. The term is also used to describe plasticizer migration, where the adhesive on a label is degraded by plasticizer from a plastic film or container to which the label has been applied.

The degraded adhesive then strikes through the paper label causing a discoloration of the label surface and eventual failure of the adhesive bond.

Bleed may also refer to the migration of plasticizers from an adhesive film into the face stock or substrate on to which it is coated.

BLEED (PRINTING) – An image or color that extends to, and slightly beyond, the trim edge of a printed label, is called bleed. The printing of an image beyond the edge of a page so that there is no margin visible.

BLISTER PACKS, CONSUMER GOODS – Blister packs which consist of carded packaging in which goods such as toys, hardware, pens, batteries and electrical items are contained between a specially made printer card and a clear pre-formed plastic such as PVC. These types of blister packs allow the consumer to visually view and examine the product through the transparent plastic before purchase.

Depending on the item(s) being packed, the card will be designed and printed around the product it contains. The plastic shell is vaccum formed around a plastic mold so that it will contain the item snuggly. The mold is fixed to the card using both heat and pressure, which activates an adhesive or heat-seal coating on the pre-printed blister card. The mold/card bond needs to be strong enough to hand on a display, but weak enough so that the package can be easily opened.

BLISTER PACK, MEDICAL – Blister packs used in medical applications can be commonly found as unit-dose packaging for pharmaceutical tablets, capsules or lozenges. They provide barrier protection for shelf-life requirements, as well as a degree of tamper resistance. A series of blister cavities may also be called a blister card or blister strip as well as a blister pack.

Strip packs are formed around the tablets at the time they are dropped into the sealing area between sealing molds. The lidding foil is brittle and enables the products or tablets to be pushed out when the lidding foil is broken. The foils may be printed or narrow-web analogue or digital (inkjet) presses or left unprinted.

The main advantages of unit-dose blister packs over other methods of packing pharmaceutical products are being able to assure the product/packaging integrity (including shelf-life) of each individual dose and the possibility of creating a compliance pack or calendar pack by printing the days of the week above each dose. Blister packs are created by means of a form-fill-seal process at the pharmaceutical company or designated contract packer.

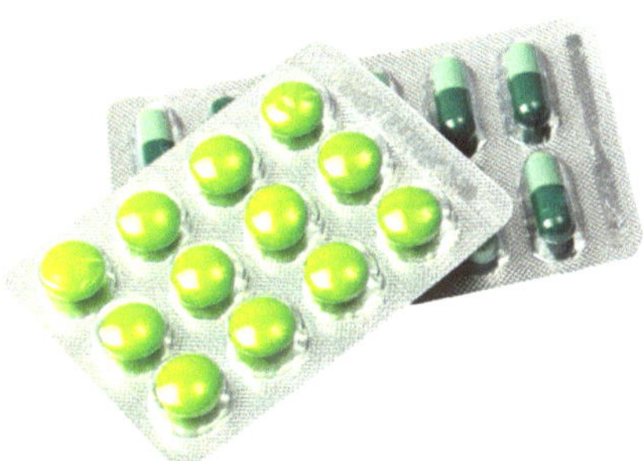

Blister pack. *Source: Tekni-Plex*

A form-fill-seal process means that the blister pack is created from rolls of flat sheet or film, filled with the pharmaceutical product and closed (sealed) on the same equipment. Such equipment is called a blisterline.

BLOCKING (ADHESIVE) – Caused by adhesion between adjacent layers of gummed, heatseal and other coated materials in both sheets and reels. This usually occurs when they are subjected to adverse pressure, temperature or humidity conditions. It is sometimes found with pre-siliconized release liners when the silicone coating is insufficiently cured prior to the web being wound into a roll.

BLOW MOLDING – A process used to produce hollow plastic bottles and containers. In blow molding a heated and softened plastic material is squeezed through a die so as to form a thin tube (a preform) within a hollow mold. Air, or occasionally nitrogen, is used to expand the preform so that the plastic is forced against the sides of a female mold cavity where it then solidifies

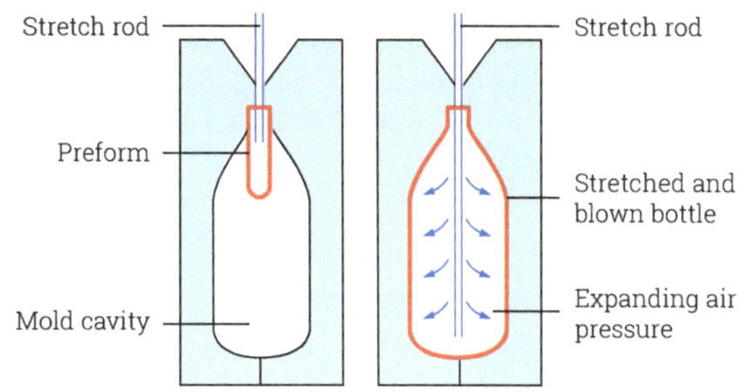

A thin pre-form tube is placed into a hollow mold and air then used to expand the heated plastic against the sides of the cavity to form the bottle.

to produce the end product.

There are three main different blow molding process methods in use: injection blow molding; extrusion blow molding; and stretch blow molding.

In blow molding applications where the label is inserted into the female mold cavity and held in place by vacuum during the forming process the label will become an integral part of the bottle wall. No further labeling is required. This is known as in-mold labeling (IML). *See also IML.*

BLOW UP – An enlargement of artwork or a photograph.

BLOWN FILM – Blown film is produced by the extrusion of plastic melt through a ring-shaped die. The

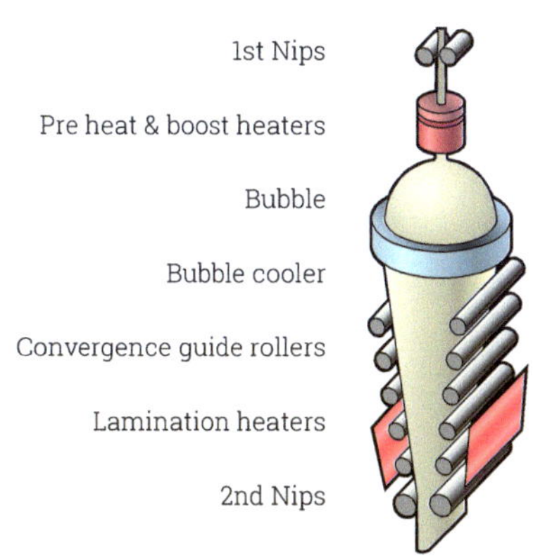

Principle of the bubble process used for blown film.

extruded tube is inflated with air to form a large bubble, which is subsequently collapsed and laid flat. Thickness tolerances are not as narrow as in the case of cast film. Blown film substrates and liners are used in the self-adhesive label industry.

BLUE ANGEL MARK – labeling system for recycled paper; this requires a minimum of 51 percent printed and domestic/office waste and also requires companies to report the energy used in the manufacturing process.

BODY – A term which refers to the viscosity, or consistency, of a printing ink. An ink with too much body is said to be stiff. Regarded more an ink-making term rather than a printing term.

BODY – The recessed area of a printing cylinder.

BOLD FACE – Name given to type that is heavier than the text type with which it is used. A heavy typeface, in contrast to a light typeface. Used for emphasis in captions, headings, sub-headings, etc.

BOND – To attach or fix materials together using adhesives.

BONDING STRENGTH – In paper, the force with which the fibers adhere to each other. In surface coatings, such as inks, varnishes and adhesives, the strength with which the dried coating adheres to the surface of the substrate. Also refers to the degree of adhesion of a pressure-sensitive face material to any surface.

BOOKLET LABELS – Developed to provide a means of adding additional text, diagrams, competitions or other material to labels. There are now a wide range of booklet labels, leaflet labels, fold-out labels, fold-in labels, concertina labels and other extended text multipage label formats. They are widely used in the agrochemical, DIY, pharmaceutical, chemical and promotional labeling sectors for information, instruction and marketing applications.

Typical uses for multiple page labels include the need to incorporate information covering:

- Instructions for use (including dilution or mixing tables) and correct handling
- Environmental, health or safety precautions
- Storage and disposal information
- Conditions of sale or limitations of liability/ warranty
- Multiple language labeling
- Dosage information
- Recipes, cooking guidelines and ingredients
- Other information required to meet legal or regulatory standards
- Promotions, competitions or coupons.
- The need for all-important information to stay with the product – not as separate information that might be lost.

Examples of extended text booklet/leaflet labels.

Extended text and multipage booklet/leaflet labels can be produced in a wide variety of formats and designs to suit virtually any container type, in any number of colors, with glossy or matt varnishing or transparent lamination to protect the information, with permanent fixture to the container or pack, with perforated or peel-off removable options, and with a re-seal capability if required. Apart from providing additional space for descriptive information, warning symbols and diagrams, multipage solutions can incorporate barcodes, holograms and tamper-evident options, or can be overprinted with batch numbers, sequential numbers, date or use-by information, manufacturer codes or machine-readable data.

Concertina and multipage leaflets and booklets are pre-printed – usually by sheet-fed offset – and folded before being combined with a self-adhesive labels supplied on a reel for use with standard label application equipment. A wide variety of permanent and removable pressure-sensitive adhesives are available. Maximum number of pages for booklet labels is usually 96 pages.

Some variations of extended text multipage labels

can be manufactured on special narrow-web presses or back-to-back presses that incorporate printing, silicone or adhesive application in patches, stripes or patterns, folding and finishing solutions.

BOPET – Biaxially orientated polyester. *See also Biaxially orientated.*

BOPP/BIOPP – Biaxially orientated polypropylene. *See also Biaxially orientated.*

BOTTLE DECORATION – The labeling of wine and spirits bottles, beer bottles, champagne bottles and liquors has been carried out in a traditional way using glass bottles for many years. Many of these products are recognised by the distinctive shape or size of the bottle, the specific label position or label shape, the positioning of the labels on the bottle and, with some products, by an additional neck foil.

In more recent years, there has been a proliferation of soft drinks, waters, carbonated beverages, fruit juices, sports and health drinks, which are now not only found in glass bottles, but also in a wide variety of shapes and sizes of plastics bottles. Bottles shapes have become more adventurous, as have the methods and techniques of labeling – which now not just encompass the traditional wet-glue applied label, but also self-adhesive labels, wrap-around film labels, full-body sleeving, and even newer solutions such as patch film or Roll-on Shrink-on (RoSo) labels. Neck collars and tamper-evident seals may also be used.

With such a wide variety of glass and plastic bottle types, label technologies, label materials, marketing demands and end-usage considerations, it has become more common to talk about bottle decoration or 'bottle dressing', rather than just labeling. The dressing of the bottle has a significant and direct effect on its sales success and should provide instant 'product recognition' at the point-of-sale.

The most common bottle label and/or dressing options today includes front and back body labels, shoulder labels, neck-around labels, champagne-bands, deep-cone wrap-around labels, medallions, wrap-around labels (paper or film), full body sleeve, tamper-evident strips, neck foils and tag labels. Generally, the names of the various label or dressing types are self-explanatory and based on the area of the bottle to which they are applied.

Examples showing different styles of bottle decoration.

Front body labels. The most important label on the bottle as it provides the main advertising or sales potential for the product, including the product name. The first label that consumers see on the shelves. Should be as large as possible to attract attention, identify the product and get the message across. May be rectangular, oval or shaped and in paper, metalized paper, paper/foil laminate, or film

Back body labels. May be of similar size to the front label and carries important additional information for the consumer in terms of sales promotion, contents, name and address, or other legal material. Also used to carry a barcode for supermarket check-out and stock control purposes. Printed on the same materials as the front body label.

Wrap-around body labels. Provide a 360 degree label that wraps-around the body of all types of cylindrical and non-cylindrical bottles. Combines the functions of the front and back body labels into one label. May be paper, metalized paper, paper/foil laminates or various plastics films.

Shoulder labels. Used on all types of glass or plastic bottles that have a truly conical shoulder/neck area. Printed in the same materials as the body labels.

Deep-cone wrap-around neck label. Commonly used for beers to provide all round brand recognition. The label may be from two- to five-times larger than conventional neck around labels.

Neck-band labels. An essential part of the labeling of champagne and sparkling wines.

Neck foiling. Used on a large scale for premium brand beers (in either gold or silver) to convey an impression of high quality, and for champagne bottles. Neck foils are totally malleable and can cover curved and shaped areas of bottle necks and the closure.

Full-body labels. Printed on heat-shrinkable film, formed into a tube, applied over the bottle and shrunk to fit to give complete 360° decoration of the whole bottle. Maximizes shelf promotion and image. Can also provide a tamper-evident shrink over the top of the bottle for additional security.

Tamper-evident labels. Used to ensure that products are 'tamper-evident' when displayed at the point-of-sale. There are wide range of tamper-evident solutions now used in the bottle sector, including strip-stamps or seals, tamper-evident closures, specially designed neck or shoulder labels which incorporate a strip, PVC shrink sleeve over the closure and neck, etc.

With the latest labeling lines being modular systems it is possible for manufacturers to incorporate flexibility of bottle decoration into their bottle filling and labeling plants with equipment that can label with wet-glue, self-adhesive, wrap-around or sleeves, all on the one machine.

The decoration of bottles can also be undertaken with direct printing or through the use of heat transfer decoration technology.

BPIF LABELS – The history of BPIF Labels began back in 1932 when the British Seal Printers Association (BSPA) was formed. This early organisation was fairly informal with activities being limited by the amount of time which the Hon. Secretary (one of the members) had available. It was re-launched (for the second time) in 2010 and has been steadily building awareness of BPIF Labels and organising events for the members and at the same time promoting the label industry in the UK.

BRAILLE LABELS – Raised image tactile warning labels to alert blind or partially sighted people to the potential dangers of the contents of drugs, medicines, hazardous chemicals, aerosols, etc, and distinguish them from harmless products. Tactile labels for the blind have also been developed and used for wines and spirits (providing information on alcoholic strength), groceries and other consumer products, while electronic embossing machines are available to create Braille labels for the home or business environment.

Some countries have also introduced legislation (e.g. BS EN ISO 11683: 1997) that states that all products sold to the public which are labeled 'very toxic', 'corrosive', 'toxic, 'harmful' or 'extremely flammable' must by law have a tactile danger warning label to alert the blind or partially sighted that they are handling a dangerous product. Such products also include any aerosol propelled by butane. These warning labels consist of a raised equilateral warning triangle or 3 raised dots forming the points of an equilateral triangle. The labels may be either incorporated onto an existing label design or applied as a separate label.

New EU requirements for Braille labels became effective from October 2010. These demand a uniform presentation of Braille alphabetical characters, including conformity to a standard height. They apply to small folded cartons, labels, leaflets, extended text labels and booklets for pharmaceutical products. In North America the voluntary Can-Am Braille set of guidelines and recommendations resemble those used in Europe,

For mass production of tactile warning labels on consumer products the label industry has developed a number of solutions. Rotary screen printing units incorporated into a roll-label press can print deposits of up to 300mm in thickness to provide the required tactile effects; others are using a tactile varnish to print dots of precise dimensions (125 microns high and less than a millimeter in diameter).

Speciality paper consisting of millions of micro-capsules (containing alcohol) in a paint-like formula that can be imaged by photocopier or special marker pens and then passed through a fusing heater to produce Braille labels, raised image graphs, charts, symbols, pictures, etc, has also been developed. Another technique developed involves a two-step process in which an image is first printed with a liquid ink; a polymer is then applied to the liquid ink and the printed image raised in a processing unit.

Table-top electronic Braille labelers are available to produce professional-looking Braille labels using a display screen and keyboard and transparent single line self-adhesive labels that will stick to most surfaces. They can be used for labeling individual cassettes, CD cases, vending machines, telephones, etc.

BRAZILIAN PRESSURE SENSITIVE LABEL

CONVERTERS ASSOCIATION – *See under ABIEA.*

BRC/IOP TECHNICAL STANDARD – A technical standard established by the British Retail Consortium (BRC) and the Institute of Packaging (IOP) which provides a common basis for accreditation for the audit of companies supplying food packaging and labeling materials for retailer branded products.

The Technical Standard assists retailers and food manufacturers to fulfil their legal obligations and to demonstrate 'due diligence'.

BREAK – A term used to denote a tear in a roll of face material or release liner. Such defects are generally spliced together and marked by a protruding 'flag' or stamped with a rubber stamp on the side of the roll.

BREATHABLE FILM LABELS – Labels that react to increasing rates of oxygen consumption by product respiration and selectively permit passes of oxygen and carbon dioxide through the pack. *See also Biomonitoring systems.*

BRIGHTNESS/WHITENESS – The strength of white or color of a substrate; the reflectivity of a substrate for blue light measured under standardized conditions on a measuring instrument designed and calibrated specifically for the purpose. Brightness/whiteness increases with bleached and coated papers and decreases with greater amounts of recycled fiber. Can be improved by incorporating optical brighteners during the papermaking process.-

Brightness/whiteness of polymer films is determined by the opacity of white film. Clear films require the use of a white printing plate.

BRITISH SHORT RUN LABEL ASSOCIATION (BSRLA) – *See under BSRLA.*

BSRLA – The British Short Run Label Association was inaugurated in 2004 in response to continual pressure on small/medium sized label printers who operated in a very competitive UK market.

The current membership stands at 67 (July 2013). Membership is only open to UK and Irish companies which provide a short run labeling service as a major part of their business and have a turnover of less than £10 million. Suppliers may not become members. Members operate a closed, private discussion group.

BRITTLENESS – The property of a material that causes it to break or fracture when deformed, e.g. through bending. Some 100% solids silicones become fairly brittle when cured, If their coating thickness is excessive, cohesive or interfacial fracture under stress can cause failure of adhesion to film substrate.

BRONZING – A method of creating a metalic image (usually gold) on a label by printing a special adhesive or gold size on to the sheet and then, while the image is still wet, applying a metalic (bronze) powder to the surface, cleaning away the excess powder and burnishing the powdered image to give the appearance of gold. Mainly used for high-added value sheet-printed glue-applied labels for liquors and cosmetics and produced on a single-color press with an added in-line bronzing machine. The process is relatively slow and expensive. A metalic image can also be produced by printing with a metalic ink or by hot or cold foil printing.

BSI – The world's international standards, testing, registration and certification organization. Founded in 1901, BSI has 65 offices worldwide and serves over 70,000 clients in 150 countries. It was the world's first national standards body, and a founding member of the International Organization for Standardization (ISO), and also facilitated and published the first commercial standards to address quality management systems, occupational health and safety management systems, environmental management systems, and project management systems, many of which are used by the global label industry. *See also ISO.*

BS 8900 – British Standard 8900 is designed to help organizations to develop an approach to sustainable development that will continue to evolve and adapt to meet new challenges and demands

BSI MANAGEMENT SYSTEMS – The North and South American hub of BSI. Established in 1991 to meet the growing demand in the North American market. It offers a wide assortment of management registration and assessment services covering virtually every industrial and commercial sector. *See BSI.*

BUBBLE JET PRINTING – A drop-on-demand non-impact ink jet printing process branded by Canon, which uses hot droplets of a liquid ink, are blown onto paper (or film) in the shape of the font character or image as required. The process is generally regarded as a variation of ink-jet printing. *See also under Ink jet printing, Impact printers, Non-impact printing and Laser printing.*

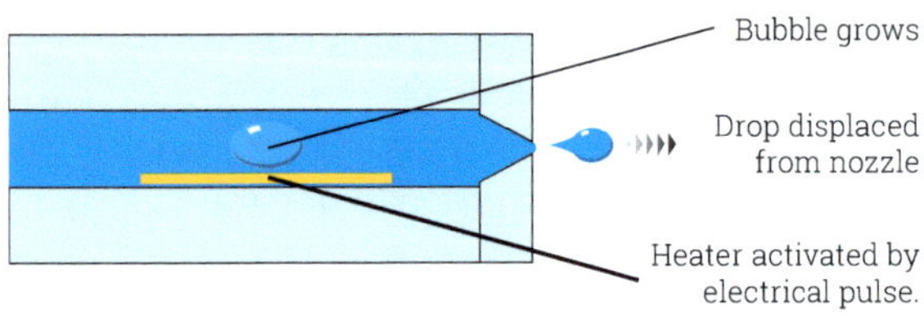

How the bubble-jet works

BULK ACTIVATOR (DE-ACTIVATOR) – Equipment for collectively activating or de-activating simultaneously multiple electronic article surveillance (EAS) tags or source tagged products/packages, such as a loaded pallet or transport packaging containing multiple retail packs.

Such equipment can be installed at various points within the supply chain, i.e. at the product/pack manufacturing stage, in the distribution chain and/or at the retailer, according to optimum logistics requirements. Typically this type of equipment is designed to be mounted readily on to existing conveyor systems.

BURSTING STRENGTH – The strength of paper or plastic film when subjected to multi-axial stress. The film or paper to be tested is stretched over a flexible rubber mat, which can be inflated. The bursting strength is the measured pressure at which the test specimen tears.

BUSINESS FORM – A generic term relating to documents which are printed with fixed data, text and boxes and then overprinted by typewriter or computer printer so as to enable a business to function and issue invoices, statements, purchase orders, etc. May be sequentially numbered, made up of several carbonless copy parts and can sometimes incorporate one or more gummed or pressure-sensitive labels. *See also Multi-part sets and Form/label combinations.*

BUTT CUT LABELS – Rectangular labels in a continuous web, which are separated by a single knife cut through the label and/or liner across the web. No matrix is removed between the labels, as is the case with die-cut labels. Separated individual labels on a backing liner are applied by hand.

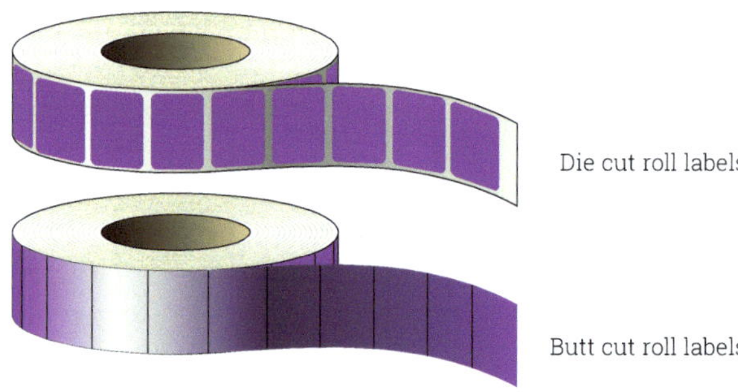

Die and butt cut roll labels

BUTT SPLICE – The end-to-end joining of two similar materials without overlapping so as to give continuity of a reel of label stock for printing or of a roll of labels for subsequent automatic application.

The butt splice process employs a knife that moves across the web to make the cut. The two ends are joined by a single-sided splicing tape applied to one or both sides of the web. *See also Flying splice.*

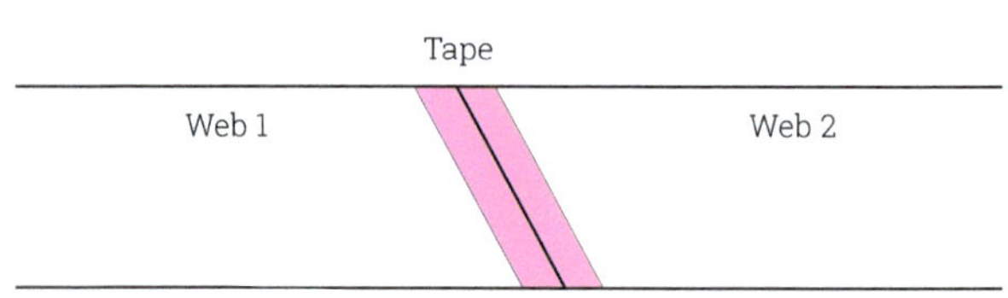

Butt splice showing the end-to-end join of two similar materials with tape.

BY-PASS SYSTEM – An arrangement whereby electronic article surveillance (EAS) detector gates at the retail store checkouts are by-passed when merchandise bearing live tags which cannot be de-activated are presented for purchase.

BYTE – Computer terminology. A unit of digital information, equivalent to one character. 8 bits = 1 byte.

C

C1S PAPER – Abbreviation used for coated one-side paper.

CAD/CAM/CAE – Computer Assisted Design/ Computer Assisted Make-up (or Manufacturing) and Computer Assisted Engineering. CAD systems are specialized workstations or high-performance personal computers that employ CAD software packages and input devices, such as graphic tablets and scanners, so as to create sophisticated designs, drawings or plans.

CAKING – The collection of dried ink on printing rollers or plates.

CALIBRATION – The setting or pre-setting of equipment to a standard measure in order to produce reliable and repeatable results. Establishing a datum or reference value and scaling for a device or process.

CALIBRATION BARS – A strip of tones on a negative, proof or print used to check print quality.

CALIPER – Caliper or thickness of a paper or film, measured by a dead weight dial micrometer, usually expressed in thousandths of an inch (mils or thou) in the USA, or in one millionth of a meter (microns) in countries using the metric system. May also be an instrument to measure thickness.

CAMERA READY/CAMERA-READY ARTWORK – Copy or finished artwork which is ready, without further preparation, for photography. *See also Artwork.*

CAPS – Abbreviation commonly used for capital letters in typesetting or on a printed page or label.

CARBON TRUST – a non-profit UK consultancy that works with national government and consumers to educate on the impacts of product usage and carbon emissions.

CARBONZERO – Also known as zero carbon footprint or carbon neutrality, CarbonZero refers to balancing the amount of carbon dioxide emissions with offsets or credits in areas such as transportation, energy production/consumption and industrial.

CARD – A paperboard material used for business cards and other applications where a stiffer, thicker, material than paper is required.

A plastic board with chips (small circuits which run a computer) attached. A printed circuit board.

CARE LABELS – Garment care labels are an essential element in telling consumers how to maximize the life of the apparel they purchase. Care information is provided with a garment so that the consumer knows whether the garment should be machine washed or hand washed, bleached, tumble dried, ironed, professionally laundered or dry cleaned in order to keep it in good condition and prolong its life. Indeed, research studies have shown that more than 60 per cent of consumers indicate that care label information is important in their decision to purchase an item of clothing, with the majority checking the garment care label before purchasing the item of clothing.

Care labels, showing the required text and/or symbols, are required on most textile clothing, must be attached firmly to the garment, be easy to find and be readable throughout the life of the garment. In addition to providing information on safe cleaning, care labels must list any necessary warnings about the cleaning method, i.e. the label must say whether any step in the care process – bleaching, washing drying, ironing, or dry cleaning – could harm the garment, or any other items that may be cleaned with it.

Care label guidelines and symbols have been developed and published in Europe by GINETEX, the body that regulates the use of garment care symbols throughout Europe – and accepted by a majority of countries of the world as an international care labeling code – and by the Federal Trade Commission's Care Labeling Rule for apparel sold in Mexico, Canada and the United States. Although the FTC care symbol system is not identical to that used in Europe, it is very similar and differs primarily in its comprehensiveness, as well as being a more user-friendly approach.

Symbols used in the GINETEX care labeling system (also subject to an international standard ISO 3758

A washing symbol for home laundry usage (although the industrial launderer can use it as additional advice to supplement his own experience). The washing symbol may also include temperature information indicated by a number inside the symbol, a hand wash symbol, and a bar under the symbol to indicate a very mild or gentle wash cycle.

A chlorine bleach symbol, with Cl inside if bleaching is allowed in a cold or dilute solution, or a cross if chlorine bleach in not permitted

An ironing symbol, with 1, 2 or 3 dots placed within the symbol to indicate maximum temperatures in degrees Celsius for ironing textile articles.

A dry-cleaning symbol (which is a little different to the American type) which incorporates the letters A, P or F to refer to the solvents that may be used by the dry cleaners. An underline indicates that a specific process should be used.
A tumble drying symbol, completed with one or two dots for different temperature levels.

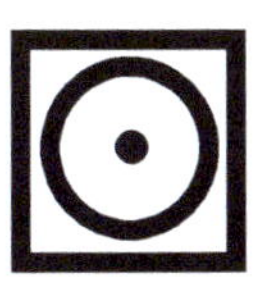

A tumble drying symbol, completed with one or two dots for different temperature levels.

'Textiles – care labeling code using symbols') – and protected by an international trademark – include:

The appropriate GINETEX, ISO or FTC guidelines should be followed and understood when producing or using care labels. *See also Garment labels, GINETEX.*

CARRIER – Sometimes used to refer to the siliconized liner material or liner layer of pressure-sensitive labels.

CAST COATED – A high gloss coated label stock on which the clay coating has been dried in contact with a highly polished chromium casting drum or drying cylinder to give an almost mirror smooth surface.

CAST FILM – Plastic sheeting or film manufactured by the casting process, as opposed to extruded or blown film. In the production of cast film the melt is extruded directly onto rolls which control its cooling and maintain its flatness. A cast film may consist of several different layers, each one having different properties. Thickness tolerances are extremely narrow. Cast film displays greater dimensional stability than blown film, for example.

CAST VINYL – Vinyl sheeting or film manufactured by coating a liquid vinyl acetate or similar ester onto a casting paper and curing in a heated oven.

CATALYST – A substance which has the capability of initiating or accelerating the speed of reaction – such as polymerisation or crosslinking – between two or more substances when introduced into their presence. An example would be the chemical (a photo-initiator) that is in Ultra Violet (UV) cured inks to make them react to exposure to UV-light and so cause a catalytic reaction and the ink to 'cure' or set.

In the manufacture of pressure-sensitive labelstocks catalysts of organo-tin are used with condensation cure silicone release coatings. Addition cure silicone release coatings, based on complexed platinum or rhodium catalysts, are also used.

CARTONS, DIGITAL – Carton manufacturers looking to produce fast turnaround, high-quality and streamlined multi-SKU production on-demand and in short runs now have the option to print digitally using the latest larger format digital presses from suppliers such as HP Indigo, Xeikon, Screen Europe and, in the pipeline, Landa nanographic press technology.

Using digital printing undoubtedly enables carton producers to manage their brand customers better – improving sell-through with targeted design, while lowering capital expenditure, time to market and obsolescence. At the same time, brand owners using digital printing for their packaging work gain the ability to affordably launch new products and gain better visibility on the shelves.

In addition, brand owners have taken to reducing

Digitally printed folding cartons. *Source: Xeikon*

their stockholding and introducing just-in-time inventories – all leading to label and package printers having to accommodate increasingly shorter runs and faster turnaround times.

In the folding carton market – a market dominated by offset – such trends mean more frequent make-readies of both presses and finishing equipment, all leading to a cumbersome, time-consuming and expensive process.

Digital technology, however, can make the folding carton supply chain more efficient and cost-effective. With digital technology, every print can be different. Because set-up and make-ready costs are minimal, small runs can be produced cost-effectively. Digital printing is extremely flexible in that it allows last-minute changes and updates to be made in short turnaround times.

In short, digital technology is a great complement to offset printing, and can also make for a profitable standalone business. That is, if the technology meets all the requirements of the folding carton market in terms of print quality, color accuracy and consistency, lightfastness, food safety, and brand protection.

While the market for personalized packaging whether for folding cartons or flexibles is undoubtedly quite small at the moment, it seems certain that it will rapidly grow to become a much bigger market in the future.

Complementing the latest digital carton printing capabilities are a number of companies that have introduced units that convert short-run cartons and boxes using semi-rotary cutting and creasing. This semi-rotary solution means that the converter doesn't need to change cylinder sizes.

Compatible folding carton boards for the new generations of digital carton presses are already becoming more widely available from suppliers such as M-real, Iggesund, Stora Enso, International Paper, Sappi and Tullis Russell.

Illustration courtesy of Delta Industrial Services shows carton finishing from a narrow-web digital press

Put together, carton producers – or label converters investing in wider format digital can now offer their brand-owner customers a complete web-fed in-line or off-line carton production capability using all the benefits and new added-value opportunities provided by digital printing.

CAVITY – Any enclosed shapes on a rotary/flexible die.

CCC MARK – The CCC Mark (China Compulsory Certification Mark) was introduced in 2003 as a means of marking a wide range of manufactured products before being exported to or sold in the Peoples Republic of China market. It integrates two compulsory inspection systems, one to check contents of products for import or export and the other for quality control, into a single procedure as part of the country's commitment for entry into the World Trade Organization. Since May 2003, designated products without the CCC Mark can be held at the border by Chinese Customs and subjected to other penalties.

The CCC Mark is required for products in 19 groups divided into a total of 132 product categories. For those wishing to know whether their products fall

The CCC Mark

within the scope of the Mark details of the various groups and categories can be located on the www.ccc-mark.com web site.

CCD – Charge Coupled Device; a photon-to-electron charge converter usually arranged in single or two-dimensional arrays that emits an electrical signal proportional to the amount of light striking it. The CCD light sensor commonly replaces film in digital cameras. Used in some scanners and consumer video cameras.

CELL – A small engraved or etched depression in an anilox roller that carries the ink to the printing plate.

CELL ANGLE – On an anilox roll, the measure of a line drawn through the centre of a row of cells and a line drawn parallel to the shaft of the anilox. The angle at which cells are arranged on an anilox roll. Typical anilox cell angles are 30°, 45° and 60°. A 60° angle allows for more complete ink transfer, and is the

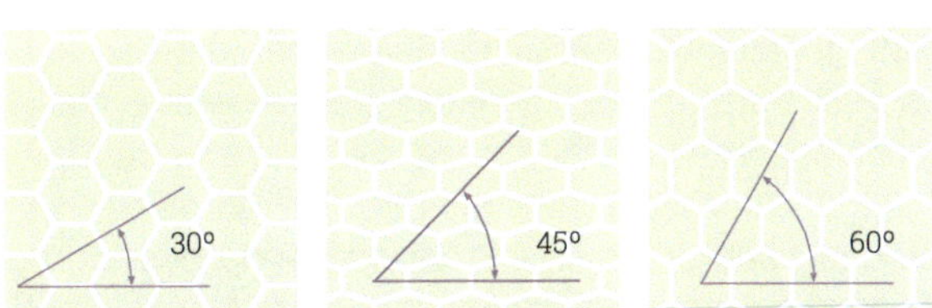

Anilox roll cells are engraved at one of three angles: 30°, 45° or 60°.

preferred cell angle. Note that the screen angle of the printing plate and the anilox cell angle can combine to cause a moiré pattern, even with one-color halftones. Moiré is avoided by angling separation screens.

CELLS PER INCH (CPI) – On an anilox roll, the number of cells in a linear inch, or cells per centimeter (CPC). As cell count increases, ink delivered to the plate decreases. As line screen resolution increases, CPI should also increase. Range is usually from 140 to 1,200 CPI (55 to 472 cells per cm).

Examples of different cell counts on an anilox roll.

CELL VOLUME – A measure of the capacity to carry ink of a square inch or centimeter of anilox roll. As CPI (or CPC) increases, cell volume decreases. Cell volumes can be calculated as theoretical volumes, or measured scientifically using an instrument. Cell depth is generally measured in microns and, when related to the cell's opening, produces a depth-to-opening ratio; the most efficient openings producing a depth-to-opening ratio between 20 and 35 per cent.

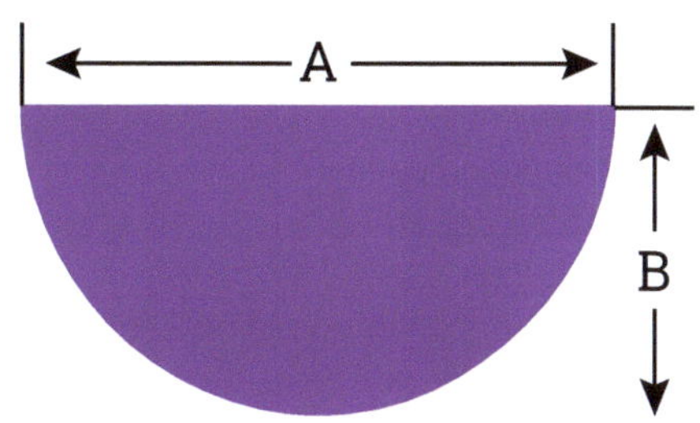

Calculation of cell volume

$$\frac{\text{Cell depth (B)}}{\text{Cell opening (A)}} \times 100\%$$

Calculation of cell volume.

The depth and shape of the cells is a key factor in the efficient delivery of a uniform ink film to the printing plate. Cell depth ratios are critical in choosing the right anilox roller for a particular application. *See also Anilox roller.*

CELLOPHANE – Transparent cellulose film material that is manufactured by chemically regenerating a dissolved cellulosic compound into a continuous thin web. Uncoated cellophane has only a few packaging uses, such as decorative wraps. It does have some industrial applications, such as release sheets, pressure-sensitive tape base, and as substrate for roll-leaf coatings. Specific coatings for moisture proofing, heat-seal ability, etc, give cellophane a range of packaging functions. Cellophane may also be metalized or colored.

CELLULOSE – Fibrous substance of wood, cotton or other vegetable matter used as components in the manufacture of paper or cellulose film.

CELSIUS – The metric scale of temperature where '0' represents the freezing point of water and '100' represents the boiling point of water at sea level. Also commonly referred to as Centigrade. To convert from Celsius to Fahrenheit multiply the Celsius figure by 9, divide by 5 and then add 32, i.e. 130°C = 266°F. *See also Faherenheit.*

As both methods of temperature measurement are used within sectors/countries of the global label industry a quick comparison table showing some of the more commonly used equivalent temperature measurements can be found under 'Converting metric to imperial'.

CEN – European Committee for Standardization (CEN). CEN had some existing packaging and environment standards that ISO 14021 packaging standards is loosely based on, as well as the EU Packaging and Packaging Waste Directive

CE MARKING – CE Marking, sometimes called the 'CE Mark', is a mandatory marking requirement for something like 70 per cent of products sold on the European Union (EU) market. CE Marking is the manufacturer's claim that proves to the buyer – or user – that the product meets the essential safety and environmental requirements of all relevant European Directives.

EU legislation makes EU importers liable for the products they import, including the machinery they provide to their employees for work. Non EU exporters may find that a prospective EU importer may not risk importing non-conforming products (those without CE Marking) which, in case of an accident or health problem, may generate legal action against them.

In terms of labeling equipment or label printing machinery, this CE Marking relates, in relatively simple terms, to: an assembly of linked parts or components, at least one of which moves, with any appropriate control and power circuits, etc, joined together for a specific application; to an assembly of machines which are arranged and controlled so that they function as an integral whole; and to interchangeable equipment modifying the function of a machine and which is placed on the market for the purpose of being assembled with a machine or series of machines.

CE

The CE Mark

The CE Mark is sufficient to allow exporting to all EC member countries, although there may be a requirement to translate some part of the operating and/or safety instructions into the language of the country the machinery is exported to. The CE Mark must be adhered to the equipment and its size must be 5mm (0.2") or higher.

CE Marking requirements may vary from Directive to Directive, and even within Directives. Third party testing, systems assessment and technical file assessments can be mandatory, but sometimes the manufacturer's unverified claim is all that may be asked for. In cases where the Directive requires products and/or systems to be independently tested, certified or inspected, this must be done by a 'Notified Body' or 'Competent Body', an organization that has been notified to the European Commission by a Member State.

Manufacturers may achieve CE Markings in different ways, although one often preferred by EU importers is to designate an Authorized Representative in the EU who will take care of the CE Marking issues.

CENTIGRADE – *See Celsius.*

CENTIPOISE – One hundred of a poise; a unit for measuring viscosity.

CENTRAL IMPRESSION/COMMON IMPRESSION (CI) PRESS – A roll-label printing machine with a number of printing units placed around a large cylinder, which serves as the impression cylinder against which the label substrate rides. May be four to eight colors. Limited to one side printing, but offers hairline register at high speeds on stretchable materials. *See also Common impressions.*

CEPS – Color Electronic Prepress System. A high-end computer-based system that is used to color correct scanner images and assemble image elements. CEPS are device dependent systems.

CERAMIC ANILOX ROLLER – Engraved ceramic roller used to meter ink in the flexographic printing system. *See Anilox.*

CERAMIC BAR – A small ceramic cylinder. Flexible cutting dies for envelope machines cut against the ceramic bar rather than an anvil.

CGM – Computer Graphics Metafile. International file format for graphic images.

CHALKING – A form of coating deterioration or dusting which is caused because the surface coating or printing ink was insufficiently bound to the base paper during manufacture or printing. Can be seen or will rub off as a chalk-like powder on or from the surface.

CHAMBERED DOCTOR BLADE – *See Dual doctor ink chamber.*

CHARACTER – A letter, typeface, figure, digit, group of bars or spaces that is used as part of the organization, control or representation of data.

CHARACTER BITMAP – Describes the arrangement of black and white dots that make up the printed image. *See Bit-mapped fonts.*

CHARACTER SET – The repertoire of characters that can be displayed, printed or understood by a computer system, or that makes up a font.

CHARACTERS PER INCH – The number and kind of characters that fit horizontally into one inch of printed data on a line. The number of characters that will fit in an inch is related to the size of the font – the smaller the font the more characters per inch.

CHARGE – Usually refers to the degree or type of electrical property carried by a substrate. *See also Static electricity.*

CHECK CHARACTER/CHECK DIGIT – A character or digit included within a barcode symbol in order to perform a mathematical check to ensure that a barcode is correct and accurately read. Its value is calculated from the other characters in the code. *See Barcode symbol.*

CHEMICAL/AGRO CHEMICAL LABEL PRINTING – Often only printed in 2 – 4 colors, with most label printing processes used – including screen. Frequently a need for performance from inks and varnishes, including three months immersion in sea water. Chemical resistance required, as well as long life and durability and, maybe, outdoor or harsh storage conditions. Must meet chemical warning label requirements, including remaining legible over the period of the label storage and usage. Both paper and film substrates used.

End-user printing and ink requirements include: printing on paper or film, a need to understand legislative demands, both long and short-run applications, some special colors and matches, and a need for performance. May be a need for variable image overprinting. *See also Chemical warning labels.*

CHEMICAL CURING – The setting or curing of an adhesive, coating or sealer brought about by the addition of a catalyst, accelerator or photo-initiator.

CHEMICAL RESISTANCE – The resistance of a pressure-sensitive label to the deteriorating effects resulting from exposure to chemicals under specified conditions.

CHEMICALLY-RESISTANT INKS – Inks used in label applications where a resistance to acids, alkalis, solvents, oil, grease, detergents, etc, is required. Labels for motor oils require resistance to oils and oil additives; wines and spirits a resistance to alcohol;

fruit juices to acids; automotive engine compartment labels to grease and detergents.

The ink maker should be advised if printed inks are likely to come into contact with chemical substances that may be affected by container or bottle contents or in areas of potential contamination.

CHEMICAL WARNING LABELS – Requirements for the labeling of chemicals are laid down by international, national or federal authorities around the world. The requirements are for each container of hazardous chemicals to be labeled, tagged or marked with the identity of the hazardous chemical that it contains. In addition, there are appropriate warnings with words, pictures and symbols – or a combination of these – which provide at least general information regarding the hazards of the chemical.

The most important information on chemical warning labels is a single word describing the level of hazard. Danger means that it is the most hazardous kind of chemical; Warning is less hazardous, and Caution is the least hazardous. However, chemicals with the Caution label may still be harmful if not handled properly or follow the proper procedures

Basic information that is normally required on chemical warning labels from around the world includes:

- The chemical's name
- Whether or not it is flammable
- Whether it has dangerous fumes
- The name, address and telephone number of the manufacturer or distributor
- A list of the chemical's ingredients
- Target organs which could be affected
- The most common types of symptoms that may result from improper use (breathing problems, skin burns, vomiting)
- How reactive the chemical is (when it catches fire, explodes or becomes a dangerous gas)
- What type of fire extinguishers should be used in case of fire
- Storage, disposal, handling and usage procedures (protective clothing, gloves, etc)
- Critical first aid instructions
- How to handle spills and leaks
- Protective equipment to be worn when using the chemical

It is important to follow the current relevant legislation, standards and guidelines when labeling chemicals, producing warning labels or when using chemicals. Other elements of chemical warning programs are Material Safety Data Sheets and training in the safe usage and handling of chemicals.

Examples of chemical and hazard warning label symbols from around the world.

CHESHIRE – The name of a specific machine designed to affix labels onto pages and/or envelopes, which has become a generic name for this type of labeling equipment.

CHILL ROLL – Metal roller or drum that is cooled internally with water, etc. Often used after the press dryer to cool the printed web prior to die-cutting, rewinding, etc. May be required where temperature-sensitive films are being coated or printed.

CHIP – Tiny piece of silicon with an integrated circuit embedded in it. This includes microprocessors, which make a computer work and memory chips, which hold the information until it is sent to disk.

CHIP CARD – Cards with one or more microchips (integrated circuits) in them. *See also Memory card and Smart cards.*

CHIP/CHIPLESS TAGS – Low-cost RFID tags can be split into two groups: those that contain a chip, and those that do not. Radical differences in cost and performance exist between these two categories, although they may overlap in some applications. In some applications chipless tags are much more suitable than chip tags.

Chip tags will cost more and can do more. They hold larger amounts of data but, due to the price of the silicon and the processes involved in making the chips, they are relatively expensive.

Chipless tags and chipless smart labels are much cheaper because their price is only limited by the price of the materials used, which are in most cases basic and readily available in large quantities. Chipless tags therefore have a price advantage over chip tags, but currently cannot hold as much data without becoming unacceptably large.

CHOKE – A specific adjustment or distortion of an image whereby the perimeter, in total or in part, is slightly pulled in (choked) towards the centre. Used when a darker element appears against a lighter background. The image area remains essentially the same except for a narrow strip or reduction around its perimeter. Choking of an element may be performed in a number of ways: manually, photomechanically or digitally, with various computer programs.

Choking of an element is normally used in conjunction with the spreading (*see Spreads*) of a neighboring element to ensure that color registration standards are achieved and that there are no color fringes or white borders around image detail due to slight misregister – from tension problems on the press, irregular plate elongation, inaccuracies in plate mounting, plate bounce, etc – during printing. Both choke and spread create a small overlap of ink colors, called the trap width. *See also Trap and Trapping.*

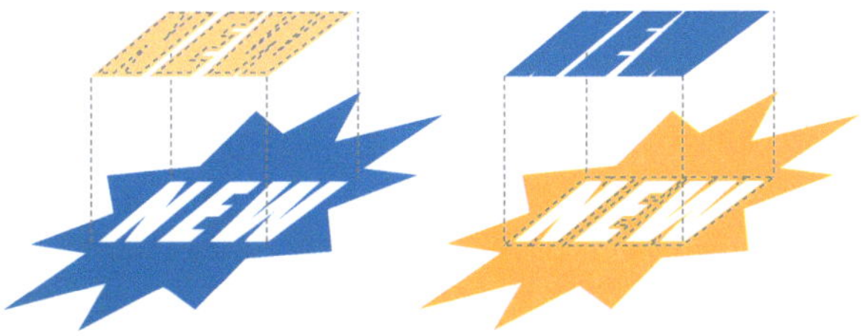

Choking and spreading of light and dark colors against adjoining colours

CIP4 – CIP4 is an international, worldwide operating standards body located in Switzerland. The purpose of the association is to encourage computer-based integration of all processes in the graphic arts industry, in particular the specification of standards.

The International Cooperation for the Integration of Process in Prepress, Press and Postpress (CIP4) is the successor to CIP3, which started in 1995 as a joint initiative of vendors for the graphical arts industry. Since then, CIP3 has developed the Print Production Format. CIP4 is about to develop and promote vendor independent standards for the graphic arts industry, such as the new Job Definition Format (JDF). *See under JDF.*

CIRCULAR PITCH, P – Distance from one face of a tooth to the corresponding face of an adjacent tooth on the same gear, measured along the pitch circle.

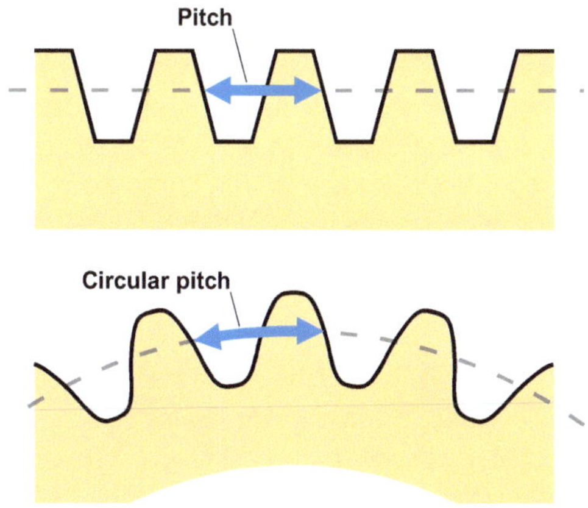

CLAMSHELL BLISTER PACKS – Blister packs with a hinged blister are known as clamshell. They are used to provide security and anti-pilferage for small high-value products, such as consumer electronics. They consist of one sheet folded over onto itself and then heatsealed, so making them difficult to open by hand.

CLAY COATED – A paper that has been coated with a clay coating on either one or both sides to produce a high quality smooth surface.

CLEAR AREA – An area that precedes the start character and follows the stop character of a barcode symbol. Also known as the Quiet Area or Zone. It must contain no machine-readable markings or extraneous spots, etc, and is quiet in terms of scanning signals.

CLEAR COAT – A surface coating that protects both the printing and the surface of a pressure-sensitive label from abrasion, sunlight, chemicals, moisture, or a combination of these.

CLIP ART – Stock line drawings, diagrams, fonts, photographs or images that are not restricted by copyright.

CLOSED-LOOP MANUFACTURING – Closed-loop manufacturing or cradle-to-cradle manufacturing minimizes a product's environmental impact through a recycling process at the disposal stage to create identical products with the used material. It employs sustainable production, operation and disposal methods, and relies heavily on the involvement of multiple value chain stakeholders.

CLOUD COMPUTING – Data storage systems which allow large amounts of data to be accessed remotely from any location on a range of devices. The Cloud also offers a platform to run applications without having the software physically installed on the user's equipment (whether this is a mainframe computer, a PC or a mobile device).

CMYK – Abbreviation for the cyan, magenta, yellow and black values, which are needed for color reproduction by process color printing. CMY + K are the primary colors of subtractive color reproduction systems. The K comes from the traditional printing description of the black separation as the 'Key', which is added to enhance color and contrast.

CNC – Computer Numeric Control.

CNC MILLING – Multi-axis machining is a manufacturing process where CNC controlled tools move in 4 or more ways to manufacture parts out of metal or other materials by milling away excess material. Multi-axis machines also support rotation around one or multiple axes.

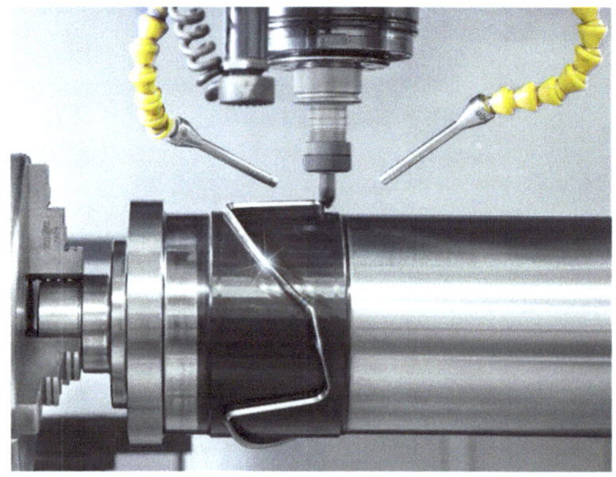

Picture shows CNC milling. *Source: Spilker.*

The number of axes for multi-axis machines varies from 4 to 9. Each axis of movement is implemented either by moving the table (into which the work piece is attached), or by moving the tool.

COATING/COATED – A paper which has received a coating of clay on one or both sides so as to give a smoother texture and a brighter finish, which may be dull (matt) or glossy depending how they are finished. It also makes them more receptive to printing inks. Coating amounts are controlled – either on the papermaking machine or off-line – by rollers, a blade or brush, or by an air knife. May also refer to polyethylene or polypropylene coated papers, which are produced by extrusion coating.

Coating also refers to the light-sensitive polymer applied to the surface of a metal printing plate, and to varnishes or lacquers applied over printed surfaces to enhance the image or to provide greater surface protection.

COATING WEIGHT – The weight of coating, e.g. adhesive or silicone, per unit area of a substrate. This is preferably expressed as grams per square meter.

COATING WIDTH – The width at which an adhesive or release coating is applied to a substrate.

COCKLE/COCKLING – A defect showing as wrinkling or puckering of a web or sheet of paper caused by unequal shrinkage, by environmental conditions or during manufacturing.

CODABAR – A barcode symbology of 16 encoded data characters; 0 – 9, 6 special characters and 4 unique start/stop characters. Each character has 7 elements (4 bars and 3 spaces).

Codabar symbol

Originally designed for use in retail price labeling, it was one of the original proposed symbologies for the UPC. However, it found a niche in diverse non-retail applications. It is the code required by the US Department of Defense.

Other applications include inventory control, libraries, blood banks, air bill applications and photo finishing.

CODES – Codes used in the label and related industries can be found in a wide variety of types and formats and are used in applications as diverse as warehousing, storage, shipping, asset tracking, security printing, cheque reading, point-of-sale, authenticity, inventory control, product and parts identification, stock taking and other areas. Most common of the code systems used in the label industry are barcodes in alphabetic, numeric or alphanumeric formats. *See Barcode and Barcode symbol.*

Other codes used by the label, forms and document markets include, OCR, MICR and OMR codes, optical codes, magnetic encoding, multi-dimensional codes or other systems used for control.

CODE 1 – A special code which can encode ASCII data, error correction data, function characters or binary encoded data. The symbol is based on a finder pattern of horizontal and vertical bars crossing the center of the symbol. Applications for Code 1 include healthcare and pharmaceutical labels and containers within the recycling industry.

CODE 27 – Code 27 (2 of 7) is a numeric only barcode symbology, which consists of seven modules, of which two are wide.

CODE 39 – Code 39 (3 of 9) is an adequate secure barcode symbology, which encodes 43 data characters. Its major advantage is that both alphabetic and numeric characters can be encoded; 0 – 9, a – z, 7 special characters and stop/start character. Any length of data can be encoded.

Code 39 symbol

Code 39 is generally the preferred symbology when alphabetic and numeric characters are required. It is the most widely used code within commerce and industry. A limitation is that the code is not especially compact, which can lead to practical dimensional constraints on small items, labels or packaging.

CODE 49 – The first truly two-dimensional barcode to be developed, responding to strong demand to condense a high volume of coded data into a symbol with a small surface area. The code is built up from a series of one-dimensional barcodes stacked one above another. Each complete Code 49 may comprise between 2 and 8 rows, according to data volume storage requirements.

CODE 128 – Code 128 is a continuous barcode in which any length of data can be encoded. It encodes the entire 128 ASCII character set.

CO-EXTRUSION – Films of more than one layer produced by more than one extruder through a common die. Some label face or liner materials are co-extruded in three or five layers to provide specific 'engineered' film constructions with enhanced characteristics of dispensability, flexibility, squeezability, stiffness, etc. The layers of polymer may be of different colors and/or types. Such films are more uniform and dimensionally stable than blown films. *See also Engineered films.*

COHESION – The internal strength of a label adhesive; the resistance to flow, or resistance to failure in the adhesive when labels are removed.

COHESIVE STRENGTH – A measure of the property in an adhesive that resists forces parallel to the surface; the time required for a standard area of pressure-sensitive material, when applied to itself, to slide apart under specific conditions of shear.

COLD AND COOL UV – Used with heat-sensitive substrates, where it is possible to use UV systems with dichroic reflectors, cold mirrors and IR filters so as to remove substrate-destroying heat from the press.

COLD FLOW – The tendency of a pressure-sensitive adhesive to act like a heavy, viscous liquid over long periods of time or when subjected to pressure or placed under load. Such phenomena as 'oozing' or 'increases in adhesion' are the results of this characteristic.

COLD FOILING – The cold foiling process is a more recent development of foil blocking in which a print unit is used to print a special adhesive on the label web in the area where the metallic effect is required. When the metalic foil is brought into contact with the adhesive it adheres to it to produce the printed foil design on the label.

The idea behind cold foiling is quite simple: using a photopolymer plate, an image is printed onto the label substrate with the use of a special UV-curable cold foil adhesive. UV light from a UV curing unit then activates the adhesive. The foil material is bonded to the adhesive and substrate, and later removed. The extracted foil is affixed to the printed adhesive and an image is created. Once printed, the surface of hot or cold foil images may be varnished, over-laminated, or encapsulated in order to provide a hard wearing, durable surface. There are two types of processes used for cold foil printing; wet or dry. Each process uses its own specially-formulated adhesive, which is not interchangeable with the other.

Cold foiling is a less expensive means of achieving foiling than the hot process and is today used in the labeling of cosmetics, wine, pharmaceutical, beverage, and food products.

COLD FUSION/COLD FIXING – The fusing – bonding or fixing – of toner applied by laser or ion deposition printing onto a label surface by cold pressure and/or with gases, thus permitting the printing of coated papers and films which are sensitive to heat.

COLD TEMPERATURE ADHESIVE – An adhesive that enables a pressure-sensitive adhesive to adhere or stick well when applied to cold surfaces – often in cold ambient temperatures.

COLOR BALANCE – The correct combination of cyan, magenta and yellow to reproduce a photograph without a color cast; produce a neutral gray; or to correctly reproduce the colors in an original object.

COLOR CHANGE INKS – Color change inks are quite widely used in security printing applications. Some change color when in contact with certain chemicals. These are used for authentication of security print and other types of secure documents. Others bleed a different color when wetted with water. Because of the nature of security printing, details of many of the highly specialized color change ink systems are not disclosed, except to authorized users.

COLOR CONTROL BARS – A colored strip printed on the margin of a sheet, which enables the platemaker and printer to check the print quality and density of each ink layer by eye or instruments.

COLOR CORRECTION – The process of adjusting an image by masking, dot-etching, re-etching and/or electronic scanning so as to correct color errors in process inks. Changing the color of pixels in an image to achieve a desired effect. May also be used to compensate for scanner deficiencies or for the characteristics of the output device.

COLOR FASTNESS – The ability of a pressure-sensitive labelstock, ink, varnish or coating to retain its original color in normal storage or to resist changes in color when exposed to light, heat or other deleterious influences.

COLORIMETER – Any of various instruments used for measuring or specifying color in the same way that the eye sees color. Colorimeters measure the spectral reflectance of a color and compute numeric values for color intensity, hue and purity

COLOR MANAGEMENT SYSTEM (CMS) – A software system used to ensure color consistency among different input and output devices so that printed results match originals. The process used to ensure that the colors on the monitor are matched by the finished result on the press. Software for this is becoming increasingly sophisticated with the introduction of new digital printing equipment.

COLOR MATCHING – The duplication or exact matching of a given color sample in terms of hue, value and intensity. In inks this is done by blending the appropriate elements of an ink or inks to achieve the color match required. Each color should be verified under correct lighting conditions. When appropriate use color measurement instruments to verify an acceptable match.

COLOR PROOF – A printed or simulated printed image of each process color using inks, toners or dyes to provide a simulated impression of the final printed reproduction. Color proofs are now most often generated by computer. Provides a visual indication that color separations will produce the required results.

COLOR SEPARATION – When required for multi-color process printing a full colored picture or illustration is normally first divided into four separate subtractive color components by either electronic/ laser scanning or through the use of color filters in a process camera prior to the production of the printing plate.

A color image separated in this way will provide films for yellow, cyan, magenta and black (CMYK) plate production. Each color on the film is represented as lines of dots set at specific angles which, when overlaid, will combine as layers of dots forming tiny repeat rosette patterns that simulate shades of color when seen at a distance.

If an image is required for line printing only, the separated films still represent each of the colors, (which in some printing processes may be more than four colors), but there is no dot formation or rosette pattern involved.

COLOR SPACE – Color spaces are used to define color measurements for different types of output device. The most commonly used are:

- **RGB:** Red-Green-Blue
- **CMYK:** Cyan-Magenta-Yellow-Black
- **HSL**: Hue, Saturation and Luminosity
- **CIE Lab:** L* is the luminancy, a* and b* are respectively red/blue and yellow/blue chrominancies. This color space is also defined with regard to the CIE XYZ color.

COLOR STATIONS – Each printing section of a printing press or set of rollers, which is used to print each individual color of a label.

COLOR TRANSPARENCY – A full-color photographic image on a transparent support from which color separations can be produced. Common formats are 35mm, 4" x 5" and 8" x 10". Transparencies can be viewed with the aid of a light box or back-lit transparency viewer.

COLOR WORK – Refers to the printing of more than one color onto a web of labels, usually where close register is concerned.

COMBINATION PRESS – A label printing press in which the printing heads or color stations can be, or are, made up of more than one printing process. A common type of combination process press is one in which the first one or two units are rotary screen printing, followed by rotary letterpress or UV flexo. A hot foil stamping unit may also be added for, say, cosmetics labels. Combination presses may additionally include a UV-flexo varnishing station.

Other combinations of processes on a press now tend to include rotary screen and UV-flexo, or combinations including offset printing and, more recently, gravure printing. Some combination process presses offer interchangeability of processes on the one line. An example might be a flexographic press that has a foil stamping unit that can be removed and replaced with a rotary letterpress unit.

Combination process presses provide the benefits of more than one printing method into one production line, offering for example, the solid color coverage of screen with the halftone printing quality of letterpress, flexo or offset.

COMMON IMPRESSION CYLINDER PRESS (CI PRESS) – A label printing press that positions all the color stations around one large central impression cylinder. The CIC press configuration offers web support throughout the printing operation and is said to provide optimum registration capabilities. May be found in rotary letterpress, flexographic and offset presses, of which the rotary letterpress CIC is most commonly found in the narrow-web label industry. May also be referred to as a Common Impression Drum or Central Impression Drum press. *See also Central Impression.*

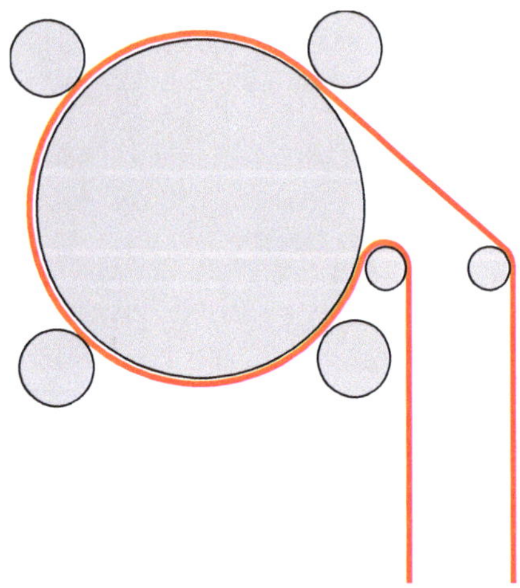

Arrangement of printing units around one large common impression cylinder.

COMPASS – Compass stands for Comparative Packaging Assessment, a web-based tool developed by the SPC in collaboration with EskoArtwork to allow brand owners, designers and converters to compare environmental footprints of various packaging constructions based on LCA and standardized data

COMPATIBILITY – The ability of ink, film, paper, substrate and/or over-varnish or over-laminate to function together in an acceptable manner without any deleterious effects to the finished printed label.

COMPRESSION – The reduction of a computer file size for the purpose of more economical storage or for a decrease in the time it would take to transmit the file over a data communication link. In the case of pictures, compression is achieved by representing the image by a smaller quantity of data by recognizing and exploiting the repetitions and inherent order that exists in a useful image.

COMPULSORY CARE LABELING – Most countries have laws that govern the labeling of goods with instructions to help the user avoid accidents or work safely with the product. Examples include aerosol cans, which have to be labeled with a warning to keep them away from heat; motorcycle crash helmets must have a label attached to them advising against painting them or using any chemicals or solvents which may weaken or damage the helmet; electric blankets have to be labeled with instructions to help them to be used safely; ladders and steps must carry information concerning safe handling; oil heaters must carry warning information on them, and also have warning instructions to accompany them. Electrical appliances and equipment carry wiring color code or safety information warnings, while a wide range of other products need to have some kind of compulsory care labeling.

It is important that all such product labels contain the proper instructions for use because manufacturers are under a duty to supply safe goods. Relevant regulations, recommendations or guidelines should be assessed and followed.

COMPUTER-AIDED DESIGN (CAD) – Computerized electronic equipment used by the package design industry to layout and design in three dimensions the structure of a package. The final result may incorporate the label design as part of the pack image.

COMPUTER ARCHITECTURE – The structure of a computer system – the chips, circuits, busses, expansion slots, etc. Open architecture allows third party developers to create add-on hardware that will work on a computer and even allow other computer manufacturers to build clones. For example, the IBM PC has open architecture, which explains the vast proliferation of PC clones; whereas Apple Mac has partially-closed architecture, which allows others to build add-in boards and peripherals, but not Mac clones.

COMPUTER GENERATED PRINTING – Any printing that is controlled by computer input, whether impact or non-impact printing processes.

COMPUTER LABELS – Labels designed for overprinting by a computer-driven printer. May be supplied either as continuous perforated paper, remoistenable gummed or heatseal paper, or more commonly as self-adhesive labels on a continuous, fan-folded backing containing sprocket-punched holes repeated down each side of the web. Computer labels may be pre-printed or un-printed.

Self-adhesive labels on a continuous backing containing sprocket-punched holes.

Pre-print and areas left clear for subsequent computer overprinting are laid out according to a grid.

COMPUTER-TO-FILM – Computer-to-film systems utilize UV beams passing through a film mask to change the surface characteristics of a plate surface prior to a mechanical washout process, drying and plate hardening.

COMPUTER-TO-PLATE TECHNOLOGY (CtP) – The manufacture of a printing plate from a computer using laser imaging. The computer-to-plate process uses a photopolymer plate coated with a black (ablation) layer that is sensitive in the infra-red range. After removal of its protective film, the plate is lasered by a YAG-laser. The laser images the information onto the pre-coated black mask. The black layer evaporates where the laser beam hits this layer and the actual photopolymer is laid bare in these areas, open for the main exposure to follow. After main exposure, the plate is conventionally washed out, dried, post-exposed and after-treated.

The use of CtP introduces digital pre-press into the pressroom. Replacing conventional step-and-repeat technology, it requires adequate digital color proofing and tighter process control. Front-end software and workflow tools must also be adapted to the CtP environment.

The printing characteristics of computer-to-plate (CtP) technology include less dot gain, greater contrast latitude and higher resolution. With CtP it is possible for flexography to come close to the quality of offset and gravure printing.

CtP technology now also applies to offset and letterpress plates. *See also Ablation layer and CtP.*

CONDITIONING – The correct storage and handling of material at specific temperature and humidity conditions for a stipulated period of time before processing. For example, all paper materials should ideally be stored in good environmental conditions in which heat and humidity are controlled. This applies to both the warehouse storage and again when the material is moved into the press or finishing room. Paper should be stored in its original wrapping and not stacked in such a way or height that will cause a subsequent web or sheet-handling problem.

CONDUCTIVE INKS – Conductive inks have been used in the circuit board industry for many years, but it is the development of such inks for high-speed printing processes and aimed at reducing the unit cost of RFID, that is now creating considerable interest.

Conductive inks are special inks that allow electricity to flow through them. This means they can act as wires, antennas or resistors and provide conductive patterns on both rigid and flexible substrates. The inks are made up of either finely dispersed particles or materials such as conductive polymers. When used in RFID applications, conductive inks serve as antennas that receive the wireless flow of information from an RFID-enabled computer.

Ink antenna compares favorably to copper antenna at ultra high frequency and microwave frequencies and are a cost-effective solution to the unit cost of RFID. This is because conductive inks are lower in cost than traditional metal antennas, and because the application of a conductive antenna or circuit on a high speed printing press is cheaper and faster than alternatives.

Traditionally printed by screen process, conductive inks are now also being developed for printing by the offset, flexo and gravure processes.

Conformability – The ability of a pressure-sensitive label or material to conform, or yield, to the contours of a specific shape or package.

CONTINUOUS CODE – A barcode or code symbol in which the space between the characters is part of the code.

CONTINUOUS LABEL – Labels manufactured from a continuous web of label stock, which are not cut into individual units prior to passing through a subsequent processing or overprinting operation. Usually produced in flat, continuous, fan-folded (Zig Zag) format and supplied in boxes. Continuous labels are mainly used in data processing applications. *See also Zig Zag and EDP labels.*

CONTINUOUS INK JET – Ink jet printing in which drops are continuously generated and electrically deflected so as to form an image, or are diverted into a gutter and recycled. *See Ink jet printing, and Drop-on-demand ink jet.*

CONTINUOUS TONE IMAGE – Often shortened to contone, it describes photographic images that contain an apparently infinite gradient or range of tones, shades and colors, which are smoothly blended together to provide a faithful reproduction of a natural image. A photographic image which has not been screened and contains gradient tones, e.g. a normal photograph.

CONTROLLED RELEASE – A release level that is greater than that which would be provided by an unmodified release coating.

CONTROLLED RELEASE ADDITIVE – A material that can be added to silicone release formulations so as to produce a range of release levels in the cured coatings.

CONSUMER PRODUCTS – Retail packaged/labeled products ranging from DIY products through food and beverages to health and beauty. Often characterized by high turnover with shelves needing to be replenished on a daily basis; such products are also known as FMCGs, or fast moving consumer goods.

CONVERTING/CONVERTING OPERATIONS – In package or label production, converting covers any process performed to manufacture a complete package or label from a raw material or an unfinished material. The process of transforming rolls of self-adhesive material into labels on a release liner (carrier) presented in rolls or sheets, so as to enable end users to apply them to products, packages or surfaces.

This process includes slitting, die-cutting and matrix stripping operations, perforating, embossing, varnishing, laminating, sheeting, and may also include printing.

CONVENTIONAL SCREEN RULING – Selection of the proper screen ruling is critical to the success of the four-color process, and is commonly dictated by the type of label substrate to be printed. The smoother the substrate, the higher the screen ruling that can be used.

CONVERTING METRIC TO IMPERIAL – Both metric and imperial measuring systems are found in the label industry, depending on the country or region. This means that from time to time it may be necessary to convert from one system to the other. Most conversions are relatively simple when the conversion factor is known.

Some of the more common conversion factors that may be required in the global label industry are shown below:

Length. To convert length measurements from metric to imperial, divide by the factor shown or, to convert to metric from imperial, multiply by the factor shown:

Inches - millimeters 25.4
Inches - centimeters 2.54
Feet - meters 0.3048
Yards - meters 0.9144

Area. To convert to metric from imperial, multiply by the factor shown or, to convert from metric to imperial, divide by the factor shown:

Square inches - square millimeters 645.16
Square inches - square centimeters 6.4516
Square feet - square centimeters 929.03
Square feet - square meters 0.0929

Square yards - square meters 0.83612

Weight. To convert weight measurements from metric to imperial, divide by the factor shown or, to convert to metric from imperial, multiply by the factor shown:

Ounces - grams 28.3495

Pounds - grams 453.592

Pounds - kilograms 0.4536

Thickness. With both microns and Mils (and sometimes gage) used for measuring the thickness of materials used in the global label industry it may be necessary to convert from one form of thickness measurement to another:

25 microns = 1 Mil = 0.001" = 100 gage

COPIER LABEL – A self-adhesive sheeted label material having butt-cut symmetrical die-cuts which has been designed for overprinting by a photocopier. Usually found in sheet form (e.g. A4) and may also be printed by an office laser or ink-jet printer. *See also A4.*

COPY – Any material supplied which is to be used in the production of a printed label. Copy may be supplied in the form of a manuscript or typescript, as a sketch, visual or layout, as artwork in the form of drawings, pictures, graphics or logos for scanning, as finished artwork, as a previously printed product, as a proof, in the form of color-separated films or, increasingly today, in the form of computer-generated artwork supplied on disk, tape or by digital transmission.

COPY PREPARATION – The process of taking copy and following instructions relating to size, colour, illustrations, position of elements, etc, so as to arrange the copy into its proper position in the various parts of the label being prepared for reproduction. Work which is being carried out for the preparation of artwork for reproduction.

CORE – A tube, usually formed from wound paper or plastic, on which paper, film, or foil labels are wound or re-wound after die-cutting, inspection and for shipment.

CORE HOLDER – A device used to affix a core to a shaft; a core chuck

CORE PLUGS – Metal, wood or compressed paper plugs', which are driven into the paper core of a finished roll to prevent crushing or other damage to the core and/or the material. May also be known as bungs.

CORNER RADIUS – A corner radius (CR) describes how rounded the corners are on a rectangular die-cut. The term 'corner radius' refers to the radius of the circle created by extended the corner arc to form a complete circle.

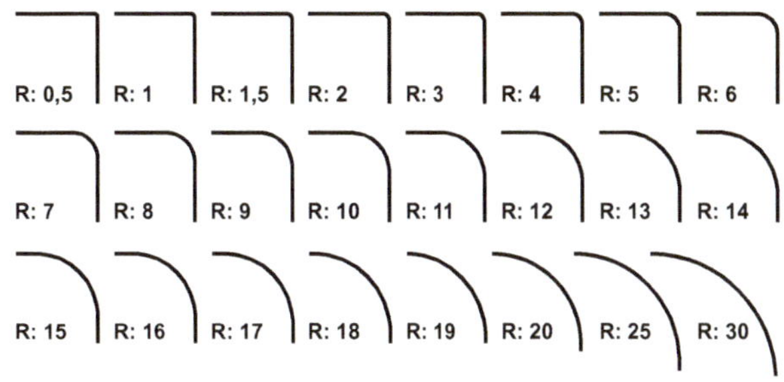

A corner radius guide. *Source: Electro Optic.*

The need for high speed automatic label application led to pressure sensitive labels – labels with adhesive already on them adhered to a liner (carrier). The liner is pulled across a sharp edge, causing the label to dispense onto the container. A square cornered label does not release as easily from a liner – thus corner radius is used to help ease automatic application.

Graphic designers will often choose rounded corners or square corners to enhance a design concept. For example, many wine labels have square corners to evoke the look of old-school cut and glue labels. Other designers use large radius rounded corners to differentiate their products on the shelf against traditional 2 mm corner radius labels.

CORONA TREATING SYSTEMS – A corona treating system consists of two major components: a power supply and a treater station. The power supply accepts standard utility electrical power and converts it into single phase, higher frequency power that is supplied to the treater station.

The treater station applied this power to the surface of the material, through an air gap, via a pair of electrodes at high potential and roll at ground potential, which supports the material. Only the side of the substrate that faces the high potentail electrode will show an increase in surface tension. The voltage build-up ionizes the air in the air gap, creating a corona, which then increases the surface energy of the substrate passing over the electrically ground roll. *See Corona treatment.*

CORONA TREATMENT – High-frequency corona treatment has been widely accepted as a process for improving the adhesion of printing inks, lacquers, adhesives and other coatings onto plastic films, paper and metal foil, providing good results, controlability and easy handling. Used as a treatment method for almost 40 years, the techniques and the effectiveness of corona treatment have been gradually improved over the years so as to keep pace with developments in coating machines and printing presses.

In corona treatment, the surface of film or other material to be treated is bombarded with electrons. These leave the electrode source and are accelerated under high tension towards the passing web material. In doing this they collide with air molecules which transmit light and react in part to ozone and nitrogen oxide. When the electrons come into contact with, say, a polyethylene film, they have so much energy that they can break the bond between the carbon-hydrogen or carbon-carbon. Reactions with the corona gas take place at these free radicals, mainly towards oxidation, with the polar functional groups thus formed providing the basis for adhesion of applied printing inks, adhesives, lacquers, etc.

Ease of operation, ease of maintenance, operational reliability and simple handling are the key criteria for selecting corona equipment. Equipment should also be easy to install within a short period of time and must include practical functions and control features. *See corona treating systems.*

CORRECTED ANVIL ROLLER – If required, an anvil roller can be altered into a plus or minus anvil roller. On the bearer position the diameter of a plus anvil roller is less than the body of the cylinder and on the bearer position of a minus anvil roller the diameter is bigger than the body of the cylinder. A corrected anvil roller is used to compensate for wear of the cutting edge or if the backing material has changed. *See Plus /minus anvil roller.*

COSMETIC LABELING – The labeling requirements for cosmetics/toiletries are set out in the respective international, EU, national or federal regulations around the world. Compliance with consumer product safety and information standards is mandatory and enforceable under the relevant Acts. This means that all suppliers of cosmetics/toiletries goods, including manufacturers, importers, wholesalers/distributors and retailers must ensure that their goods comply with mandatory standards.

Cosmetics and toiletries are generally defined in regulations as substances or preparations intended to contact any external part of the body for use in cleansing, improving or altering the complexion, skin, hair or teeth, and include deodorants and perfumes. Cosmetics therefore include lipsticks, mascara, eye-shadows, nail polish, shampoos, conditioners, moisturizers, cleansers, hair dyes, epilatories and depilatories, and foundations.

Manufacturers of cosmetics must print certain information on the labels of each product and, if the product is of a size, shape or nature that prevents ingredients labeling, then information must be provided by using swing tags, pamphlets, brochures, panels, leaflet or booklet labels, charts or similar which are attached to, provided with, or prominently displayed near the product.

Generally, the requirements provide that cosmetics labels must contain:

- The identity of the product
- A list of ingredients
- The net quantity in metric units (or other specified units depending on country)
- The name and address of the manufacturer or distributor
- Any warnings or cautions necessary for the safe use of the product

Regulations may require that the names of ingredients may be in a specific language or as in the International Nomenclature Cosmetic Ingredient (INCI) names. The INCI names are listed in the International Cosmetic Ingredient Dictionary published by the Cosmetic Toiletries and Fragrance Association (CTFA) of America.

In all cases the appropriate current Regulations, Standards, Directives, etc, should be followed when labeling cosmetics products.

COSMETIC LABEL PRINTING – Labels for the

cosmetics and toiletries sectors may only be printed in 2 – 4 colors, but may incorporate hot or cold foiling, embossing, and overlaminating. Often require special brand or house colors.

Generally longer and more flexible lead times than, say, food labels, as there is frequently a need for a development and testing phase to create unique image and/or ink requirements. Cosmetics labels use a very high percentage of filmic label stocks (around 70%) and metallic foils or metallized substrates.

Inks and varnishes used for cosmetics/ toiletries labels may need to provide resistance to products or, in the case of certain bathroom toiletries, water or moisture resistance. Printing may be screen process, rotary or semi-rotary letterpress, offset – or combination process presses.

Key end-user requirements are for a quality image and impact from substrate and inks, high usage of special house or brand colors and for color matching, product, chemical or moisture resistance, both paper and film substrates, with the sector historically tending to be screen and letterpress orientated. Combination process printing, offset and even digital are used today.

COUNTERFEITING – The making of an imitation, copy or forgery of a genuine bank note, credit card, document, product, label or package with the intention to deceive or defraud is counterfeiting. The imitations, copies or forgeries produced are counterfeits.

Historically, counterfeiting mainly related to pieces of paper that had a value, such as currency, checks or bonds. Passports too, which also have a 'value', have been traditional markets for counterfeiting. More recently, counterfeiting has rapidly spread to pieces of plastic that have a value, such as credit cards or debit cards. All such paper or plastic items with a high value are now generally protected by a variety of overt and covert security features to deter the counterfeiter.

Nevertheless, counterfeiting of currency, passports, credit and debit cards, etc., is still being undertaken by the determined counterfeiter and often as part of an organized crime operation, or as part of an illegal immigration network. Even if a passport itself is not a fake it may have been obtained with an application using counterfeit documentation.

Counterfeit brands. In more recent times, counterfeiting has significantly extended beyond currency, passports and credit cards into the illegal copying and selling of all kinds of branded goods – clothing, sportswear, shoes, watches, toys, computer software, music, videos, cosmetics, car parts, and much more – together with all the associated labeling and packaging of those goods, so as to make the goods appear as if they were genuinely produced by the legal trade mark or brand owner.

The growth of street markets, car boot sales, the Internet and other outlets has undoubtedly helped to spread the growth of counterfeiting. The manufacture of counterfeit goods has additionally become a means by which organized crime gangs can launder money obtained through illegal trading activities, such as dealing in drugs, guns or pornography.

Counterfeiting has also become far easier to perpetuate in recent years. Relatively inexpensive computers, software, scanners and table-top color printers – as well as digital cameras – have made it far easier to reproduce a brand owner's label or packaging. Similarly, linking a computer to a small programmable embroidery machine means that almost any logo, design or mark can be quickly and easily reproduced on clothing – often with relatively good profits for the criminal. The bigger and better known the brand and the more the promotion and marketing of it, so the possibility of it being counterfeited generally grows.

Quite simply, counterfeiters set out to make money by illegally copying or duplicating a reputable manufacturer's products, but without having all the normal brand and trade mark registration procedures, without having to invest in expensive capital equipment, without participating in the extensive brand marketing and global promotion, or having to develop comprehensive brand protection procedures.

Dangers of counterfeiting. While many counterfeit products only damage a person's finances by obtaining cheap or inferior goods under a brand name, there are many well-documented cases of counterfeit goods causing health and safety problems, which, in some cases, have led to serious injury and/or death – particularly of young children. The World Health Organisation for example, has estimated that some 7% of the world's medicines are fakes. These fakes lead to many cases of unnecessary illness and death each year – as well as losses to the pharmaceutical industry of many $ billions, with a good part of these losses going into the hands of organized crime syndicates and terrorist organizations. Not only is there a loss of revenue from the brand, but the credibility of the drug company and the brand is put at risk.

In the medical field there have been cases of counterfeit medical tubing kits used in open heart surgery, of counterfeited eye drops containing contaminated water – which could cause blindness if put into an infected eye – as well as ineffective antibiotics and potentially lethal injectable drugs.

Life and death issues of counterfeiting. Outside of the medical/pharmaceutical world, another area where counterfeiting has been found to be a danger to life is that which comes from counterfeit aircraft and automotive parts. Certainly, causes of both air disasters and road crashes have been traced to the failure of sub-standard counterfeit components, ranging from bolts to counterfeit brake linings, while components like counterfeit air filters or oil filters may have a very adverse effect on the performance of an engine.

In the electrical goods and electrical components markets there are again very real dangers that have been found in counterfeited products with sub-standard wiring, plugs or other components. Electrocution and loss of life may be the end consequences of such goods.

Even comparatively low cost items have been counterfeited with consequent safety hazards. Fake washing powder for example, which was essentially caustic soda flakes, has caused skin problems for those putting the flakes into hot water, followed by their hands.

Inferior counterfeit toys too, may not be manufactured to the correct standards of safety or health for use by children and can therefore be a serious hazard for both unsuspecting parents and children. Problems may come from swallowing items, from the use of non-safety paints, or from sharp or rough unfinished edges.

When it comes to children's clothing there have been some disturbing cases where the necessary flame-proofing or flame-retarding properties have not been used. Such garments can then become a real fire hazard that may lead to a child being burnt.

Not for consumption. Wines and spirits is another area where there have been safety or health issues. Contamination, raw alcohol, inferior ingredients, etc, are just a few of the potential problems, while bottles of drink that have be re-filled and re-corked with inferior quality drink may not have been adequately cleaned or washed.

There have also been cases of fake fertilizer that have caused farmers' crops to wither and die off. A real problem when you depend on farming for your living – or a country that depends on the farmer and the food he grows to survive.

Such a list of counterfeited goods that have, or may, cause health and safety issues for the, often unsuspecting, consumer can extend to cigarettes and tobacco, spectacles and sun glasses, sun block cream, fireworks, oil heaters, some food products, household chemicals, and the like. Indeed the list can be almost endless.

Certainly it can be readily seen, that the dangers to health and safety can come from many different types of counterfeit goods and products. Health and safety issues may be minor or very serious, sometimes even leading to loss of life. The only safe recourse with any suspected counterfeit products is not to buy them.

Products most counterfeited. Any highly visible, well-promoted, branded product is a possible source of attack for the counterfeiter – and it is not necessarily always very expensive or highly coveted items. It may simply be everyday quite mundane products that are being counterfeited, such as paper clips, ball point pens, disposable lighters, instant coffee, plastic garbage bags, washing powder, copier paper or labels.

Although the initial cost of such day-to-day products is quite small – and the return for counterfeiters is low per item – the sheer volumes that come onto the market make them legitimate targets for the counterfeiter. Enforcement expenditures and/or the legal penalties on those caught producing or selling this type of product may also be minimal, making them low risk targets for the counterfeiter.

However, it is the well-known brand name products that are most likely to be the ones that reach the headlines. Some of the most vulnerable products or sectors currently counterfeited are:

- Branded clothing, footwear
- Toiletries and cosmetics
- Cigarettes and tobacco
- Household consumer products
- Wines and spirits
- Drugs and other pharmaceutical/medical products
- Computer and leisure software
- Music and films
- Automotive spare parts
- Sports and leisure goods
- Textiles

- Other branded goods

It is generally in these sectors that most effort is expended by global brand owners in detecting, authenticating, seizing and prosecuting the criminal perpetrators – and where the opportunities are for brand protection technology providers.

COUPON – A label or part of a label, which can be used to supply information or have a redeemable value. Coupons may be pressure-sensitive, non pressure-sensitive, dry-peel or adhesive blocked.

COURTAULDS COMMITMENT (CC2) – An agreement aimed at improving resource efficiency and reducing the carbon and wider environmental impact of the grocery retail sector. Within this scheme, the carbon impact measurement is to be based on 'an estimate of the carbon impact of the (packaging and label) materials, effective recyclability and recycled content'.

COVERAGE – The surface area covered by a given quantity of ink or coating material; ink or coating mileage on a surface.

COVERT IMAGE – A hidden image not visible with the naked eye, e.g. an image printed in invisible fluorescent inks. Encoded or latent images. Conversely, an overt image is one that is visible with the naked eye under normal viewing conditions.

COVERT TAGGING – Application of a fully functional electronic article surveillance (EAS) tag to a product or packaging in such a way as to be hidden from the consumer. Conversely, overt tagging means the application of a tag in such a way as to be visible.

CPU – Central Processing Unit, one of the core components of all computers. The chip that runs a computer. CPUs consist of three parts – the control unit which handles the interpretation and execution of the program instructions; the arithmetic and logic unit which carries out all the arithmetic calculation and logic operations, such as comparing the contents of two stores; the immediate access store which stores the program, data and the current processing results.

CRADLE-TO-GRAVE MANUFACTURING – looks at the life-cycle of a product from extraction through to disposal in landfill or incineration. There is no recovery or reuse stage. This how manufacturing has traditionally produced products.

CRAZING – The appearance of a network of small cracks in a varnish coat or a plastic facestock.

CREEP – The movement of an adhered pressure-sensitive label or tape under tension on a surface due to its low cohesive strength.

CROMALIN – A one-piece off-press color proof for four-color process printing using separations to construct an image by the successive exposure and application of adhesive polymer-dry powder toners. Manufactured and marketed by DuPont. Originally used by the platemaker to make color adjustments but now more widely used by printers, designers and customers as an alternative to mechanical or on-press proofing.

CROP – To eliminate portions of the copy or artwork by masking off or cutting out or indicating (with cropmarks) unnecessary portions of the label. The process of reducing the areas of an image by cutting pieces off it rather than re-sizing the complete image.

CROP MARKS – Marks made on the outer edges of artwork to indicate the size of a printed label after die-cutting or trimming.

CROSS CUTTING – Cutting across the web.

CROSS DIRECTION – The grain direction at right angles to the machine direction in a web or sheet of paper. Papers often have a greater tear strength in the cross direction, but a weaker tensile strength, and are affected more by changes in relative humidity in the cross direction than in the machine grain direction.

CROSSLINKING – The development of a three-dimensional structure in a polymeric material, such as a silicone release coating, adhesive or ink. In adhesives, crosslinking usually results in increased shear resistance, temperature resistance and solvent

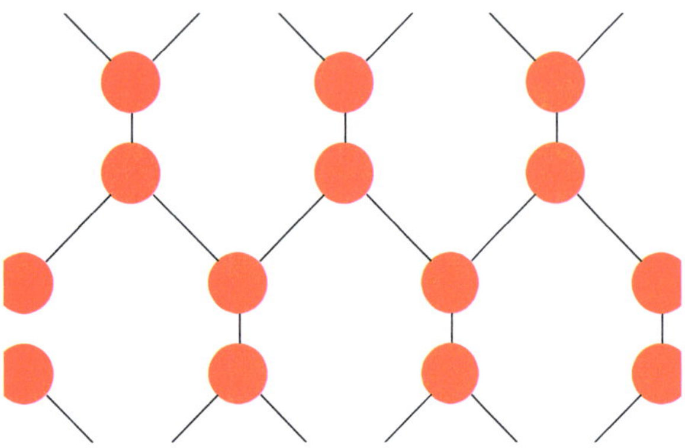

Cross-linking structure of a thermosetting plastic.

resistance. Crosslinking transforms release coatings from liquids into solids that are firmly bonded to the liner substrate. Crosslinking is also a characteristic of thermosetting plastics.

CROSS PERFORATIONS – Perforations across the width of a continuous web for easy separation of individual sheets and/or fan folding of continuous labels. Also any perforation applied at right angles to a label or page, depending on its printed format.

Examples of horizontal cross perforation on continuous labels.

CRT – Cathode Ray Tube. An electronic vacuum tube used as a device to display images, mainly in phototypography.

CRUSH CUT – A cut made by a rotary blade in contact with an anvil or base roll.

CRUSHED CORE – A core that is damaged or gives way and becomes out-of-round from too much tension, dropping or bumping.

CTP – Computer-to-Plate. There are now a growing number of systems for preparing litho and flexo plates directly from the computer, so eliminating film and related chemicals used in traditional origination and colour separations and plate preparation. *See Computer-to-Plate technology.*

CURE – The process of changing the structure and properties of an adhesive, silicone, coating or ink by chemical reaction. This change may be effected by crosslinking or polymerization initiated by catalysts, thermal treatment or by UV or EB radiation. The curing of inks uses high intensity UV lamps; the curing of rubber requires considerable heat and pressure. *See also Accelerated drying.*

CURE TEMPERATURE – The temperature at which the cure reaction is carried out.

CURE TIME – The time taken to effect a cure.

CURL – The tendency of a paper, either by itself or in a laminate construction, to bend or partly wrap around the axis of one of its directions. Frequently caused by humidity changes or improper tension control. It may sometimes be defeated by using a de-curling bar set-up on the paper unwind of a machine.

CUT AND TIE – The term used when describing a perforation. The cut penetrates through whilst the tie remains to hold the material together.

CUT AND STACK FILM LABELS – Cut and stack wraparound labels have been used for many years to apply labels to glass bottles and cans. Historically always used with paper labels, the technology began to evolve rapidly when the requirement to label plastic bottles – and use filmic labels – became necessary. Today, the latest cut and stack wraparound labelling systems can work with either paper or film labels.

Printing of cut and stack film labels would normally

be with flexo (or UV flexo) or gravure. Although some offset litho printers are now printing on film. Narrow to mid-web presses using combination printing, including screen process and hot foil techniques, could also be utilised.

In use, pre-cut rectangular film labels are placed in a hopper or magazine in the label applicator. A non-aqueous hot melt glue strip is applied to the bottle and, when the bottle passes by the magazine, the glue strip picks up the label, which is then wrapped around the bottle. At the same time, adhesive is applied to the trailing edge of the label, which secures the end of the label to itself on a small overlap.

Application speeds are moderate at around 300 labels a minute/18,000 an hour. Changeover time from one bottle type or from one label to another should take no more than 15 minutes, which simply involves changing a base plate and the labels. A servo-driven bottle positioning system is also available.

For bottle size or label application height changes the glue roller and glue scraper are simply exchanged without any need for tools. For label size changes, the entire label magazine is replaced. If the diameter of a bottle changes it is easy to adjust the timing belt without the use of tools.

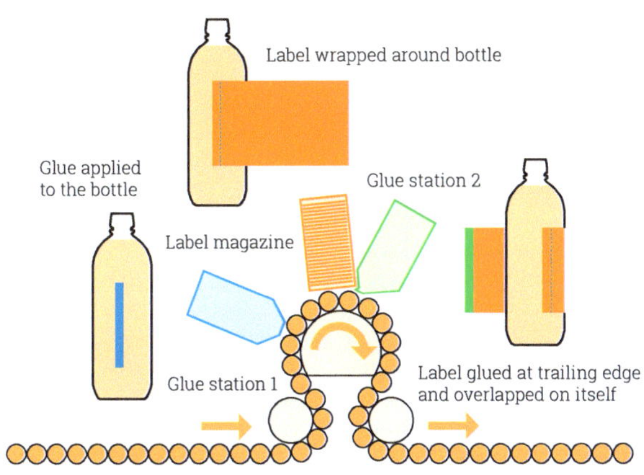

Application of pre-cut film labels using two glue stations.

Cut and stack wraparound film labels can only be used on parallel-sided bottles or cans, or those with a very slight contour. The system cannot apply separate front and back labels (only one wraparound label) nor a neck label on a tapering neck. If a neck label is required with a body label, then an additional self-adhesive labelling head would be required on the application line.

CUT AND STACK LABELS – Glue applied paper labels which are cut to size after printing and piled in stacks of 500 to 1,000 labels ready for gluing and application by the user.

CUT-OFF – A closing time or date when no further input can be accepted and the production process must commence. In web printing, the cut or print length corresponding to the circumference of the plate cylinder and/or die cutter; repeat length.

CUTS – The number of rolls slit from a master roll.

CUTTER BEVELS – In die-cutting, the angle of the material supporting the peak of the cutter. The smoothness of the bevel sides directly affects the amount of pressure that is required to penetrate the laminate surface.

CUTTING STATION/UNIT – The part of a label press that contains the equipment to cut a shape or pattern into a given material. A cutting unit includes an anvil roller and/or support roller rotatable on a machine frame about an axis of rotation. The anvil roller has an anvil surface. A cutting tool is mounted on the machine frame for rotation about an axis of rotation, with a cutter interacting with the anvil surface and with supporting rings which are held on the

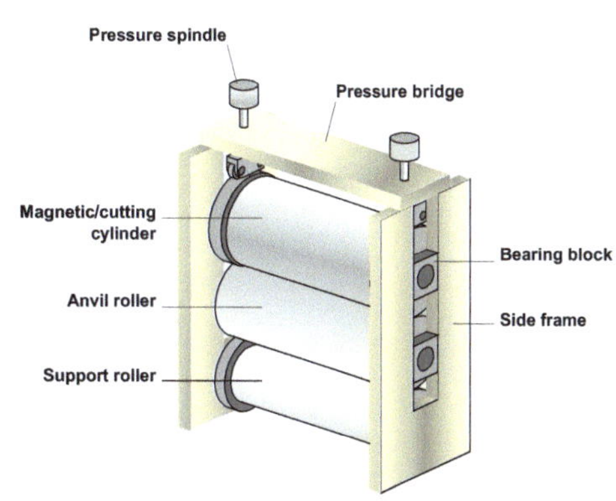

Illustration shows a label press cutting unit.

cutting tool and support it relative to the anvil roller with their supporting ring surfaces and/or vice versa.

CUT-TO-PRINT TOLERANCE – Variation in size on labels between the edge of the label and the edge of the printed image on the label.

CYAN – A subtractive primary color, which reflects blue and green light and absorbs red light.

CYLINDER – In flexographic, offset litho, gravure, semi-rotary and rotary letterpress printing the term cylinder usually refers to the rollers on which the printing plates are mounted (plate cylinders), and the one that receives the impression (impression cylinder). In offset litho there is additionally a cylinder that carries a blanket (blanket cylinder) onto which the plate prints and then transfers or 'offsets' the printed image onto the printing substrate.

Other examples of cylinders used in the labelling industry are: magnetic cylinders, anvil rollers, solid dies, sheeters, foiling dies, etc…

CYLINDER GAP – Gap or space in cylinders of a sheet-fed press where the plate or blanket clamps and grippers are housed.

CYREL – DuPont's trademark for photopolymer plate material.

D

2D, 2D/3D AND 3D – Classifications of Diffractive Optically Variable Image Devices (DOVID's) in relation to their 3D image content. 2D relates to flat artwork with no depth; 2D/3D holograms are flat artwork in two or more two-dimensional planes; 3D refers to completely three-dimensional holograms with actual or computer-generated three-dimensional objects. *See also DOVID and Holograms.*

2D BARCODES – Unlike conventional barcodes, 2D barcodes are able to carry information in both horizontal and vertical planes, enabling more information to be printed in a smaller amount of space. Many 2D barcode symbols can carry 2,000 characters of data in a single symbol, as compared to a regular or linear barcode capacity of 10 to 22 characters. They can also carry binary data, which can be made to run a computer program or generate photographs or drawings.

2D barcodes are being used in applications where there is a limited amount of space and for the coding of high cost sophisticated devices where the code may need to contain a list of components, serial numbers and/or manufacturing history.

2D MATRIX CODES – Matrix codes have a checkerboard appearance in which each uniformly spaced square or polygon-shaped cell corresponds to a binary bit. They are made up of a mosaic of light and dark elements that must all be read before any characters can be recognized. They only have one width, so the scanner must read the center of each bar to interpret the code. *See also Data matrix code.*

DANCER BAR/DANCER ROLLER – A control system used at various points along a roll-label press to control the feed rate of label material, under the correct tension, into the next unit on the press. Also required to detect web breaks and the condition of the reel of material being fed into the press.

DANISH LABEL ASSOCIATION – *See under DLA.*

DAT – Digital Audio Tape. A small cassette type cartridge that can hold vast amounts of data. Ideal for backing up data.

DAMPENING – The use of a liquid in the offset litho process so as to allow the image and non-image areas of the plate to be defined and ink to be placed in the correct places. The dampening liquid (solution) is transferred to the printing plate during printing through a dampening system and dampener rollers. These are cloth-covered, parchment paper or rubber rollers that distribute the solution.

DANCER BAR/DANCER ROLLER – A control system used at various points along a roll-label press to control the feed rate of the web of label material, under the correct tension, into the next unit on the press. Also used to detect web breaks and the condition of the reel of material being fed into the press.

DATA MATRIX CODE – A proprietary dynamically variable 2-D (two dimensional) matrix code, which provides a visual representation of a machine executable electronic binary code. The code can range in size from .001 inch square (0.65 sq. mm) to 14.00 inches square (9032.24 sq mm) and has a capacity of between 1 and 2,000 characters in any language and can be read at any angle of rotation.

Two-dimensional matrix code.

For security applications, the code can be printed on labels or products with invisible ultra violet sensitive ink that can only become visible when illuminated with 'black light'. Invisible codes can be read with the same speed and accuracy as visible data codes.

DATASTRIP2D – A variable-size, variable-density two dimensional barcode symbology offering a very high data capacity. The symbology has a unique combination of matrix density and variable rectangular aspect ratio and handles large data capacity in a single symbol, typically 2800 bytes and above.

AIM, the Association for Automatic Identification and Data Capture Technologies has approved the Datastrip2D symbology as a new work item for development into an AIM International Technical Standard.

DEACTIVATABLE LABEL – A self-adhesive electronic article surveillance (EAS) label, which can be cancelled (deactivated) at the point of sale checkout by means of a deactivator. *See also Deactivation.*

DEACTIVATION – The process by which a deactivatable EAS label is passed through a deactivating field. This field is generated by a deactivator control box and is emitted by a deactivator pad or similar device. EAS tags and labels of all types ideally need to be cancelled, de-activated or removed at the point-of-sale to enable the customer to leave the store without setting off an alarm.

De-activation or removal should not impede throughput at the checkout and ideally should be 100% effective in order to avoid false accusations of theft. De-activation scanners may be hand-held, flat-bed or vertical, pen or concealed types.

DEAD TAG – An electronic article surveillance (EAS) tag that has been deactivated permanently. This involves destroying electronic components, or at least substantially re-aligning magnetic properties for the tag.

DECODE – To interpret and verify the signal from a code.

DECODER/DECODING DEVICE – A piece of equipment or software which is able to accept data in one format and output in another; an electrical or electronic device which is able to translate the signals from a scanner and convert them to a recognizable form or into computer data.

A decoder performs checks on an electronic signal so as to validate it as meaningful and then processes the signal through a decode algorithm designed to detect errors in the signals. Some decoders are microprocessor controlled and can carry out limited data processing operations or store data for batch transmission.

DECORATION OF PLASTIC BOTTLES – The methods and techniques of labeling and decorating plastic containers have been evolving for more than 30 years. Originally, paper labels applied with a wet glue were the most commonly used solutions. Pressure-sensitive paper labels have also been used for many years, particularly in the toiletries, cosmetics, healthcare and industrial markets.

By the early 1980's, companies such as Procter & Gamble – in association with Owens Illinois and Multicolor – were pioneering the in-mold labeling process for decorating plastic containers. Again, this was with paper labels, and primarily for blow-molded containers for toiletries and under-the-sink products. Today, synthetic papers and films are also used.

Towards the mid 1980's, shrink sleeving of twin packs began to be used – initially un-decorated but soon evolving for printed, decorated product labeling using shrinkable sleeves. Toiletries and pharmaceuticals were often the key market applications for such technology, but this has now extended into the drinks sector.

Direct printing of plastic bottles and containers by the screen process has also been a long-used process – again for under-the-sink products, as well as for the labeling of industrial products and applications like milk bottles. Offset printing may be used for some applications.

Other long used techniques for decorating plastic containers include transfer decoration. Developed originally by Dennison Manufacturing, transfer decoration has been used for more than 30 years. The technology has undergone further evolutions in recent years, with others now also developing the transfer decoration technology.

The impact of carbonated beverages It was the evolution of PET bottles being used for carbonated beverages that began to bring change to the methods and techniques of labeling and decorating high-quality plastic bottles for the drinks markets. Originally using paper labels, it was soon found that expansion of the carbonated drink content would split or tear paper labels. Methods and techniques for labeling using extensible film labels soon began to evolve. Many newly introduced and high added value drinks labels – wine, spirits and speciality beers – are now clear self-adhesive or high quality coated paper labels.

During the 1990s and into the 2000s, much work was done to develop materials and application solutions where cut-and-stack film labels and wrap-around labels rapidly evolved. Today, such technologies are amongst the fastest growing of all labeling possibilities for PET bottles containing carbonated beverages, soft drinks and bottled waters.

Stretch sleeving was a further bottle labeling technology that evolved in the 1990's and this has found applications for certain drinks markets and other products.

The packaging waste and environmental revolution In the late 1980s pressures began to build on the volume of packaging waste, the nature and type of waste, and the impact of waste on the environment. The use of PVC film for product labeling soon came under growing pressure and other plastic film types – such as OPP, BOPP, polyethylene and PET began to grow rapidly.

By the early 1990s compatibility of label material and bottle/container material began to grow in importance so that the bottle and label could, potentially, be recycled together. The impact of packaging waste and environmental legislation still continues to play a major role in the selection of label and bottle decoration solutions today.

Present and future In the 2000s the range and variety of PET bottle decoration and labeling solutions is more diverse than ever before. Certainly all the methods of labeling or direct decoration already outlined are currently being used to some degree on plastic bottles – although not all have gained any widespread use for the decoration of PET bottles.

It is the potential of PET bottles in the beer market that is stimulating further development and evolution in decoration technologies. Currently much of that development relates to speciality and niche beer products, mainly shorter runs or smaller volumes. Here, the labeling solutions may be somewhat different to the high volume mass beer markets.

Most recently, development, trials and interest in direct printing onto plastic bottles using inkjet technology is gaining ground and may well become a viable growth plastic bottle decoration option in the coming years. Already used for product trials, test marketing, extended proofing, some personalization, etc, direct digital printing, now also becoming known as direct-to-shape printing, is achieving good quality printed results and ever higher production speeds. *See also Bottle decoration and Direct-to-shape printing.*

DEFOAMER – A substance or mixture of substances which, when added to foaming solutions causes small bubbles to collect into large bubbles which then rise to the surface and break (burst). Defoamers are used in papermaking and paper coating and in flexographic inks.

DEGRADABLE – Capable of being broken down (degraded) into natural elements or simpler compounds under normal environmental conditions. In the label industry it is a term most commonly used in relation to papers, laminates or films, which will degrade or biodegrade in, say, landfill sites (not preferred disposal today).

DE-INKING PROCESS – The process of removing printing inks from paper fibers during the paper recycling process.

DELAMINATION – The splitting of a material into layers in a direction approximately parallel to the surface, e.g. the separation of a self-adhesive print carrier from its release liner (silicone backing paper) in the printing machine.

DELPHAX – Previously incorrectly described as an Ion Deposition printing process, this was subsequently corrected by Delphax and is now described as an E.B.I. (Electron Beam Imaging) printing system. The print engine is capable of printing variable information text and barcodes at good resolution, very quickly. It has been used for labels and tags and successfully incorporated into those production lines.

The toner is heated and becomes sticky while the material to be printed is passed over a pre-heated platen before the image is transferred onto it under cold pressure, which ensures a good toner bond. Some Delphax machines use flash fusion to bond the toner onto the material being printed, which is less demanding.

DENSITOMETER – A photoelectric instrument that measures reflected or transmitted light on photographic film, colors or printed products. Densitometers vary in sophistication and in the number of features provided, such as black and white, color, data display, read-out memory and computer printer-out.

Densitometers used for the measurement and control of color inks on a substrate are reflection densitometers. They use red, green and blue color filters to measure and control color proofs and sheets printed with special color bars, which consist of small blocks of each color, graded halftone tints and overprints of the colors. Measurements taken are ink strength, dot gain, neutral gray balance and ink trapping on the press.

DENSITY – In photography and printing, density is the degree of light absorption or opacity of a photographic or printed image. In computing, density is the degree of packing of data onto a disk, tape, CD, etc. In artwork, it is the closeness of substance in the image and non-image areas of the black and white artwork.

DESIGN FOR SUSTAINABILITY – (D4S) or sustainable product design, is a globally recognized method for companies to improve profit margins, product quality, market opportunities, environmental performance and social benefits.

DESILICONIZE – Industrial process to separate silicone coating from a non-filmic release liner so both materials can be effectively recovered in a recycling system.

DESKTOP PUBLISHING – The ability to produce printed images or pages on a computer for reproduction. Frequently referred to as DTP.

DESTRUCTIBLE LABEL – A type of pressure-sensitive label that, once removed from the release liner and bonded to the intended surface, cannot be removed without damage, such that any attempt to remove it is readily evident. May also be referred to as a tamper-evident or security label. *See also Tamper-evident labels.*

DETACHER – A device which is used to release hard electronic article surveillance (EAS) tags – at the retail checkout – from items to which they have been affixed, commonly clothing.

DETECTION AISLE – The area covered by EAS detection gates after point of sale and prior to store exit.

DETECTION FIELD – The zone of radio signals (created by EAS radio transmitters/detector gates) of a required radio frequency to detect tags. The zone is designed to respond at that frequency. *See also Detector gates.*

DETECTOR GATES – To be effective all electronic article surveillance (EAS) systems require detectors containing radio antennas to be placed at or near the store exits or check-outs – whether for customers or staff. These create or receive radio signals at a required frequency, producing a measurable response in an electronic tag or label (which has not been deactivated) as it passes through the gates. The detector system emits an audible and/or visible alarm if tagged items have not been de-activated on payment for goods at the checkout, which alerts staff or security personnel to the theft. The system may additionally be linked to security cameras.

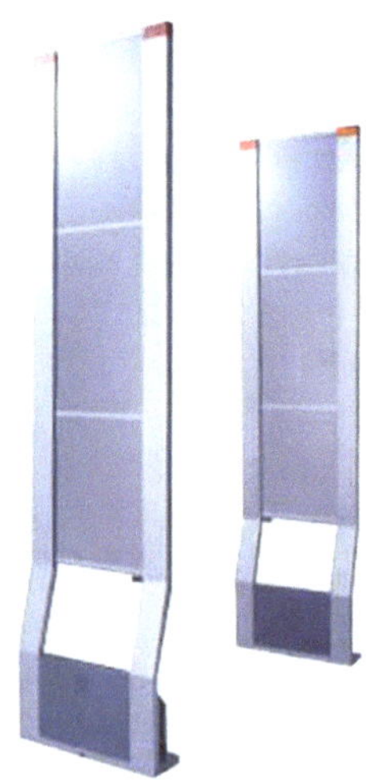

Detector gates placed at or near store exits.

Dector gates are available in different shapes and sizes, which can be customized or colored in line with the decor of the store. *See also Antenna.*

DIE – A tool or device used to cut a desired shape, form or finish in any material; most commonly to produce a label or a specifically-sized sample. In the case of a self-adhesive label the die is placed within the converting line or press and is used to only cut through the face materials of the pressure-sensitive laminate, but not the release liner. The matrix waste can then be removed to leave the desired label shape on the web. Dies may be constructed from flat steel strip cutting rules or as solid rotary tools or made as flexible dies for mounting on a magnetic cylinder.

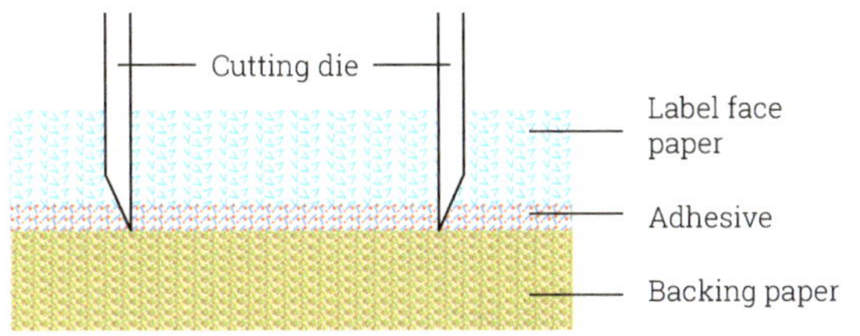

Die cutting through the label face material and adhesive, but not the backing paper.

DIE-CUTTING – The process of cutting a label shape with a die. Most self-adhesive labels and some wet-glue and in-mold labels have to be die-cut to shape as part of their manufacturing and finishing procedure. Depending on the type of label and the printing and/or die-cutting requirement, the operation may be performed using high or hollow dies, flat dies, rotary dies or, most recently, with digital die-cutting.

High or hollow dies used for cutting glue-applied labels to shape are made of cold-rolled steel, which is forged and welded to create the required shape and height. The inside of the dies are parallel for about 25mm (I inch), after which they flare out. In use, the stack of labels is either stationary – with the die moving – or the die is stationary and the stack is pressed against it. In either case, sufficient space must be allowed between each label to ensure clean cutting. Sharpness of the die is critical, as are finished cut label tolerances.

Flat dies are most commonly produced by bending lengths of accurately fashioned steel rule, which has been finished to a cutting bevel along one edge. This rule is around 0.4mm (0.0157") in thickness and nominally 12mm (0.472") in height. To form a cutter, the rule, once bent to shape using a special bending tool, is placed in a base into which the shape or shapes of the label(s) has been cut. In this way the rule is supported during use on the press and retains a high degree of accuracy.

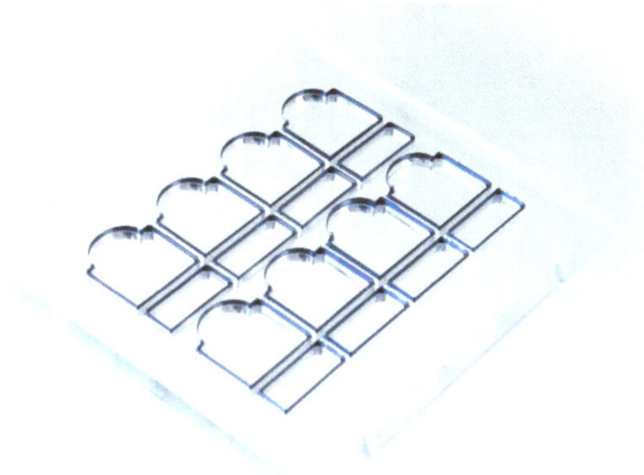

Illustration of a flat rule die. *Source: Wink.*

Rotary dies are engraved by electronic discharge from a cylinder of solid steel so as to leave the cutting edge standing proud around the cylinder circumference.

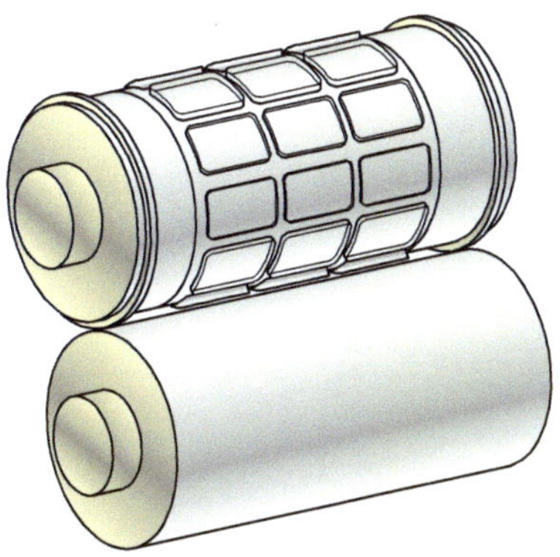

Rotary die cutting cylinder in contact with the anvil to allow space for the thickness of the liner or backing.

Each of the types of rotary die require some form of final finishing following on from the machining or etching, which is undertaken using computer-guided equipment and which sets the seal of quality on the die.

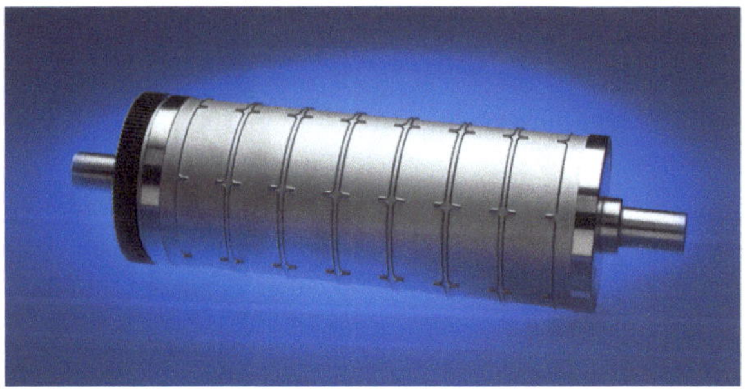

An electronic discharge machined (EDM) rotary dies. *Source: Gerhard/Rotometrics*

Flexible dies are thin plates made from spring steel that have the die configuration etched over the surface and milled to the right dimensions. They are then mounted for use by wrapping the thin steel around a magnetic cylinder.

Digital die cutting of labels is a relatively new development, which has evolved from laser cutting technology used to cut out die base boards prior to inserting the rule. As with variable imaging or digital printing, the required shape and size of label are programmed by computer, with a laser beam and lens system used to direct the beam in cutting-out the label shape. Both paper and filmic labels are today being cut with a degree of success, although the technology is still a somewhat expensive method of cutting labels to shape.

Flexible die. *Source: Electro Optic*

Digital label printing is now increasingly using laser die-cutting and finishing for printing self-adhesive labels in small quantities and on-demand when printed on UV inkjet presses. *See also laser die-cutting.*

DIE HEIGHT (FLEXIBLE DIE) – The height of a flexible die (base + height of cutting line)
Market standard: 0.48 mm (cut through plates)
0.44 mm (kiss cut/glassine backing)

DIE LIFE – Meterage expected from a new die or that expected following a re-sharpening of a die. Estimates of die life depend on machine setting, type of labelstock, type of adhesive and on operator handling. Estimates of meterage from a die may vary considerably from company to company, machine to machine or job to job.

DIE LINE – A blueprint, drawing or computer-generated layout of the cutting shape or shapes of a die. May be supplied with artwork as an overlay, as a blue line on the base artwork or supplied as computer-generated data on disk, CD or transmitted electronically.

DIE STRIKE – The impression left on a backing liner after being converted by a cutting die.

DIE WIPE – A test to check the evenness of a rotary die cutting pattern on a backing liner. A solvent pen/marker or fine colored powder is wiped over the silicone surface where a die impression is evident. Where the die has penetrated the silicone coating the ink or powder stains the backing, highlighting unevenness in the cutting depth which can cause the waste stripping matrix to break.

DIFFERENTIAL RELEASE – A two-sided silicone release coated paper or film whose release value for an adhesive on its face side is different to that on its reverse side. One side has easy release while the opposite side is tighter, such that the adhesive stays with the tighter side during winding and other subsequent converting operations. This provides two levels of release on two release surfaces, so that for example, when unwinding a transfer tape the adhesive remains as one release surface only, so allowing it to be applied to a substrate and for the backing to be removed.

DIGIFLEX – A computer-to-plate system that eliminates films and film exposure while still using affordable analogue printing plates.

Creation of the image area on the plate is achieved using a reactive inkjet ink that produces a UV opaque mask onto a polyester film substrate. This chemically reacts instantly with a second reactant on top of the plate to freeze the ink droplets without allowing the ink to spread. The outcome is high resolution, with zero dot gain on the plate, and the capability of achieving a two percent dot on the press.

The reactive layer is then transferred from the polyester substrate to the top of the plate using a lamination process. The primer layer, which has no reaction with the plate surface, is washable during the plate development process and can be used with all standard water-washable plates, with solvent washable plates and with rotary screen plates. Effectively, after the image has been created on the plate, the rest of the plate-making process remains unchanged, so providing minimal adaptions to an already familiar workflow.

DIGITAL – Pertains to the representation, manipulation or transmission of information by discrete, or on-off, signals.

DIGITAL ARTWORK AND REPRO – Instead of producing finished artwork for labels by the more conventional and traditional methods, digital artwork and repro (DAR) enables electronically-generated artwork to be taken straight to the final film, plates or press from the computer, so reducing lead times and often, lowering overall costs.

In operation, line artwork, logos, color illustrations or transparencies are scanned in and dropped into position – or digital camera images are imported – text boxes and any other elements added. All necessary adjustments are then made to compensate for printing variations (spread, choke, allowances for dot gain, etc). Digital color proofs can be produced, modifications made and the final electronic artwork made ready for output onto film, or straight from digital artwork to plate production, enabling even faster turnaround times and more accurate output. *See also Digital workflow.*

DIGITAL COLOR MANAGEMENT – Each device in the digital workflow works with its own color space and therefore has its own color representation (e.g. RGB versus CMYK). Colors from the same input file may look different on different monitors and the colors on the printed output file will look different from the colors displayed on a computer screen.

Color management is therefore necessary to ensure color consistency and predictability among the different devices, i.e. users will know what a given yellow on a computer screen will look like once it is printed on paper

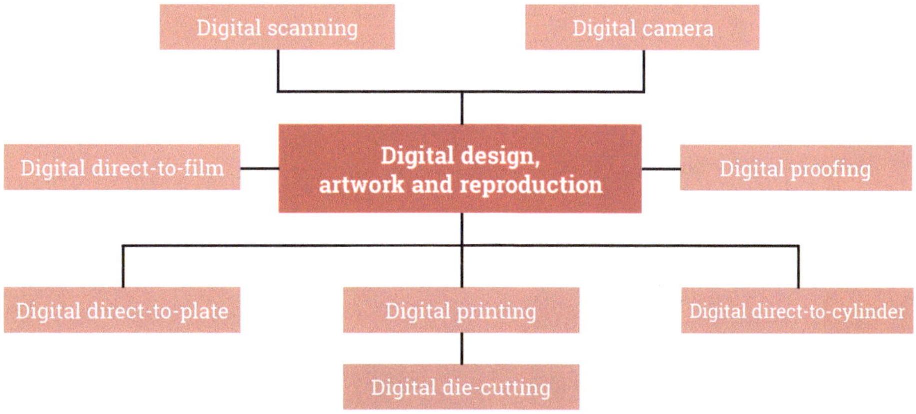

The digital design, artwork, reproduction and workflow revolution.

DIGITAL COLOR PROOF – An off-press color proof that has been produced from digital data without the need for color-separated films. *See also Remote digital proofing and Soft proof.*

DIGITAL FONT – An electronically stored type fount (font) in which the characters are stored as a series of digital signals.

DIGITAL FRONT-END – The digital front-end in the digital printing process is a combination of hardware and software, encompassing the entire suite of software modules necessary to process the files to be digitally printed and to control the print engine. Depending on the press manufacturer, digital front end modules may be Esko, Xeikon-800, Fiery XF, Prinext, Screen Equios, etc. The modules or processes typically impacting output quality are calibration, color management and screening.

Calibration. The calibration of the imaging device in order to increase the accuracy of its output, i.e. to ensure its output matches certain criteria, is closely linked to all aspects of process stability.

Color management. Each device in the digital workflow works with its own color space and therefore has its own color representation (e.g. RGB versus CMYK). Colors from the same input file may look different on different monitors and the colors on the printed output file will look different from the colors displayed on a computer screen.

Color management is therefore necessary to ensure color consistency and predictability among the different devices, i.e. users will know what a given yellow on a computer screen will look like once it is printed on paper.

Screening. Most digital printing systems cannot print continuous tone images as they can only deposit dots with a limited number of bits per process color, which is why for printing purposes continuous tone images need to be converted to a pattern of dots containing the right number of bits per process color. The process used to do this is halftoning.

DIGITAL IMAGING AND PRINTING PROCESS – The digital imaging and printing process takes place in three stages: input, imaging and output. These stages are examined below:

Input. This is a typeset page, label or package design created with, for example, Adobe Indesign and made up of images (contones), text and graphics (line work). This page or label/pack is typically device independent and is basically a file of instructions describing line work in the form of vector graphics, text, fonts, and images in a page description language which specifies all elements of the layout,

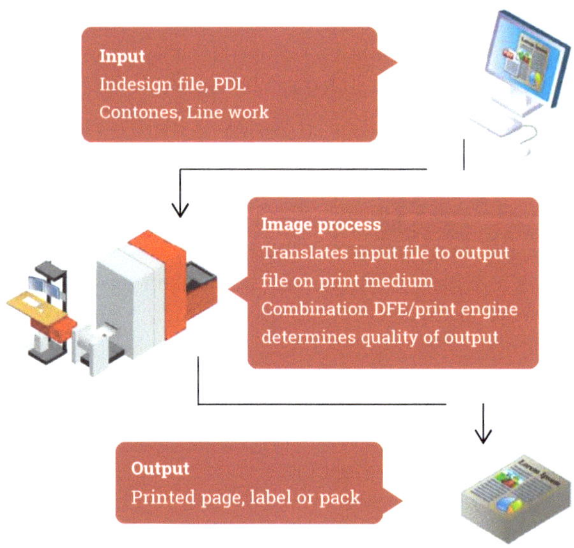

Illustration shows the digital imaging process.
Source: Xeikon

Imaging process. Digital print engines typically print interfacing lines of dots. Contone images or line work can therefore not be printed without converting them to a series of dots. This is done by the RIP (raster image processor), a component of the digital front end (DFE) that interprets the instructions of the page description language and translates them into a bitmap, i.e. the actual pattern of dots, producing one bitmap for each process color.

Typically every dot in the bitmap contains 8 bits per process color. The amount of information contained in such bitmaps is generally too large to be handled by the printing process. This is why, in order to print this bitmap it has to be 'halftoned'. Halftoning or screening is a process that converts a bitmap containing dots of 8 bits per process color into a bitmap containing dots with fewer bits per color. These 're-sampled' bitmaps are then printed by the print engine.

Output. This is the reproduction of the electronic

original (input) printed by the print engine. The quality of the output is influenced by the combination of the DFE and the print engine. Although the design parameters or specifications of these components do not by themselves define the level of perceived (image) quality, they do have a direct influence on the ultimate quality of the output, and as such they can be manipulated to maximize image quality.

DIGITAL INSPECTION – With digital presses running ever faster it has become increasingly important to consider fully-automated universal inspection and verification throughout the entire digital printing process, enabling even the shortest of runs to be automatically inspected. A number of key industry suppliers now partner together in providing digital print process control and quality assurance solutions for both label and package printing.

DIGITAL PLATEMAKING – Digital platemaking uses a laser inside a plate processor. The plate itself is covered with a mask that is ablated during the laser imaging process. The plate is then exposed to UV light, placed on a thermal plate processing device which removes the unwanted material from the plate surface using a dry method before finally ejecting the plate for mounting.

Used to achieve high quality output of water or solvent-wash digital photopolymer flexo plates, sleeves, digital letterpress and dry-offset plates. *See also CtP.*

DIGITAL PRINT FINISHING, IN-LINE OR OFF-LINE – Digital print-on-demand color label printing today is carried out with liquid toner, dry toner and inkjet engines which print on a web of, predominately, self-adhesive labelstock in four or more colors. Once printed, the web needs to be die-cut to produce the individual labels for re-winding and subsequent label application. This die-cutting operation may be undertaken in-line as part of the digital label press or off-line as a stand-alone digital finishing operation.

In-line or off-line. Working offline gives the opportunity to have much more flexibility. The more complex the label job, then the more the converter to-date has looked at an off-line finishing scenario.

If jobs predominately just require flexo varnishing and die-cutting, then in-line may well be the best option. The more processes required in the finishing line (foiling, foiling, embossing, etc), the more it is likely to slow the digital press speed down, which negates the need to have a press running at 50-60 meters or more a minute. If changeover and set-up on the finishing line is longer than the press changeover time, then off-line finishing probably makes the most sense.

In-line finishing however, does offer key advantages. The converter will get from the beginning to the end of the job in one operation. With the new generation of faster digital label presses the ability to print and ship without a separate finishing operation is a definite advantage – providing the press and output is not substantially slowed down by the finishing requirements.

Laser die-cutting in-line has also become more attractive as inkjet speeds have risen, and also when jobs can be batched or 'Ganged' on the web. Doing everything in-line also means that it is possible to modify the pre-press or image to match the die-cutting or the hot stamping, or to a screen edge.

The choice of in-line or off-line may also depend on how many digital printing machines a converter operates. With multiple digital printing presses it may be more feasible to have one or two off-line finishing systems servicing 3, 4 or 5 digital presses – even providing finishing for conventional presses where more complex finishing operations are required.

DIGITAL PRINT FINISHING TECHNOLOGIES – Die-cutting is a primary requirement for finishing on almost all label presses, whether carried out in-line or off-line. This operation may be undertaken using flat-bed, rotary, rotary cylinder or, increasingly, with laser cutting. The great majority of digital presses will also require varnishing (usually UV flexo) in the finishing line.

Almost all the digital finishing lines installed to date have die-cutting and varnishing options as standard, as well as web-handling, edge trim, matrix strip, slitting and rewinding. In essence, a basic off-line finishing machine for self-adhesive labels will likely include:

- An unwind
- Web handling
- UV flexo varnishing
- Semi-rotary die cutter
- Liner score/edge trimming
- Matrix stripping and matrix rewinding
- Slitting and rewinding into finished rolls.

Depending on the market sectors that the converter is involved with, so other finishing operations will be required. Hot or cold foiling for the cosmetics or wine and spirits sectors; possibly embossing for those same sectors as well; maybe sequential coding or numbering for pharmaceuticals; perhaps a vision system for web inspection; sometimes over-laminating.

One or more conventional printing processes may also be incorporated into the finishing line to add-value or special effects for, say, high-quality cosmetics or wine label applications. Options may include full rotary or semi-rotary flexo with UV or infra-red drying; an optional spot varnishing station with servo-driven register control; flat-bed or rotary screen module (the cost of the mesh, frame, pre-press and set-up will be cheaper with flat screen). Rotary screen may well be considered if large volumes and increased speed are required.

DIGITAL PRINTING – A process involving the creation, storage, transfer and reproduction of printed images in a digital format so as to allow fixed or completely variable information to be printed on a label without printing plates. The main plateless digital printing technologies today are:

- Electrophotographic (EP) liquid toner (HP Indigo)
- Electrophotographic dry toner (Xeikon, iSys)
- UV inkjet – e.g. Domino, Durst, EFI Jetrion, Stork Prints
- Water-based inkjet (Epson, Memjet)
- Dye-based inkjet (Swiftcolor)
- Laser (Primera)
- Thermal Transfer (Matan
- Nanographic offset (Landa)

It is the more recent development of print-on-demand (POD) halftone digital color printing using powder and liquid toner digital presses and color ink jet printing equipment that has become of most interest to the label industry – particularly in the narrow- to mid-web sector.

The means to print labels digitally in up to four or more colors direct to paper or synthetic materials using these methods began to emerge in the early 1990s and has now sufficiently developed to become a label printing technology in its own right. The majority of the 2,500 or so digital label presses installed worldwide by the end of 2014 will be predominately using dry toner (Xeikon) or liquid toner (HP Indigo) technologies.

In the toner-based digital label printing process, a latent image of one of the color separations is created on a sensitive drum. That image is then developed for printing with the dry powder or liquid toner before being transferred to the label substrate as a printed image. Each color is printed in turn.

In the HP Indigo web-fed machine each image is first transferred to a rubber blanket where the liquid ink becomes tacky. Each color is added to the hot blanket in turn and the finished image is transferred to the substrate complete.

In the Xeikon dry toner digital press printing is from five stations using a dry powder toner. Each print station consists of a central electrophotographic drum. During each print, the drum is cleaned, charged, images and developed. The developed toner image is then transferred to the substrate using heat and some pressure.

The most recent digital label press development is that of color ink jet printing, where there are now some 40 to 50 vendors selling inkjet color press ranging across UV inkjet, UV LED inkjet, water-based inkjet, solvent-based inkjet and dye-based inkjet.

All digital label printing systems may be controlled from a computer keyboard with the copy and images to be printed having been prepared on a VDU screen for accurate composition and assembly. *See also Digital front end, Digital imaging, Indigo, Xeikon, Inkjet.*

DIGITAL PRINT RESOLUTION – Digital print resolution, although not the only measure, is one of the key measures of digital print quality. For those more used to flexo printing this is commonly quoted in the form of lines per inch (lpi) or lines per centimeter (lpc). Ink jet, including the desk top versions that readers' may have at home, as well as the numerous UV-curable ink jet-based digital presses that are now rapidly growing market presence, will all tend to quote dpi or dots per inch.

Dots-per-inch defines the density of dots that can be printed in an inch. The more dots per inch available, the higher the print resolution of the image; consequently, the more smaller dots that can be utilized, the better the image definition is going to be.

Consequently, the print resolution depends on the dot size and therefore, in turn, on the hardware of the print engine – and on its imaging device in particular.

Resolution	Dot size
600 dpi	42 µm
800 dpi	32 µm
1200 dpi	21 µm

For example, in the table above, a 1200 dpi print engine can place 1200 discrete dots per inch, i.e. dots with a size of 1/1200 inch = 21 μm dot size. A 600 dpi print engine could place 600 dots per inch, with each dot being 1/600 inch = 42 μm. Using LED imaging technology this would correspond to a LED array with 1200 and 600 LEDs respectively. The number of LEDs on a diode array being fixed, resolution, as it is defined here, is a hardware specification.

In some inkjet systems resolution can also be affected by print speed – higher resolutions are achievable but at lower speeds and vice versa. So buyers should always check what speeds a quoted resolution refers to. Similarly, the highest quoted print speed may only be achievable with a reduced resolution.

DIGITAL PROOFING – Producing proofs directly from stored digital information without film or plates, e.g. laser, ink jet, digital halftone or thermal transfer dye sublimation.

Digital proofing can refer to both soft proofing (viewing a proof on a screen) or hard copy remote digital proofing remotely in a customer's office. *See Remote digital proofing and Soft proofing.*

DIGITAL SLEEVE TECHNOLOGY – The production of flexographic printing sleeves using laser-exposable technology. There are two possible routes to producing a sleeve: producing a sleeve with a seam on the basis of laser exposable plates; or sleeves without seams, produced by continuous, endless, laser exposure.

More than 90 per cent of flexo orders are printed from plates mounted on cylinders or sleeves, and it is only in applications where continuous image patterns are required for products such as wallpaper, floor covers, coffee packaging and cigarette packaging, that seamless sleeves are required.

DIGITAL WATERMARKING – Technique for building covert anti-counterfeit and ownership information into a digital file. The digital watermark can be read by a smartphone or other scanning device with the appropriate software application.

DIGITAL WORKFLOW – The use of the Portable Document Format (PDF) has provided the printing industry with a predictable, platform-independent, page-independent entity which solved many of the limitations encountered with PostScript. Add to the success of PDF, the computerization of areas such as pre-flighting, trapping, imposition, digital proofing and digital-to-film, plate or press, and the route to completely digital workflow has become a reality. Digital scanning, digital cameras and digital direct-to-plate technologies also add to the capabilities for full digital workflow.

Previously, most of these functions were carried out manually, often requiring a different server to perform a function or different functions. Some of the possibilities, such as the sophisticated digital cameras and digital die-cutting, that are now available, are relatively new developments. Now, with automated and more comprehensive and cohesive systems, fully digital workflow is widely used in the packaging and labeling sectors. *See also Digital artwork and repro and Pre-flighting.*

DIGITIZE – The process of converting analog information into digital format so that the computer can process it. For example, scanners divide images into a grid of small dots and records the intensity of light reflected from each dot as a number for the computer to re-assemble as a picture.

DIMENSIONAL STABILITY – The degree to which the dimensions of a material remain constant, particularly in relation to external influences,such as moisture and temperature.

The ability of a material to retain its size and/or consistency of dimensions under varying atmospheric or environmental conditions.

DIN CONNECTOR – Deutsche Industrie Norm. The German standard for cable connectors.

DIRECT COAT – The coating of an adhesive direct on to a face stock, as distinct from transfer coating in which it is coated on to the release liner and then laminated to the face stock.

DIRECT THERMAL LABELS – Labels that allow a direct thermal printer to create an image by applying heat directly to a label material which has a heat-activatable coating. No ribbon is required in the printer. Originally referred to as chemi-thermal labels.

DIRECT THERMAL PRINTING – An electronic

printing process in which a special heat-sensitive surface coated material is turned selectively black by the direct application of heat through the contact apertures of an electronically-controlled thermal print head. The print head consists of numerous pixels (needles) in the form of a grid or matrix that are heated and cooled selectively by a microprocessor controller.

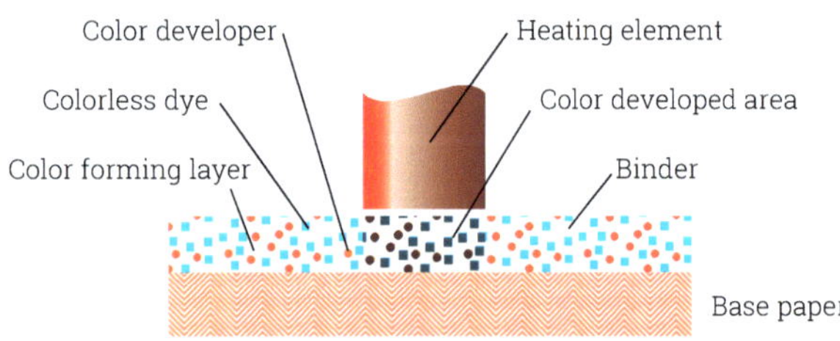

The special heat-sensitive chemical coating on the label substrate darkens under the action of the heating elements.

A simple test to determine whether a substrate is a direct thermal image material or not is to hold it close to a radiator, light bulb or other hot surface where direct thermal materials will rapidly darken and turn black. Accelerating a finger nail across the label surface will produce a black friction line.

Direct thermal label paper is available in two basic qualities. The first is used for well-controlled, short-term labeling where there is very little chance of the image being contaminated. This known as 'Unprotected' or 'Non top-coated thermal'; most in-store printed price-weight labeling of fresh produce – fish, delicatessen, self-weigh fruit and vegetables – use this quality.

The other direct thermal label material is called 'Top-coated' thermal paper. This may also be barrier coated on the reverse to give protection on both the print and the adhesive surfaces. These protected coatings are designed to prevent contamination of the thermally-printed image from any condensation, fats, grease, solvents and plasticizers from packaging films that the labeled product may come into contact with during transportation and storage.

In the supermarket environment, top-coated grades will be found on fresh and frozen products that have been packed and labeled in a factory then transported in refrigerated trucks to the supermarket. Principally pre-packed meat and cheese that requires an individual description and barcode.

A premium grade of top-coating to give better protection is also available, while

direct thermal film on direct thermal paper/film laminations are used where greater protection or label strength is needed, e.g. airline baggage loop labels.

For thermal printers that print faster, users should specify thermal paper that is more sensitive. Thermal labels with a faster thermal response can print at speeds up to 12 inches (305 mm) per second, or at slower speeds where the print head temperature can be reduced, allowing a longer print-head life.

For applications that require a shelf life longer than six months – unless it is very well controlled or dark environment – thermal transfer labeling is used. This is because the thermal transfer printed image has very good longevity. *See also Thermal transfer printing.*

DIRECT-TO-PLATE – The process of using digital information to laser image a printing plate or design roll thereby bypassing the film stage of production.

DIRECT-TO-PRESS OR DIRECT IMAGING PRESS – A type of offset press in which the printing plate is imaged directly on the press, using a built-in laser exposure unit, from instructions received via data from PostScript RIPS stored on a computer. This eliminates many of the manual steps of conventional pre-press, with preparation time per job being greatly decreased.

The concept was invented by Presstek and first seen on the Heidelberg GTO-DI.

DIRECT-TO-SHAPE PRINTING – Direct digital printing onto bottles, cans, tubes and other types of packaging without the need to decorate the shape using printed labels.

While full-color inkjet printing onto shaped surfaces has existed for several years, it is still in its infancy in terms of full-scale production for printing volume shaped products. Most systems to date are primarily used for prototyping, test marketing, ultra short runs, etc.

The potential of using inkjet printing for producing personalized bottles (particularly plastic bottles) and cans, shorter runs, multiple versions and variations, etc, and at the same time reducing time to market, having greater production flexibility and the opportunity to adding value, are very attractive and at least 10 or more companies currently offer or are developing direct-to-shape inkjet solutions for bottle or can decoration.

Inkjet printing direct on to pre-formed bottles

With bottles and cans normally being cylindrical, they provide a convenient potential market for direct digital printing. A cylinder turning on a mandrel presents a consistent surface for an array of inkjet nozzles. If only a small percentage of the total bottles and can market turns to direct digital printing it still represents a significant target market for solutions providers, as well as an interesting opportunity for brand owners to add value.

There are also other companies that currently make systems for digital printing of cosmetics and other tubes – another potential added-value market opportunity.

Depending on the particular application, speeds of direct-to-shape printing solutions currently range from as low as 800 or so pieces per hour (tubes) up to as high as 36,000 (plastic bottles) per hour. There are other companies looking to use direct-to-shape digital printing for products such as cell phone and soccer balls.

There seems little doubt that UV inkjet technology will have a growing future for direct-to-shape printing.

DISCRETE CODE – A barcode in which the spaces between characters are not part of the code as each character begins and ends with a bar. The spaces can vary in width within defined tolerances.

DISPENSER – A device that feeds pressure-sensitive labels, either manually or automatically, presenting them ready for application by hand or mechanical means.

DISPERSION – A uniform distribution of solid particles in a liquid vehicle.

DISPROPORTIONING – More commonly known as dispro, disproportioning is the calculation that ensures that a printing plate produced in the flat will print an image of the correct size once it is wrapped around a printing cylinder. It involves reducing a plate-ready film in overall image size to compensate for the known distance that the photopolymer plate will stretch or distort on the cylinder.

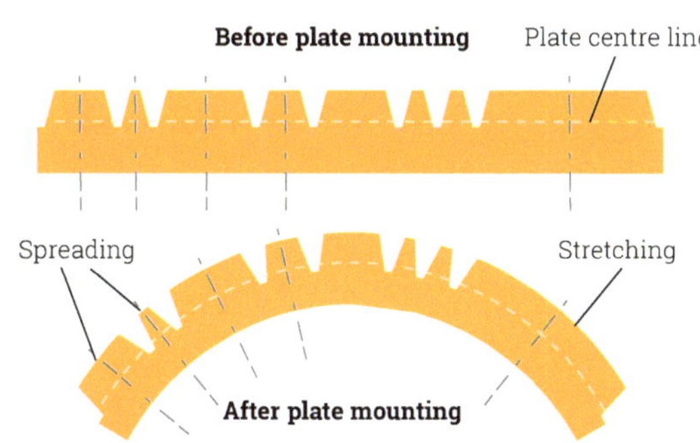

The printing plate produced has to be compensated for the known distance that the plate will stretch or distort on the cylinder.

The amount of distortion will depend upon the thickness of the plate and the thickness of the tape used to mount it on the circumference of the cylinder. Generally, thicker plates and shorter repeat lengths will increase elongation.

Direct-to-plate imaging on to a curved plate, or to a plate already mounted on the cylinder, avoids the need for disproportioning. As the image is applied to a curved surface, no stretching occurs. *See also Plate elongation*

DISTANCE ACROSS – Horizontal gap between labels on a plate or web. *See also Gap across.*

DISTANCE AROUND – Vertical gap between labels on a plate or web. *See Gap around.*

DISTORTION – A change in the dimensions of an object to compensate for change in length when a flexible die is wrapped around a magnetic cylinder.

DITHERING – A system or technique by which a computer achieves an optical illusion by blending or simulating colors from adjacent dots on the screen to make the eye believe there are more colors there than there actually are.

DLA – Danish Label Association. Founded under this name in 2002. Although an association had existed in Denmark since 1983, called the Danish Label Producers (DLP) (and before that the Danish Label Society) this original label association was part of a more general label & packaging association. It was the membership of this association that decided to introduce a dedicated label association in 2002, using the same logo as the DLP but with the emphasis on label production.

Today the DLA membership has 24 printers and 13 suppliers covering over 90 percent of the label production in Denmark.

DNA TAGGANTS – Interest in the use of DNA taggants for the provision of hidden or covert markers used in brand protection applications has been growing in the past few years, with a number of companies now supplying DNA taggant materials. In application, specially processed DNA is mixed or attached to specific media, such as ink, polymer or adhesive, so as to use the unique and hard-to-replicate characteristics of the DNA to be able to tell genuine products from fakes.

Patented anti-counterfeit DNA solutions include a uniquely stable and environmentally tolerant anti-counterfeit ink, which can be authenticated by laboratory DNA analysis; a food grade version; and an instant detection ink that allows for an instant detection function for on-site, real-time authentication. The instant verification process is designed to allow sampling anywhere a product is found using a proprietary testing buffer fluid. A unique biochemical reaction occurs providing a first indication that the label is authentic.

DNA ink can be applied to garment labels or hang tags, or printed onto logos or any other surface. The labels are easily tested for authenticity with the instant detection kit.

DOCTOR BLADE – A thin flexible blade of metal or polyethylene mounted parallel to and in contact with an anilox roller for the purpose of scraping off or metering excess inks or coatings from the non-cell areas of the roller surface.

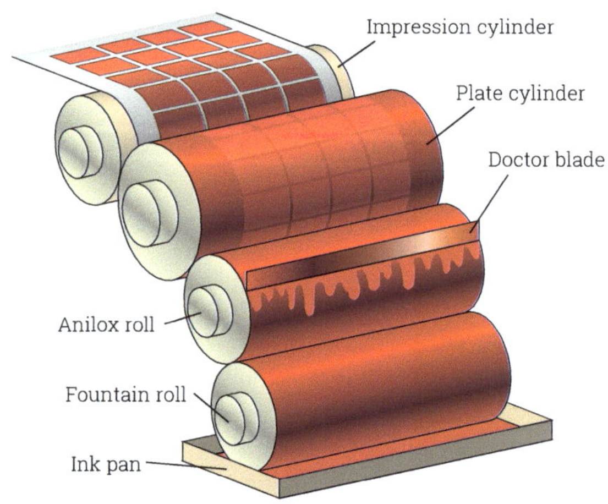

Two roll doctor blade

DOCTOR ROLL – The fountain or metering roller in a flexographic press which doctors off excess ink from the engraved anilox roller.

DOT – The individual element in the halftone printing process. The dots, visible through a magnifying lens, have equal density and spacing but vary in size so as

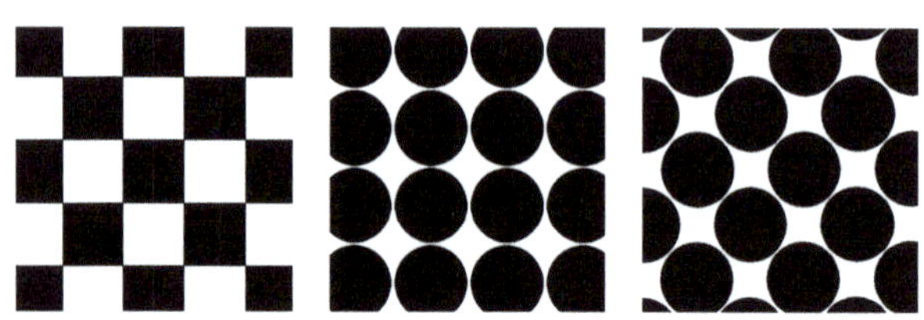

Square dots at 50% | Symmetrical round dots at 75% | Asymmetrical round dots at 75%.

Halftone dots can be generated in a number of shapes.

to control the photographic reproduction effect (in color or black on white) or intensity of the printed color.

The shape of the halftone dot used to produce a continuous tone image can significantly affect the density of the image. Dots can be generated in a number of shapes, including square, elliptical, octagonal, and both symmetrical and asymmetrical dots. With stochastic or frequency modulated dots, they are of uniform size – down to 18 microns – and are scattered in a semi-random pattern to form the printed color.

DOT GAIN – A printing characteristic or defect in which the size of a half-tone dot changes as a result of platemaking and printing by different printing processes. For example, letterpress or flexo dots print larger on a softish paper surface than they do on a harder film surface, which in turn has the effect of darkening the color or tone of the printed image as the ink is now spread over a greater area than originally intended.

Compensation for dot gain can be accommodated at the repro stage provided specific plate, substrate or press information is known. Typically, the printer will undertake a press fingerprinting analysis to obtain dot gain information, which can then be passed to the color separator or designer.

When dots are transferred from a film to a plate, they will tend to grow in size during light exposure. Ink transferred from the plate to the label substrate will again increase in size. Fluid inks and compressible plates used in the flexographic process also tend to increase dot gain, varying according to press and substrate used. Fortunately, dot gain is largely predictable and consistent, and can therefore be compensated for.

DOT MATRIX PRINTER – An electronic or computer-controlled printing machine which produces an image by mechanically firing small pins or hammers against a color or black ink coated ribbon, in turn transferring the ink onto a label surface to produce a printed image. Each pin or hammer produces a fine dot and these can be positioned in a grid to produce letters, words, numbers, codes or simple graphics. The dot matrix print head usually fires against a hard rubber roller or similar hard backing so as to provide a sharp image.

Not all dot matrix printers are capable of printing barcodes. Only those with graphics capabilities are suited to the formation of the expected range of bar widths and layouts. Dot matrix printers are able to print documents, labels, multipart forms and form/label combinations. A key advantage lies with the fact that they can print onto a very wide variety of materials including paper, board, non-paper materials, aluminum and other thin metals.

Generally speaking, dot matrix printing offers the use of low-cost label materials and multi-pass ribbons. The printing density is not high and when related to barcodes would come in the medium to high ranges. The very method of printing the image means that as a system, dot matrix is somewhat noisier than other electronic printing methods.

Dot matrix printers tend to be rugged and thus ideal for those locations where they may be required to run around the clock.

DOUBLE-SIDED LABEL STOCK – Label stock made with two face layers that are mutually supportive, each performing the role of release liner for the other. Silicone and adhesive are applied in alternate stripes to the two face layers, which are applied together so that each silicone stripe meets its adhesive counterpart – so making it possible to use the material like a normal label stock. Used for multi-functional information labeling.

DOVID – Diffractive Optically Variable Image Device. A generic term used in the security and authentication sectors. DOVIDs are a type of OVD, a term used to describe several types of authentication, anti-copy component. *See also OVD and Hologram.*

DOWNTIME – Non-productive time in a label plant which may be due to ink, roll, plate or other press changes, to plate or die problems, to equipment failure, or through accidents or employee absence.

DPI – Dots per inch. For conventional halftone printing, resolution is expressed as the number of lines of halftone dots per inch, also called screen ruling. Higher screen rulings produce higher resolution images because there are more dots per square inch used to reproduce detail. May also be expressed in lines or dots per centimeter.

DRAW DOWN – A term used in ink-making or color matching in which a small amount of ink is drawn down across a paper surface by a meyer rod or pallet knife to roughly determine the color or shade.

DRIER – Also spelt dryer. Metallic compounds or other substances added to printing inks to speed up the drying time. The term is also used for that part of a coating machine or printing press through which a web travels to dry the coating, adhesive or ink.

DRIFT – Printing which runs or drifts out of register due to mechanical or substrate imperfections.

DROP-ON-DEMAND INK JET – Inkjet printing offers the widest range of applications of all the digital printing processes, as well as the largest selection of inks, printing formats and substrates. In inkjet printing the printheads eject micro droplets of ink through nozzles so as to produce images on a wide variety of substrates (paper, plastics, glass, ceramics, textiles, etc) without any contact.

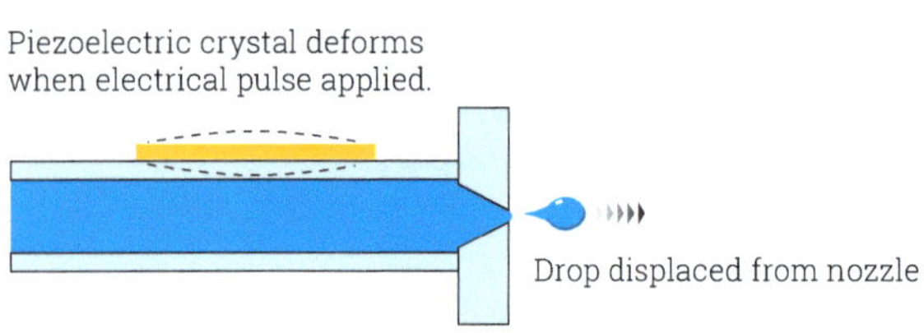

An electrical pulse is applied to a piezo-elecrtical crystal, which deforms and emits a droplet of ink.

Inkjet printing may be continuous or drop on demand. In Continuous Inkjet (CI), a continuous flow of ink is deviated to create an image pattern; Drop on Demand (DOD) creates and ejects a droplet only when it is needed. For the majority of inkjet label and package printing presses the droplets are ejected when the ink reservoir wall is deformed by an electrical pulse applied to a piezo electric crystal. Depending on the electric current applied, an ink droplet of a uniform or variable size is ejected.

The DOD process is able to handle a wide variety of inks and, being digital, it has the ability to produce variable information on demand. Two different types of drop-on-demand printers are available: piezo-electric and thermal.

Piezo-electric printers make use of a voltage pulse applied to a piezo-electric crystal to generate a pressure pulse in the imaging head, which then emits a droplet of ink. This piezo Drop on Demand process may be binary inkjet or greyscale inkjet. In the greyscale process the ink droplet size may vary, unlike binary inkjet in which the drop, when expelled, always has the same size.

Greyscale inkjet printing is able to achieve a much higher level of quality than that obtained from binary inkjet using the same resolution.

Thermal in-jet printers may be bubblejet – which use a resistive heating device to make the ink boil and so produce a droplet that will be expelled onto the substrate; or solid ink/phase change type printers that use heat to melt solid sticks of ink into a liquid, which is then converted into droplets and ejected using impulses from a piezo-electric crystal. The droplets then convert back from liquid to solid once they contact the substrate.

Modern printheads can have a large number of nozzles printing relatively small drops and producing fast coverage with high quality printed results. Drop-on-demand ink jet is already being used in a variety of printing markets, from posters to banners, documents and tile decoration and, more recently into the label and packaging fields where it can be easily integrated onto a production line without creating a significant disruption to the process.

One of the key aims of package printing is to print the label direct on to the primary packaging in-line with the filling process. Currently, drop-on-demand technology is about the only technology capable of doing this, although there are challenges to achieving and acceptable quality and performance. It is anticipated that drop-on-demand ink jet will become a major player in the field of package printing, including direct printing of full graphics quality images onto bottles or containers. *See also Inkjet resolution and Direct-to-shape printing.*

DRY EDGE – To avoid adhesive or silicone coatings flowing off the sides of reels and contaminating machine rollers during pressure-sensitive laminating, narrow adhesive-free dry edges are allowed. These are also called selvedge and are usually removed during subsequent slitting.

Adhesive strip coating patterns can be designed to allow a dry adhesive free end to a label to facilitate easy hand removal from the release liner for application. This can also be achieved by de-lamination, printing an adhesive killing pattern of varnish onto the adhesive coating, re-laminating and then die-cutting the label to register with the pattern.

DRY GUM – A label stock that needs to be moistened to enable it to stick. *See also Gummed paper or Particle gummed.*

DRY LAP – *See under Pattern coated.*

DRY OFFSET – Dry offset, also called waterless offset, is a development of the offset litho process and utilizes a coating of silicone which covers all parts of the printing plate not required to print and so eliminates the use of water as a means of keeping the non-printing areas of the plates clear of ink. The inks, being greasy, are instead repelled by the silicone and the difficulty of maintaining the correct balance between water and ink on the printing plate is avoided. The printed and non-printing areas of the plate are still at a single level and the process remains planographic. *See also Offset printing and Planographic.*

DRYER – *See Drier.*

DRYING TUNNEL – Part of a coating machine or printing press fitted with heaters (usually hot air blowers or infra-red) which enables a printed or coated web to pass through to speed up the drying process.

DTP – Desk Top Publishing.

DUAL DOCTOR INK CHAMBER – A cartridge on a flexographic press designed to supply ink to the anilox roll in an enclosed area (chamber). Ink is pumped into and circulated through the cartridge positioned on the anilox roll. Two thin doctor blades are in contact with the surface of the anilox roll; one blade contains ink within a cartridge, the other contains and meters the ink from the non-cell areas. When using these systems the fountain roll can be dispensed with.

DUAL HEIGHT – Found when two different cutting heights combined in one tool, possible in both solid cutting tools and flexible dies. In a die drawing, it should be clearly indicated which cutting lines are cutting the different layers of material.

DUCT – The trough of ink on a letterpress or offset printing machine, usually including an adjustable flexible blade, which contains the supply of ink and by means of which the ink is presented to the duct roller.

DUROMETER – An instrument or gauge used for measuring the hardness of rubber (inking rollers, etc). Hardness comparisons of rubber surfaces are provided by 'Shore' hardness readings.

DWELL OR DWELL TIME – The length of time in which pressure is applied to a pressure-sensitive label during production or application. The time that a pressure-sensitive material remains on a surface before it is tested for adhesion or removability. Dwell time is also used to describe the period during which a hot stamping die or embossing head remains in contact with the surface of a label material. Sometimes referred to as contact time.

Dwell time is also used to describe the period a heated platen and an instantaneous heatseal label are held together on a labeling surface to achieve adhesion.

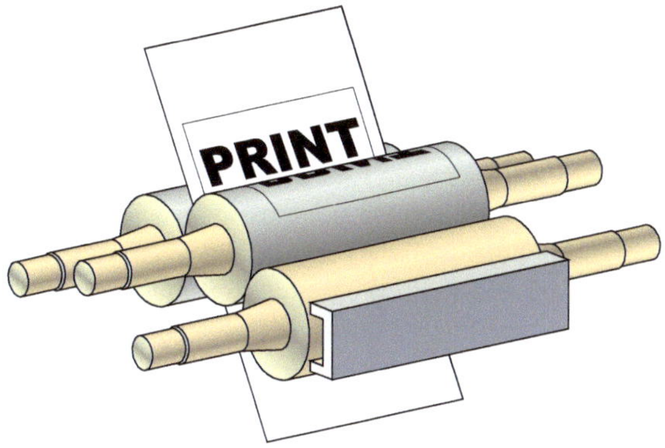

Diagram showing dual doctor ink chamber. Ink is pumped into and circulated through the enclosed chamber. Twin doctor blades are in contact with the surface of the anilox roll.

E

E. & O.E – An abbreviated form of the term 'Errors and Omissions Excepted.'

EAN – The European Article Numbering system that was administered by EAN International. The bar code standard used throughout Europe, Asia and South America. Now administered by GS1. *See EAN International and GS1.*

EAN INTERNATIONAL – Originally set up as the 'European Article Numbering Association, it was so successful in developing the EAN bar coding system in the retail sector in Europe that it led to the establishment of Numbering Organizations in many other countries and from all continents. Following this success it subsequently changed its name to EAN International and saw EAN standards adopted by other industries to meet their identification needs, including health care, packaging, transport, electronics and publishing.

In 1990. The Uniform Code Council (UCC) and EAN International (EAN) signed a co-operation agreement formalizing their intent to co-manage global standards.

In 2005, a new name for the organization – GS1 – was launched worldwide. GS1 partners with the International Standards Organization (ISO). *See GS1, ISO, GS1 EPCglobal.*

EAS SYSTEMS – Electronic Article Surveillance systems are anti-theft systems designed to protect retail merchandise and other products prior to sale. Central to the system are electronic security tags or labels attached to products/assets. These tags are used with activation/de-activation and detection equipment to control unauthorized removal of products from controlled exits and enclosed areas in retail stores or other premises. When someone passes a gate area holding an item containing a tag that has not been de-activated at the checkout, an alarm sounds.

There are several different types of EAS systems, but in each case the EAS tag or label is affixed to an item. The tag is then deactivated when the item is legally purchased (or legally borrowed in the case of, say, a library book) at the checkout. As the item passes through the exit gates, the gate senses whether the tag is active or deactivated, sounding an alarm if necessary.

EAS systems are widely used in applications where there may be a chance of theft, such as retail theft, lending libraries, etc. When the tag is built into the product at the point of manufacture or packaging it is known as source tagging. This makes the subsequent labeling of goods unnecessary, so saving time and money. *See also Source tagging.*

EAS TAGS – Miniaturized and low cost anti-theft detection devices that can be attached to retail products or incorporated into labels or packaging. The proximity of such tags is automatically sensed by EAS detection equipment, which then triggers an alarm.

ECO-EFFICIENCY ANALYSIS – Eco-efficiency analysis compares both the economic and environmental impacts that a product or process has over the course of its life-cycle.

ECOLOGO – EcoLogo, an environmental standard and certification mark that was established in 1988 by the Canadian government but is now used globally. EcoLogo meets the ISO 14024 standards for eco-labeling

EDGE GUIDE – A web edge guidance unit on a web-fed label press that keeps the web travelling straight and true as it passes through the machine. Web guides play an important part in the control of the web through a press and may consist of photo-electric, UV light or ultrasonic sensors that are focused on the edge of the web. These sensors feed data back to either an adjustable pivoting roller, which guides the web to the left or right as required, or to the infeed reel stand which can also move to the left or right.

EDGE LIFT – Occurs when the edge of a label rises from the labeled surface, most frequently on small diameter curved surfaces such as ampoules or vials. Resistance to edge lift will depend on the bonding strength of the adhesive and the flexibility of the facestock used.

Self-adhesive labels that have edge lift on one

corner are usually exhibiting a memory curl in the label material. This is usually caused by peeling the label away from the liner before application. The corner that lifts is the one which has been held to remove the label, which may have adversely affected the adhesive surface with sweat or other hand-borne contaminations.

If the backing liner is pulled away from the label, the label will subsequently remain flat. This allows the adhesive to remain in contact with the surface being labeled during the adhesive bedding down period which, depending on the type of labeled surface, may be up to 48 hours.

The use of a finger nail or spatula when touching the adhesive will avoid 'blinding' or deadening the adhesive surface.

EDM – Electronic discharge machining process for removing metal, as in rotary die manufacture.

EDM DIE – Label cutting dies produced using electronic discharge machining (EDM). Dies made using this process last longer than standard dies.

EDP – Electronic Data Processing. The processing of web format, fanfolded and sprocket punched labels – usually blank – through pin-feed computer printing equipment, or through laser printing operations, some of which no-longer require pin-feed..

EJECTION RUBBER – A variety of materials used for facilitating the flatbed and rotary cutting process. These materials used vary in thickness, structure and hardness (shore). It is important to use ejection rubber that is slightly higher than the actual height of the cutting line. This will cause compression (sufficient energy and ejection force) on the rubber thus ensuring complete ejection of the cut out part.

ELASTIC MEMORY – The tendency of some materials to try and return to their original length after they have been elongated.

ELECTRO-CONDUCTIVE INKS – Used to provide an electro-conductive coating or strip on tickets or labels.

ELECTRONIC DISCHARGE MACHINE/ MACHINING – *See under EDM.*

ELECTROPHOTOGRAPHIC PRINTING – Non-impact printing process used by the majority of laser printer and photocopier manufacturers, as well digital press manufactuers such as HP Indigo and Xeikon.

Digital data, which has been converted to raster scan patterns, is transferred onto a pre-charged electro-sensitive belt or drum by a laser beam or an array of L.E.D (Light Emitting Diodes). Toner powder (or liquid), which carries a charge which is opposite to that of the latent image, is then cascaded over the electrically-charged image and adheres to the charged areas. The toner coated image is brought into contact with the label material to which it transfers.

The correct condition of the label material is critical to the image transferring evenly. Paper materials with a moisture content of 4% are ideal, with a maximum of 6% being allowed by most OEMs (Original Equipment Manufacturers). Finally, the toner is bonded (or fixed) to the label material by heat, pressure or a combination of both.

Some flash fusion units that use Xenon and halogen lamps are also used and these are less demanding of self-adhesive label materials because only the dark toner powder is heated during flash fusion.

Toners used in electrographic printing may be powders (such as used by Xeikon presses) or liquids (as used on HP Indigo presses). Two-component powders, one-component powders and liquid toners are used depending on the machine manufacturer. Liquid toners have much smaller particles and enable higher resolution and finer details.

ELECTROMAGNETIC SPECTRUM – The complete spectrum of specific wavelengths of electromagnetic radiation which includes both visible light and 'invisible' light and which ranges from cosmic, gamma and X-rays at one end of the spectrum, through ultra-violet, visible light and infra-red radiations, to ultra-short, short and long radio waves.

Visible light, which enables colored objects to be seen, is only a very small part of the electromagnetic spectrum. Objects are seen as colored because they selectively absorb some wavelengths of light, usually white light, falling on them and reflect the rest as the color.

UV radiation is commonly used by the label industry as a method of drying (curing) UV-curable inks and varnishes and some adhesive and silicone technologies. Electromagnetic radiations are also used in some security printing applications and some types of EAS anti-theft tags, while radio frequencies are used in RFID systems. *See also EAS systems, RFID and UV inks.*

ELECTRONIC DOT GENERATION – A method of producing halftones electronically in digital imaging using scanners and pre-press systems.

ELECTRONIC PRE-PRESS – Computer-assisted design stages for creating a new label from original concept through to the separated, stepped films required for plate making. *See also Digital workflow.*

ELECTRONIC PRODUCT CODES – *See under ePCs.*

ELECTRONICS/ELECTRICAL LABELS – The electronics industry is generally regarded as a single entity, with much of the global production, distribution and purchasing centrally controlled by major multinational companies working to regional or global strategies. Components (and labels) are frequently sourced globally and many companies may not even order or pay for their products in the country of consumption. Some large OEMs have become almost virtual companies, which basically invent, design and market products – but do not actually make anything. Manufacturing is sold to contract electronic manufacturers.

Key sectors for the electronics industry, and for the use of performance, durable and/or security/authentication labels are:

- PCs, including components and peripherals
- Hand-held computers
- Consumer electronics equipment (video and audio equipment)
- Digital cameras
- Electronic games and consoles
- Mobile phone handsets
- PCBs
- Rechargeable batteries
- Power supply AC/DC cables
- Industrial components
- Electronic devices (electronic tubes, semi-conductors, integrated circuits, etc)

As can be seen, the electronics sector manufactures a wide range of products, many of which are then incorporated into other products or equipment. Consequently, the labeling requirements for this sector are also quite diverse and demanding, with durable, bar code, asset management, inventory and security labels required.

A guide to the range and variety of label requirements required in the electronics sector can be seen in the table shown below.

Label types used in the electronics and electrical sectors

- Consecutively numbered and/or bar code labels for identification
- Void, tamper resistant and tamper-evident labels
- Asset management and inventory labels
- Rating and serial plates
- Circuit board labels - both during processing and for end-usage
- Care/use instruction labels
- Warning labels
- Warranty labels
- Flexible nameplates
- Cord and plug labels
- Quality inspection/checking labels
- Warehouse and inventory control labels - shelf, rack, bin, location
- Security labels - authenticity, anti-counterfeiting, theft deterrence

---

ELECTROSTATIC ASSIST – A method of using electrostatic forces to assist the removal of ink from gravure cylinder cells cleanly during printing.

ELECTROSTATIC PRINTING – A printing process that uses a specially coated paper which accepts a charged area created by an electron beam. Toner sticks to the electron charged areas to produce the printed image. Now principally used in large format plotters and printers, but previously used for copiers that were superseded by plain paper copiers and electrophotography. *See Electrophotographic.*

ELLIPTICAL DOT – A halftone dot with an elliptical shape rather than a circular shape, which produces better tonal values. In halftone printing, elongated dots which improve gradation of tones, particularly in middle tones and vignettes. May also be called chain dots.

ELONGATION – The distance a material will stretch lengthwise before it breaks, expressed as a percentage of the original length.

EMBOSSING – The process of raising a design or image above the label surface, often through the use of a set of matched male and female dies. The female die is a depressed image, while the male die is prepared so as to push the label paper into the female embossing die to create a raised (embossed) area on the label design. No ink is used but the web may, or may not, be already printed. Where embossing is carried out on an un-printed area of the substrate it is known as blind embossing; if embossing over a printed image to convert it to a raised image it is called register embossing. Embossing may also be combined with hot foil stamping in one action.

Embossing is also used to reproduce the extremely fine holographic interference patterns in a polymer substrate provided with a reflective layer, which gives the hologram its brightness. Holographic patterns may also be directly embossed into a reflective metal as an authentication mark.

EMBOSSING DIE – Embossing dies are used to deform the surface of a substrate to create a raised (embossed) or sunk (debossed) design. It can be achieved with a matched pair of male and female dies. The female die has the required image incised into the surface; the male die has a matched raised image. An offset is applied to one or both images to enable the dies to accommodate the thickness of the material being embossed.

During the embossing process, the two dies are pressed together through a substrate to create a raised image (blind embossing).

In the process of de-bossing – which is the reverse of embossing, the positions of the relief die and the counter die are reversed.

Dies are manufactured from a range of materials – most commonly steel, brass or aluminum – depending on the following factors:

- Complexity of the image
- The nature of the material to be embossed
- The depth of relief required
- The perceived product life
- Any specific product use requirements, e.g. pharmaceuticals or food.

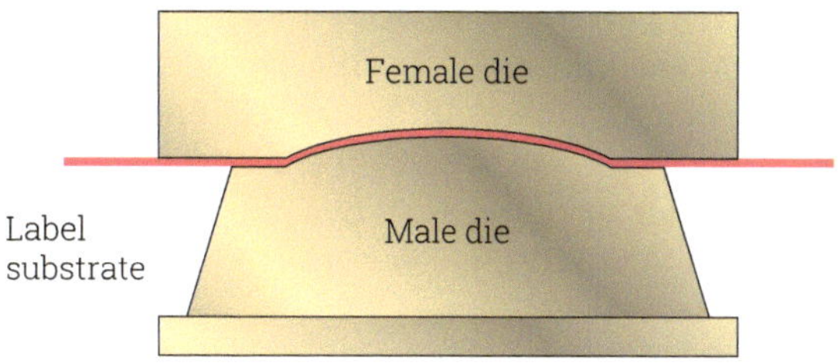

Male and female embossing dies create a raised image in the label substrate.

Picture shows male/female rotary embossing unit. *Source: Spilker*

Male/female embossing dies have their widest application in rotary embossing, although other approaches may be used for certain applications.

EMI – Electromagnetic interference. The electronic 'noise' or other extraneous signals that can set off the detector alarm in an electronic article surveillance (EAS) system, or conversely reduce the detector's sensing area and/or efficiency.

EMID TAG – Electromagnetic Identification tag. A much cheaper alternative to RFID tagging which can be used by brand owners, particularly in the pharmaceutical sector, to identify and track products in the supply chain. EMID technology consists of a tag and a reader, with each tag being considered as a magnetic bar code. The technology is similar to electronic article surveillance (EAS) systems, except that it is capable of doing a lot more.

EMID tags, which can be read without the need for an optical line-of-sight scanner, enable the remote reading of magnetic features. Simple linear magnetic tags are made up elements of soft magnetic material of less than one micron in thickness, mounted on a thin layer of PET. The system is extremely accurate.

Unlike RFID, except to the extent that it can provide each individual item with its own unique identity, EMID tags can be used as a basic authentication technology. They can be delivered as threads, labels or by hot-foil transfer, and are compatible with RFID tags.

EM TAG – Electromagnetic tag. A miniaturized and low-cost anti-theft detection device that can be attached to retail labels or packaging. First introduced to retailers during the 1980s, although their origins in lending libraries date from much earlier.

A typical EM tag comprises Transmit and Receive functions in separate antennae or combined receiver format, together with control electronics and a power supply. The transmitter generates a field of oscillating electromagnetic energy at a fixed frequency as determined by the system design. As energy saturation is reached, the tag generates a set of harmonics and produces a signal that is interpreted by the receiving antenna, thus activating an alarm.

To de-activate the EM tag, its magnetic properties must be altered to prevent it reaching saturation when entering the detection field. EM tags may be repeatedly de-activated and re-activated, depending on the material used for their construction. *See also EAS systems.*

EMULSIFICATION – The process of dispersing one liquid in another when the two liquids do not naturally mix.

EMULSIFYING AGENT – A substance used to produce an emulsion of two liquids which do not naturally mix.

EMULSION – When two or more immiscible (or un-mixable) liquids are held together in a homogeneous mixture by the action of an 'emulsifying agent'. The photosensitive layer on a piece of film, paper or plate onto which an image can be exposed. Some pressure-sensitive adhesives are emulsion adhesives.

EN 13432 – Established as the EU Directive on Packaging and Packaging Waste standard for packaging recoverable through bio-degradation and compostable packaging in 2000.

Many EU member states' national standardization bodies have adopted the standard, for example in the UK it is known as BS EN 13432 under the British Standard Institution.

ENCAPSULATION – The process of encapsulating or trapping a substance, such as a fragrance or dye, within a coating so that it can be applied like an ink on a printing press. Label and forms printers will be familiar with NCR type carbonless papers where encapsulated dyes are coated onto the reverse of a CB (Coated Back) paper so that when pressure is applied with a pen or impact printer the capsules break. This enables the dyes to flow into a receptive clay coating on the surface of the CF (Coated Front) paper or CF label below it in the form or form/label set.

ENERGY RECOVERY – Conversion of energy, such as methane gas or the heat from decomposition or incineration of waste, into electricity or other energy.

ENGINEERED FILMS – Filmic release liners demand properties not normally found in blown or cast extruded films made for other end-users. To meet the specific film liner requirements of properties such as the tensile and tear strength needed to withstand the stresses of die-cutting and matrix stripping, and the right stiffness/flexibility balance for trouble-free label application, special multi-layer customized co-extrusion films have been developed. These are known as 'Engineered' films.

Specific properties for label liner – or face materials – can be provided in each layer, e.g. a central layer that provides web break resistance for high speed dispensing, and excellent web flatness; a top and bottom co-polymer layer that provides a highly receptive surface for siliconization. The engineered constructions are free from chemical additives that could interfere with the cure chemistry of thermal or radiation cure silicones. To enhance silicone receptivity one side can be corona discharge treated.

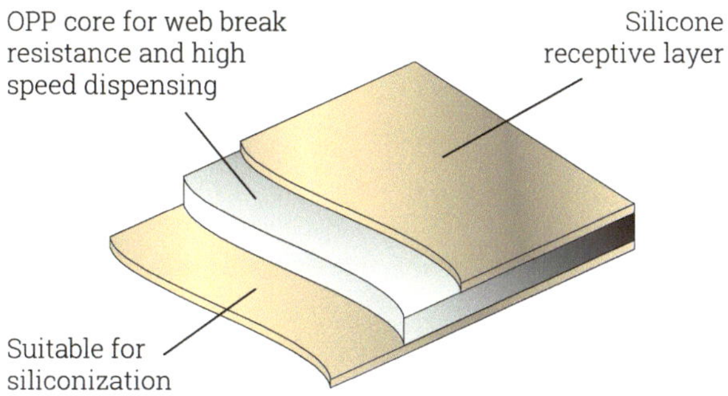

Three layer co-extruded film for a silicone release liner

ENGRAVED ROLLER – Rollers available with various surfaces, such as chrome or ceramic, which have been engraved with mechanical or laser engraving equipment to produce a pattern of tiny cells which are used to meter inks or coatings. *See Anilox roller.*

ENVIRONMENTAL MANAGEMENT SYSTEM (EMS) – An EMS should be part of a label converter's overall management system and is intended to provide a framework for the continual improvement of the company's environmental performance. It is ideally integrated with other management systems that may be in use within the company, such as ISO 9000 or Six Sigma.

Environmental performance and management systems have become of increasing interest to the label industry in recent years as new environmental and waste management legislation has been introduced by governments around the world. Implementing a successful environmental management program offers benefits to industry suppliers and label converters, including:

- The more efficient use of resources, with consequent cost savings
- Competitive advantage with customers
- Confidence in the ability to meet new legislative requirements
- A reduce likelihood of unforeseen environmental liability
- Increased environmental awareness by employees and the public.

Companies in the label industry can introduce and adopt an environmental management program through the use of ISO 14000, a system borne out of the BS EN ISO 9000 quality management system. ISO 14000 requires a company to determine its current environmental performance and to develop programs and systems to ensure that environmental performance is continuously monitored and improved. *See also ISO 9000.*

ENVIRONMENTAL PROTECTION AGENCY – *See EPA.*

EPA – The U.S. Environmental Protection Agency. Established in 1970 by the White House and Congress in response to a growing public demand for cleaner water, air and land. EPA works to develop and enforce regulations that implement environmental laws enacted by congress and, where these are not met, can issue sanctions. Voluntary partnerships with industries, businesses, organizations and the State and local governments work to set pollution-management goals; including conserving water and energy, minimizing greenhouse gases, cutting toxic emissions, re-using solid waste, controlling indoor pollution and getting a handle on pesticide risks.

EPCGLOBAL – EPC Global leads the development of industry-driven standards for the Electronic Product Code™ (EPC) to support the use of Radio Frequency Identification (RFID) in today's fast-moving, information rich, trading networks.

ePCS – Electronic product codes. The electronic equivalent of a physical bar code. Used with RFID technology an ePC enables each item in a smart tag's chip to be uniquely identified, significantly enhancing manufacturers' ability to protect their brands and trading partners from the threat of counterfeit merchandise. They do not require a direct line of sight to be read.

The ePC code is a 96-bit code, which enables components used or sold to be individually labeled with a unique serial number or even provided with a digital certificate. ePC codes are expected to one day replace bar codes.

EPS – Encapsulated PostScript (EPS). A PostScript document with additional restrictions which is intended to be usable as a graphics file format. In other words, EPS files are more-or-less self-contained, reasonably predictable PostScript documents that describe an image or drawing and can be placed within another PostScript document. Simply, an EPS file is a PostScript program, saved as a single file that includes a low-resolution preview "encapsulated" inside of it, allowing some programs to display a preview on the screen.

EPSF – A four-letter acronym used in Macintosh applications to identify Encapsulated PostScript Files.

EQUILIBRIUM MOISTURE CONTENT – The moisture content of a paper that has reached a balance with the atmosphere surrounding it, i.e. in a condition in which it will neither give up nor absorb moisture.

ESD – Etiket Sanayicileri Dernegi, the Turkish Label Association, was founded in 1998 in Istanbul by 9 founder members. The membership has since grown to 124 members. It is a fairly active association which has integrated with other associations and in particular FINAT.

ETCHING – A chemical process utilized in producing flexible cutting dies. Ferric Chloride is used to remove all unnecessary material from a plate. The cutting line remains untouched as it is protected by a photo-resist film.

EUROPEN – The European Organization for Packaging and the Environment is a multi-national European trade group that promotes the profitable and sustainable packaging for the region. It includes Coca-Cola, DuPont, P&G and others.

EVAPORATION – The changing from a liquid to a gaseous or vapor state, as when a solvent or water leaves a printed ink film during the ink drying process.

EVENT MONITORING – Refers to software-controlled recording (using CCTV/video and/or computer) of a number of pre-determined activities or events likely to be associated with retail theft. May also be called ‘critical event monitoring’.

EXIT PROTECTION – The arrangement of electronic article surveillance (EAS) devices at a store exit(s) to detect all live EAS tags which are passed through or taken near the exit devices. *See also Detector gates.*

EXTENDED GAMUT
Where CMYK print stations are augmented by orange, green and violet. Allows a press to cover up to 90 percent of the Pantone color gamut. Available for multiple print systems including UV inkjet, UV flexo and offset.

EXTENDED TEXT LABELS – *See Booklet labels.*

EXTENDERS – Any material added to an ink to reduce its color strength and/or viscosity.

EXTRUDER – A plastics processing machine used for the processing of plastics molding compound, mostly in pellet form, in preparation for downstream conversion, whereby the molding compound is compressed, mixed, plasticized, homogenized and/or chemically converted. Extruder applications include the manufacture of blown, cast and calendered film in rolls and sheets.

EXTRUSION COATING – A method of coating a web of paper with resins, plastics or similar hot-melt compounds. The coating is applied through an extruder die positioned immediately above the nip between a supporting roll and a chill roll.

F

FAHRENHEIT – A method of temperature measurement named after Gabriel Fahrenheit, who invented the mercury thermometer in 1714. Two reference points are used with the Fahrenheit scale; the freezing point of water (32 degrees Fahrenheit) and the boiling point of water (212 degrees Fahrenheit).

Body temperature on the Fahrenheit scale is 98.4 degrees. To convert from Fahrenheit to Celsius/Centigrade first deduct 32 from the Fahrenheit figure, then multiply the result by 5 and divide by 9, e.g. 70 degrees Fahrenheit minus 32, multiplied by 5 and divided by 9 = 21 degrees Celsius/Centigrade

As both the Fahrenheit and Celsius/Centigrade scale are used within the global label industry a conversion table for some of the more commonly used temperatures is provided under Converting metric to imperial.

Fahrenheit	Celcius/Centigrade
-112°F	-80°C
4°F	-20°C
23°F	-5°C
32°F	0°C
122°F	50°C
158°F	70°C
212°F	100°C
250°F	121°C
320°F	160°C
450°F	232°C
482°F	250°C

Conversion of some common Celsius/Centigrade temperatures used in the label industry to their Fahrenheit equivalents.

FABRIC LABELS – Labels consisting of printed or woven fabrics – or combinations of both – which are sewn or heat-sealed to other fabrics in garment or textiles labeling to provide brand identity, care information, descriptive details, fiber content, country of origin, manufacturer or other information.

Fabric labels are used with clothing, most household textiles (bedding, sleeping bags, curtains, carpets, tablecloths, towels) and for items such as umbrellas or flags. Printed care labels have to be applied to most clothing; other information on fabric labels may include that on flammability or flame resistance, shrinkage or pre-shrunk information, Woolmark or Seal of Cotton identification. *See also Garment labels and Care labels.*

FACE-CUT LABEL – Any pressure-sensitive label in which the face material has been cut to the liner. A die-cut label product from which the waste matrix around the labels has not been removed.

FACE MATERIAL (FACE STOCK) – A paper, film, fabric, foil or other material which is bonded to the adhesive layer and which forms the functional part of the label construction in a pressure-sensitive laminate. The top layer of a pressure-sensitive laminate, the three layers being the label, the adhesive and the backing or release liner.

FACESTOCK – Any paper, film, foil, synthetic or other material that can be converted into labels.

FADING – A gradual decrease in the intensity or brilliance of a color, which may occur through ageing or exposure to light.

FALSE ALARM – The alarm triggered by the detection equipment in an electronic article surveillance (EAS) system when no EAS tag or label is present within the detection area.

FAN-FOLDED LABELS – Labels manufactured from a continuous web of label stock and then folded

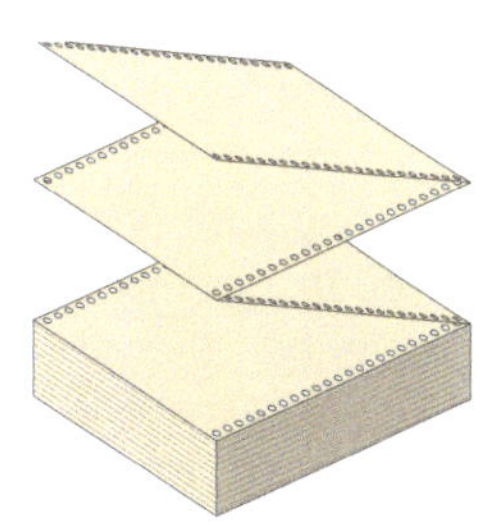

Fan folded accordian format of continuous labels

into a zig-zag accordion style format for packing flat in boxes. The fold is created by horizontal perforations in the release liner. Continuous fan-folded labels are mostly used for data processing applications. *See also EDP labels, Continuous labels and Zig-zag labels.*

FASTNESS – The property of a paper, ink or a printed label, which makes it resistant to a change in color.

FDA – The FDA (US Food and Drugs Administration) is the U.S. federal agency responsible for ensuring that foods are safe, wholesome and sanitary; that human and veterinary drugs, biological products, and mechanical devices are safe and effective; that cosmetics are safe; that electronic products that emit radiation are safe. The FDA also ensures that these products are honestly, accurately and informatively represented to the public, including issuing regulations governing the use of self-adhesive labels for contact with foodstuffs (FDA 175.125 for direct contact, and FDA 175.105 for indirect contact).

As the FDA continues to look at strengthening food labeling regulations by making food labeling clearer and stricter, as well as easier to read, it is important that the latest regulations should always be assessed and acted upon.

Some of the Administration's other responsibilities include OTC (Over-the-Counter) and Prescription Drug labeling and the issue of publications such as Cosmetics Labeling and Label Claims and The Cosmetics Labeling Manual. Current details on the FDA website.

FELT SIDE – In paper making the felt side is the opposite side to the wire mesh that is used for forming the paper on the wet-end of the paper-making machine. The top side of the paper during manufacture. The felt side has a felt like texture with more short fibres than the wire side. Historically, the smoother side of the paper for printing, although modern paper-making machines can make a very similar surface on both the felt and wire sides of paper. *See also Wire side.*

FESTOON – System for taking up a label web in order to continue feeding a press while a splice is being made.

FIBERS, SECURITY – Fibers in a variety of colors or colorless can be introduced into paper at the manufacturing stage. The fibers can be made to react to and show up under specific light sources, i.e. UV or IR. Used for the production of security documents, such as checks, vouchers and tickets.

FILE – A collection of data stored on a magnetic medium, which can be processed by a computer.

FILLERS – Materials, often clay and chalk, which are used to fill the pores in paper to improve smoothness, opacity and printability.

FILLING IN – Areas of small type, halftones or a design that have been filled with ink during printing.

FILM – A dark material (usually acetate) coated on one side with an emulsion that is light sensitive. When exposed to light an image is created in the emulsion, which can then be developed in chemicals to produce a film negative or positive for the subsequent making of printing plates or cylinders.

FILMS – A transparent or opaque polymer material used as a face material or release liner for pressure-sensitive labels, or for the manufacture of shrink and stretch labels, in-mold labels, wrap-around labels or cut-and-stack film labels.

Filmic label materials may be manufactured from a diversity of plastics, such as polyester, polyethylene, polystyrene, polypropylene, vinyl or other synthetic materials. Transparent label films can be used to provide a 'no-label' look on the labeled product. *See entries for specific film material, i.e. Polyester.*

FINAL ADHESION – The strength with which a label adheres to a surface under normal conditions. The time taken to reach final adhesion will depend on the type of adhesive, the nature of the label substrate and the conditions under which the label was applied. The minimum time is 24 hours.

FINAT – Fédération Internationale des Fabricants Transformateurs d'Adhesives et Thermocoulants sur Papiers et Autres Supports. FINAT was founded in 1958 in Paris, but since 1970 has been located in The Hague, The Netherlands.

FINAT today provides not just a European but a

global platform for self-adhesive label printers and suppliers to meet, network, exchange and enhance collective knowledge and expertise, develop and maintain standards and tests, provide the facilities for management development and education, and to promote the image of the industry as a dynamic and innovative workspace.

The association has around 600 members companies in over 40 countries worldwide. It is an international association covering almost all aspects of labeling technology and works to define credible industry standards covering product specifications and performance, best practices, quality, terms and conditions of sale, payment terms, intellectual property rights and technical recommendations.

The Association's Technical Handbook and the Test Methods that it contains are recognized as the international reference standard for suppliers and users of self-adhesive materials.

FINAT is also active in the political arena, presenting a collective industry view at EU and government levels and in providing a source of information to individual members and member associations on pan-European and international legislative and political issues – from waste legislation to registration of chemicals, food contact and the environment.

FINAT TEST METHODS AND REFERENCES – Standard, internationally approved methods developed by FINAT for the testing and handling of self-adhesive laminates and labels. The 9th edition, published in 2014 contains two new test methods (FTM28 and FTM29). The edition also incorporates a full review of sections 2 and 3 of the Handbook. Current publications available are:

Section 1:

FTM 1 Peel adhesion (180°) at 300mm per minute
FTM 2 Peel adhesion (90°) at 300mm per minute
FTM 3 Low speed release force
FTM 4 High speed release force (10 – 300 m per minute)
FTM 5 Accelerated ageing – Extended storage
FTM 6 Resistance to ultra-violet light
FTM 7 Silicone coat weight by energy-dispersive X-ray fluorescence spectometry
FTM 8 Resistance to shear from a standard surface
FTM 9 Loop tack measurement
FTM 10 Quality of silicone coated substrates for self-adhesive laminates: release force (300 mm per minute)
FTM 11 Quality of silicone coated substrates for self-adhesive laminates: subsequent adhesion
FTM 12 Adhesive coat weight
FTM 13 Low temperature adhesion
FTM 14 Dimensional stability
FTM 15 Surface tension of plastic films
FTM 16 Chemical resistance – Spot method
FTM 17 Chemical resistance – Immersion method
FTM 18 Dynamic shear
FTM 19 Recycling compatibility of self-adhesive labels
FTM 20 Fluorescence and whiteness
FTM 21 Ink adhesion (basic)
FTM 22 Ink adhesion (advanced)
FTM 23 Die strike (a. paper; b. filmic paper)
FTM 24 Mandrel hold
FTM 25 Evaluation of silicone coverage of coated papers by use of a water based stain test
FTM 26 Wash-off paper and film labels
FTM 27 Ink rub test for UV printed labels, ink surface against substrate or ink surface
FTM 28 Evaluation of bleeding through paper (NEW)
FTM 29 Scratch resistance of an UV ink film on different print materials (NEW).

Section 2: Guidelines

2.1. Recommended joins in self-adhesive laminates for roll labels
2.2. Processing recommendations
2.3. Surfaces requiring caution
2.4. Safety guide for users of self-adhesive label stock products
2.5. Handling and storage of label stock throughout printing and converting
2.6. Qualitative adhesion testing
2.7 Adhesives & Nomenclature for adhesives used in self-adhesive laminates
2.8. Unwind roll chart
2.9. Recommendations regarding static electricity
2.10 Advices about light fastness of printing inks and light exposition (NEW).

Section 3: Test equipment

3.1. Standard FINAT test roller
3.2. Automatic rolling devide
3.3. Sources of Test Equipment
3.4. Testing institutes
3.5. tesa test tapes product information and worldwide addresses.

Copies of the above FINAT publications can be ordered through the 'Knowledge' section of the FINAT website – www.finat.com.

FINGERPRINTING – The printing of a special test form or test run on a press so as to determine the registration, dot gain, distortion and other characteristics of the press. Once known, these can be compensated for at the design, film or platemaking stages.

FINISH – The surface property of a paper or film determined by its surface contour and gloss, e.g. gloss, matt or embossed, rough, smooth, also vellum (Machine Finished, Machine Glazed, Super-calendered), cast, etc.

FINISHING – The last work to be carried out on a client's label order prior to the job being despatched, i.e. slitting, rewinding, inspection, packing, labeling, etc.

FIRST IN, FIRST OUT – Stock issued or used in the order it is received. May be abbreviated to FIFO.

FIRST IN, LAST OUT – Stock issued in rotation. FILO.

FIX-A-FORM™ – Extended text, leaflet-label product created by Denny Bros in the UK by combining a self-adhesive label with a folded leaflet. Originally developed in1970 for a leading agrochemical producer. A key element was the application process in which the Fix-a-Form leaflet labels were supplied on a reel and applied using standard labeling equipment.

The benefits of an automatically applied extended text product information system were soon accepted across a broad industry, not just for information purposes but also for on-pack promotions. By the early 1980s Denny Bros had established an international licensing operation (Fix-a-Form International Ltd) to grant local manufacturing rights to leading label printers around the world.

Today, Fix-a-Form leaflet labels remain the most widely available and most flexible system of its kind. *See also Booklet labels.*

FLAG – A marker, usually a strip of colored paper or board, inserted into rolls of paper, film or pressure-sensitive materials and extending from the edge to indicate a deviation from the standard, such as a splice, specification change or defect, or to mark a specific length. The flag provides a warning to the slitter, press or inspection rewind operator during the next operation in the converting and finishing process. Flags can be inserted into rolls, sheets or fan-folded stacks on the run.

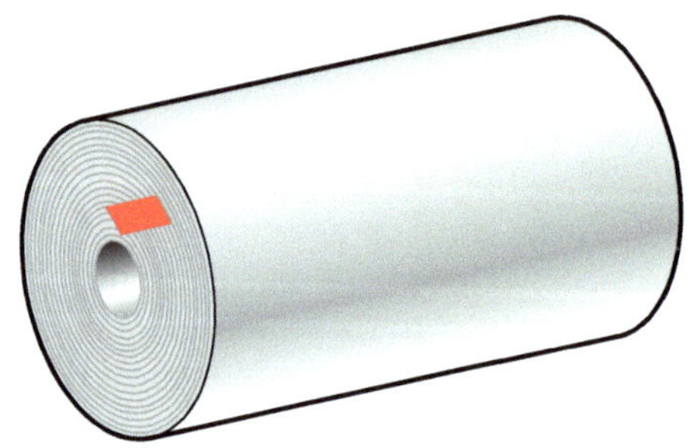

A colored marker inserted into a roll to indicate a deviation or fault.

FLAME RESISTANT PAPER – Paper which has been treated with chemicals to make it resistant to fire or flames. The paper is not actually fireproof, but will not support combustion or carry a flame, although it will char.

FLAME TREATMENT – A method of altering the surface characteristics of polymer materials to enhance the keying of inks or adhesives by passing the material through an open flame to promote surface oxidation.

FLAMING – Surface treatment of polyolefin films or blow-molded containers with an oxidizing gas flame for the purpose of improving printability or adhesion. *See also Corona treatment.*

FLATBED SCANNER – A device that scans images in a way similar to a photocopier. The original image is placed face down on a glass plate prior to the scanning action.

Also refers to a fixed scanner (like those used in supermarket checkouts), which can read barcode symbols at any angle as goods are passed in front of it.

FLAT BED CUTTING – A method of die cutting in

which material is converted through vertical movement of a die against a given material

FLATNESS – The degree to which a sheet of (self-adhesive) paper or film lies flat on the printing table and during its passage through the printing press.

FLEX – The deflection of rollers or cylinders in a printing press. Also describes the bending qualities or characteristics of any material, including printing plates.

FLEXIBLE DIE – A thin, flexible, steel die-cutting 'foil' or plate for use on magnetic cylinders, magnetic bases or other special die-cutting systems.

Flexible cutting dies are etched from specially formulated steel of around 0.5 mm in thickness and then mounted on a permanently magnetized cylinder in the printing press. Flexible dies are lower in cost than solid dies and the economics of use become more attractive as the complexity of the label shape increases. The life is much the same as for a solid die, providing cleanliness, setting, anvil condition, adhesive and label design are properly controlled.

Flexible dies are available today with a wide range of surface treatments and provide good results with a wide variety of materials. *See also Flexible wrap-around dies, and Magnetic dies.*

FLEXIBLE PACKAGING – Flexible packaging accounts for more than 30 per cent of all plastic packaging sales and has been growing rapidly in recent years, particularly replacing other forms of packaging – such as rigid plastics containers or corrugated materials. Major applications are as diverse as the multi-colored frozen food packs in supermarkets, sweet wrappers, confectionary packs (Mars wrappers, etc), garden center and horticultural packs (fertilizer, bark, grass seed), DIY packs (wallpaper paste, plaster), stand-up pouches (washing liquids, detergents, soups) and agricultural packs (feeds, seeds).

Almost 90 per cent of all flexible packaging, however, is for the food manufacturing and retail sectors, although non-food applications in pharmaceutical, agricultural, industrial chemicals, cosmetics and toiletries and household cleaning products are now growing fast. All are web-fed printed, mainly in multiple colors using a wide variety of press widths and in-line converting options. Flexo and gravure printing processes are the most widely used. Shorter runs and personalized flexible packs are increasingly being printed digitally.

Flexible packaging materials are of three main materials types:

- Plastics
- Paper or paper-based
- Aluminum foils and foil laminates

Of these materials, some 75 per cent are plastics, of which the main film materials used are Polyethylene (PE), Low Density Polyethylene (LDPE), and Bi-axially Orientated Polypropylene (BOPP).

These films are used for wrapping fresh produce, dry foods, meat and cheese, freezer packs, carrier bags, or foods on high-speed packaging lines. Aluminum foils are used where their better barrier properties give them the edge over flexible films or papers. Aluminum foils are used for confectionery, ready-meals, pharmaceuticals, soups and sauces, preserved foods, liquid foods. Paper and paper-based materials are used for a wide range of printed bags and sacks.

New developments with flexible packaging include single-use or re-fill pouches for liquids. Single-use pouches include those used for soups, while re-fill pouches are being used for liquid detergent.

Printing of flexible packaging. In terms of the printing of flexible packaging, gravure is regarded as being superior in print quality, although the quality of flexo has improved dramatically and is now considered cost-effective compared to gravure. All printing is undertaken on web-fed (reel printing) machines, including wide-web, mid-web and some narrow-web presses. Some 60 per cent of flexible packaging printing is direct onto the film surface; 40 per cent is printed on the reverse and laminated. Apart from print performance, flexible packaging films may have to withstand lamination, sterilization, pasteurization and thermo sealing and therefore may be surface coated or treated.

Inks and varnishes play a key role in not only providing the decorative and information performance, but also adding barrier properties, coatings and seals.

As most flexible packaging is used for food or for sensitive applications (body-care, cigarettes, etc.), it is important that neither the materials nor the inks should be able to contaminate the product by odor, migration or through trace particles. A small, but increasing, use of UV inks is tending to push flexibles printing towards flexo. Standard tests for flexible packaging may include sniff tests (color assessment) and flavor tests (taste assessment).

FLEXIBLE WRAP-AROUND DIES – Flexible cutting dies are etched from specially formulated steel of around 0.50mm in thickness (0.02") and then mounted on a permanently magnetized cylinder in the printing press. Flexible dies are lower in cost than solid dies and the economics of use become more attractive as the complexity of the label shape increases. Die life is much the same as for a solid die, providing cleanliness, setting, anvil condition, adhesive and label design are properly controlled. Flexible dies are available today with a wide range of surface treatments and provide good results with a wide variety of materials. *See Magnetic dies.*

FLEXIBILITY – A property of face materials, measured under specified conditions, which indicates how readily they will conform to curved surfaces such as bottles, ampoules or syringes.

FLEXOGRAPHY – Flexography is a rotary relief printing process in which ink is applied to the raised surface of an image on a flexible printing plate and, from there, to the surface of a label or flexible packaging substrate. The key differences between letterpress printing (which also uses a relief printing plate) and flexographic printing lie in the printing plate and the method of metering and applying the ink to the raised surface.

Printing plates for flexographic printing may be made from a softer, more resilient rubber or flexible photopolymer plate material than letterpress, though similar exposure and wash-out stages to that of letterpress are required. Rather than the thicker, sticky ink used for letterpress printing, the flexographic process uses a very thin solvent (declining) or water-based (growing) ink. The thin ink requires a special type of ink metering system, which uses an engraved anilox roll containing thousands of tiny recessed cells – which carry the ink – and a doctor roll or blade which removes excess ink from the anilox roll surface. The degree of fineness and the shape of the cells can be changed depending on the effect required.

Narrow-web flexographic printing is the most widely used label printing process worldwide for the production of self-adhesive labels. In the USA it makes up more than 90 per cent of the installed base, and over 50 per cent of the installed base in Europe. The process is used for the production of almost all types of labels today and has become increasingly accepted for the production of higher quality prime labels, particularly with digital plate imaging and the latest UV-curing flexographic press technology.

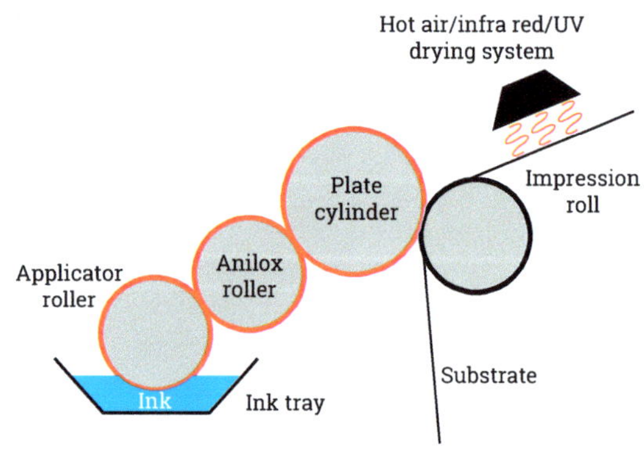

Diagram showing the components of a flexographic printing station.

There are three basic types of flexographic press configuration: in-line, common impression drum and stack. Any of these configurations may be used for printing a wide range of materials.

Quality of origination, pre-press, press maintenance and good press fingerprinting, are important factors in the production of good quality flexographic printed labels.

Apart from narrow-web flexographic presses used for self-adhesive label, sachet, pouch, blister pack and unsupported film production, the process is used in mid- and wider-web formats (500mm (20"), 600mm (24") wide and upwards for the printing of some sleeve labels and wrap-around film labels – as well as for wider-format flexible packaging and carton production.

FLOOD COAT – A method of coating an entire surface with an ink, coating or adhesive.

FLOOR COVERING LABELS – Labels for textile floor coverings may need to meet national or international standards that require that they carry recommendations indicating dimensions, nature of construction, fiber content, and instructions for laying and cleaning, as well as performance when subjected to a small source of ignition.

Label producers and users should check with any current regulations before manufacture or usage of textile floor covering labels.

FLUORESCENT INKS – A range of inks which can be printed in either visible or invisible forms and which change color when subjected to shortwave light or longwave light. Historically mainly used for fraud protection, but now used for promotional and other applications. Availability of pigments (short wave) for fraud protection is strictly controlled to preserve the authenticity of such inks.

Originally limited to screen-printing, the pigments now have sufficient strength to permit colors to be printed in one impression by lithography, flexo and gravure.

FLUORESCENT PAPER – A paper that is coated with a fluorescent pigment which not only reflects a visible wave length, but is activated by most of the remaining light to re-emit it as a color of a longer wave length which results in reinforcement of the reflected color.

FLUOROCARBON FILMS – Films characterized by the presence of carbon to fluoride bonds, which result in very low flow energies. An example is DuPont's Teflon™ based on polytetrafluoroethylene.

FLYING SPLICE – A joining together of two webs (splicing) while the web is in motion, most commonly when reels are changed from an expiring roll to a new roll automatically. This is done without stopping the machine, either at the unwind or rewind ends of a web-fed press or coating machine.

FMCG – The descriptive abbreviation applied to Fast Moving Consumer Goods.

FM SCREENING – Frequency Modulated or FM screening is a means of digitally screening dots on film using special randomizing software to provide dots which are varied. *See Stochastic screening.*

FOIL LAMINATE – A foil (usually gold, silver or silver aluminum foil which has been colored) which has been laminated to paper, board or plastic film to produce a label or flexible packaging substrate. Foils for printing are top-coated with either a clear (for silver) or colored lacquer to improve ink receptivity.

To determine that a sample is aluminum foil, rather than metalic coated or vacuum metalized material, ignite the sample. The paper support will burn leaving the foil behind, whereas other paper or plastic materials will be destroyed.

FOOD CONTACT – A label and/or adhesive that is in direct contact with food. Most countries have standards which determine which label and packaging materials may come into direct contact with food-stuffs and human skin, which products are prohibited, and those where there may be some transfer or migration of substances to food. The label user and/or label manufacturer are held liable for any failure on their part to comply with standards. *See also FDA.*

FOOD FRESHNESS – The freshness of food is a key factor in its overall quality and the combination of many factors that make the food acceptable to eat. Being able to tell when food is fresh is important to the consumer, whether at home or when eating out. Indeed, food freshness is perceived as the most helpful factor in assessing the safety of foods at the time of purchase. It is also important to the grocery store that sells the food.

The freshness and quality of food depends very much on the distribution and marketing systems used; any mishandling in the food distribution and selling chain can have a significant impact on its overall quality. Careful food storage and handling at home is also important in ensuring that food retains its high quality.

New label and packaging technologies that have been introduced in recent years are available to monitor temperature, or time and temperature, variations in the supply and sales chain, while new inks, varnishes, indicators and coatings that can be applied to, or incorporated into, packaging and label materials are now able to provide an indication (by change of color) of freshness.

Food quality indicators for example, can detect volatile amines on shellfish and other sea foods. Tiny detecting discs the size of dots are incorporated into the packaging. By seeing if the packaging has changed color consumers can make informed decisions about the quality of the sea food, particularly pre-packed products where it is difficult to use the sense of smell or touch to assess the quality of the fish.

Other recent new developments can help to stop microbial/bacterial deterioration of certain foods. Fish, poultry, meats, soft fruits and carbohydrates are some of the foods now benefiting from food quality, food freshness or anti-microbial label developments. *See also Food labeling, Food product-dating, Anti-microbial inks and Temperature indicators.*

FOOD FRESHNESS INDICATORS – Used to detect the freshness of foods and whether the product has reached the point where is starting to lose its original freshness and should be eaten. Food freshness indicators in the form of an in-package label, are designed to detect the by-product gases formed by bacterial action which are indicative of delay.

As protein-based foods spoil they give off gases which can be detected in parts per million by the indicator labels, which then change color or UV fluorescence to provide a measure of the food's freshness.

The indicator labels are designed for use with all fresh meat products, some vegetables, prepared foods and with sea foods.

FOOD LABELING – The labeling requirements for foods are set out in the relevant EU, FDA and national regulations around the world. They generally relate to most prepared foods, such as canned and frozen foods, breads, cereals, snacks, desserts, etc. Nutritional labeling may be voluntary, while other foods may have more detailed separate labeling requirements.

The aim of most food labeling regulations is to provide consumers with information so as to choose between foods; to tell them what they are buying; to protect the consumer from false claims or misleading descriptions and to provide clear guidelines about what the label can or cannot show.

As a general rule, all pre-packed food has to be marked with the following items:

- The name of the food
- A list of ingredients
- The quantity of certain ingredients or categories of ingredients
- An appropriate durability indication (e.g. 'Best before', 'Consume by', 'Sell by')
- Any special storage conditions or conditions of use
- The name or business name and address of the manufacturer or packer and/or a seller

Where necessary, the food label may also need to indicate particulars of the place of origin if failure to give such information might mislead a purchaser. Additional requirements and instructions for use should be shown if it would be difficult to make appropriate use of the food in the absence of such instructions.

Supplementary regulations are found in many countries to cover the labeling of specific foods and such things as dietary supplements, fatty acids, health claims, etc.

In all cases the appropriate specific international, national or federal regulations should be followed when determining what should appear on food labels.

FOOD PRODUCT DATING – Many food manufacturers incorporate food product dating into the packaging/labeling of specific foods, although there is, as yet, no universally accepted dating system. Historically, the commonly used definitions that help consumers to understand dates that are (voluntarily) printed on various food products include:

- **'Sell by' –** which tells the store how long to display the product for sale. Consumers should not buy the product after this date.
- **'Best before' or 'Best if used by' –** which tells consumers how long the product will retain its best flavor or quality. This is not a purchase or safety date.
- **'Use by' –** which tells consumers the last date that is recommended for using the product while at its best quality. This date is determined by the manufacturer. *See also Food labeling.*

In some countries the regulations underwent change in 2001, with guidance for food producers stating that 'use-by' labels should only be used where the food could be unsafe after that date. Most other foods should have a 'best before' date only, to indicate when the food is no longer at its best, but still safe to eat.

Food manufacturers and label/packaging producers should always check the current regulations/guidelines for their particular country to ensure they meet the correct requirements.

FOOD/SUPERMARKET LABEL PRINTING – Usually printed in 4 – 6 colors, plus over-varnish. Fifth or sixth color is frequently a special house or brand color. Primarily printed on paper using the rotary letterpress, flexo or UV-flexo printing processes. The sector is often characterized by short lead times. Inks may need chill cabinet or freezer cabinet performance and be rub resistant.

End-user printing and ink requirements for this sector include: good 4-color halftone results, special house colors, low temperature performance for cold displayed products, handling and rub resistance during packaging, transit and retail display, as well as a quick response and troubleshooting service. Generally a very cost-competitive sector.

FORGERY – A forgery is a document, packaging, label, identity card, passport, badge, etc, that contains altered or imitated information. The alteration or replication of the information with intent to defraud is forgery. Examples of forgery include:

- Utilization of a brand name and/or company logo in an unauthorized manner.
- Falsification of a package, label or container in an attempt to pass off counterfeited contents as genuine.
- Falsification of label/packaging markings to pass off damaged or unusable products as good. For example, forged alterations to use-by-date for products that have in fact passed their sell-by-date, e.g. pharmaceutical or food products.
- Signatures on documents.
- Details contained on passports or access passes to restricted areas (this may be text and/or ID photographic images).
- Changed cash amounts on payment documents.

The problems of forgery may be minimized by the addition of some form of forgery protection to documents, labels, packaging, etc.

FORM/LABEL COMBINATIONS – Developed to save business form and label end-users money, time and overall applied costs, form/label combinations are used in a range of commercial and industrial application by combining order information, invoice, inventory stock picking, parts distribution, laboratory or medical test documents, direct mail documents, etc, with one or more address, identification, shipping or handling labels.

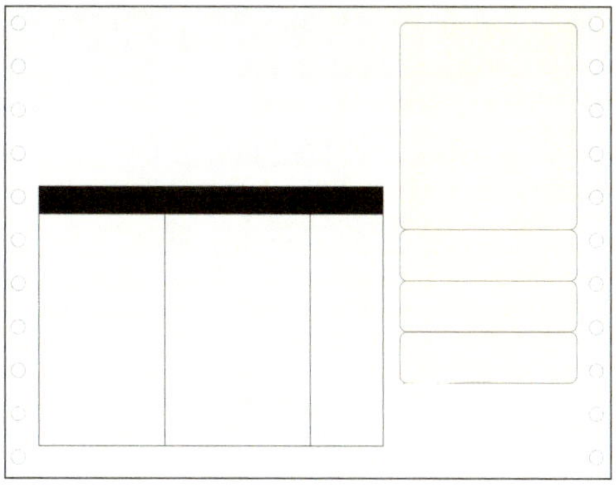

Peel-off self-adhesive labels incorporated into or on a continuous business form.

Mostly custom manufactured, form/label combinations allow a labels(s) and corresponding or related forms to be computer/electronically imprinted at the same time, offering users:

- savings in computer processing time
- accurate matching of forms and labels
- more efficient handling, distribution and shipping
- savings in total form and label costs.

The production of form/label combinations can be undertaken in several different ways: the marrying of form and label webs together on a label press using a glue line; the use of NCR type carbonless papers form sets with a CF label as the last ply; affixing pressure-sensitive labels onto a form, either by blowing or rolling; and applying a silicone and adhesive coated patch to the back of the form and then die-cutting the label from the front.

FORMAT – The size, style, layout, margins, etc., of a label or pack. Also to format a computer floppy disk – prepare a disk to receive computer-generated information.

FORM ROLLERS – Inking or dampening rollers which directly contact the printing plate on a label or other printing press.

FOUNT (FONT) – A complete set of typeface

characters of the same design, size and face which includes upper and lower case, bold, italic, numerals, punctuation marks, accents, etc.

FOUNTAIN ROLL – In flexography, the fountain roll is a rubber-covered roll in the two-roll flexographic inking system. The function of this roll is to deliver and meter ink to the anilox roll.

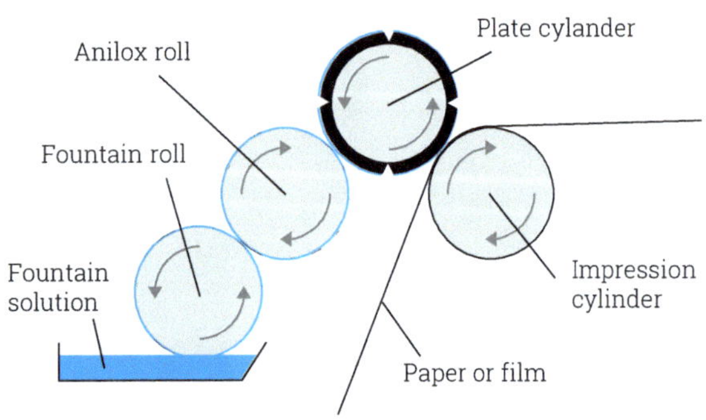

Use of the fountain roll to deliver ink to the anilox roll in two-roll flexographic sytem.

FOUNTAIN ROLLER – The roller on an offset-litho printing press that initiates the supply of moisture to the dampening system.

FOUNTAIN SOLUTION – A water-based solution containing additives of natural or synthetic gum and other chemicals, which is applied through the dampening system to the printing plate on a lithographic printing machine so as to keep the non-printing areas from accepting ink.

FOUR-COLOR PROCESS – Four-color process printing involves the use of four screens and four inks (the three subtractive primary colors of cyan, magenta and yellow, plus black) to reproduce the wide spectrum of colors found in the original. This is referred to as process printing, or CMYK printing, after the inks used.

The printing of color pictures by means of the three subtractive primary colors (yellow, magenta and cyan) and black. The original color image for printing must be separated into individual color films, together with a black film to correct flaws in the cyan, magenta and yellow inks, and to boost the blackness of the printed image. These films are used to make the four plates needed for 4-color printing.

Each plate must be printed in exact register with the others on the press to achieve the highest quality results. This is an industry-standard printing method. CMYK is a widely used color system for computer printing devices.

FRACTIONAL TAGGING – The application of electronic article surveillance (EAS) tags to a proportion of products/packs (i.e. less than 100%). This reduces the number of tags utilized and hence overall cost, but covert application of tags at source is required to make this technique effective.

FRAGRANCE BURST – A process in which a scent or smell is encapsulated within a defined area on printed matter. When scratched or opened the fragrance is released.

FREEZER GRADE ADHESIVE – An adhesive which is designed to give good adhesion of a self-adhesive label on a cold or frozen substrate. An adhesive which can be applied at, and function at, temperatures below freezing point. Some can perform down to -50C degrees Centigrade (112 degrees Fahrenheit).

Different manufacturers recipes have varying performance characteristics; some deep-freeze grades have good permanent performance at room temperature, whereas others are repositionable or even removable, depending on the surface of the material being labeled.

FRENCH LABEL ASSOCIATION – *See under UNFEA.*

FRONT END SYSTEM – A workstation or group of workstations that contain the application software for the preparation of label text and graphics.

FSC – The Forestry Stewardship Council. A non-profit organization that encourages responsible management of the world's forests and provides global standards for forest management which cover a balance of environmental, social and economic aspects

FTA – American and International Flexographic Technical Association. The FTA was founded in 1958

and is the leading technical society devoted exclusively to the flexographic printing industry. It promotes, develops and maintains the advancement of flexography and has a key emphasis on developing and maintaining standards of quality for the process. It maintains a library of technical aids, CD-ROM training programs and textbooks, which focus on every area and aspect of flexo printing, and runs seminars, workshops and conferences throughout the year.

Members come from all aspects of the flexo industry and include printers, suppliers, trade shops, consumer products companies, designers, end-users, consultants and educational institutions. *See FTA website for further information.*

FUEL PELLETS – Biomass fuel pellets can be constructed from recovered material waste from the manufacture of labels and packaging to be used as an alternative energy source to replace coal.

FUGITIVE COLORS – Inks which are not fast to light.

FUGITIVE INKS – Special inks that change color or are removed when water or chemicals are applied. Mainly used for fraud protection applications.

G

GAGE/GAUGE – A unit of measure, usually of thickness or diameter and generally expressed by a number. The thickness of paper or film expressed internationally in microns (μ). A micron is one millionth of a meter. Some converters still express film thickness in 'mil' (one thousandth of an inch) or 'gage' (e.g. 48 gage = 0.00048 inch).

GAMUT – The color gamut of a system defines the limits of the shades and hues which can be displayed or reproduced.

GAP AROUND – space between the labels in web direction.

GAP ACROSS – space between labels across the web.

GAPLESS – Refers to plates or blanket cylinders on printing presses which are without gaps.

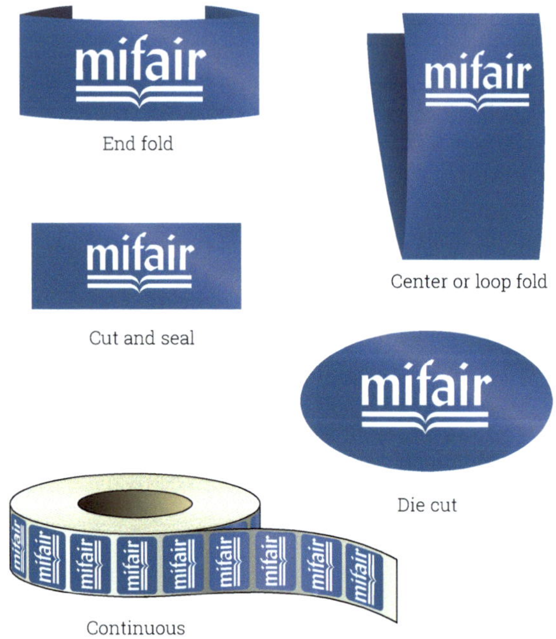

Various garment label formats.

GARMENT LABELS – Labels are an essential element of every garment, providing both legally required care instructions and content information. For most garments they also provide the primary method of brand identification. Garment labels may be woven or printed, or a combination of the two processes and, for many applications, can all be used for the same purposes. However, there are important differences: printed labels are less expensive and faster to produce than woven labels, while woven labels are more durable so as to survive commercial or home washing.

For the more highly priced fashion sector, woven labels are the main standard, offering a luxurious and sophisticated appearance – although good quality printed labels are becoming more common. Printed labels will be more legible than woven labels where there is lot of text or very small text, such as with care labels.

Printed labels are most frequently made on rotary letterpress or screen presses, or thermal transfer (with suitable durable ink ribbons). Flexographic presses are also used, especially for self-adhesive removable garment labels. There is also some use of offset on coated fabrics, as well as ink jet printing. Depending on the printing equipment and process used so printing may be carried out on one or both sides of a wide range of different materials, from satin, acetate, polyester, nylon and cotton twill, in a continuous format in anything from one to eight colors.

The labels are printed with brand, care, barcode or other information from a roll as required and, after printing, folded to end fold (the left and right edges of the label are folded), center or loop folded (folded in the middle), miter fold or single cut to size (using a cold or hot knife), die cut to shape or left in a continuous tape format. The most common material used is satin-coated acetate.

Woven garment labels are manufactured on industry-standard weaving equipment in a wide variety of weaving styles – single shuttle, double shuttle, needle loom, damask weave and twill weave – and in a range of patterns, usually using polyester, taffeta, damask or satin materials. Damask weave labels have the best life and can reproduce smaller type sizes, although not to the same clarity as printed labels. The majority of woven labels carry a brand name.

Garment label materials are selected according to the garment type and performance, from strong, durable materials for jeans to ultrasoft woven edge satin fabric for lingerie. They may have a slit edge, woven edge (generally used for any garment which has direct skin contact) or be specially treated, i.e. fire retardant or self-extinguish nylon, stone wash nylon to withstand chemical and stonewashing of workwear.

Woven labels and most printed labels are sewn in place on the garment with a sewing machine. Printed labels can be applied flat to a garment using iron-on, heat-activated or self-adhesive backings.

GAS-SCAVENGING LABELS – Labels that are able to extract and absorb oxygen consumption by product respiration, which in turn reduces pathogenic growth and extends shelf-life. *See also Biomontoring.*

GATF – Graphic Arts Technical Foundation, a member-supported, non-profit, technical and education organization serving the international graphic communications industries.

Originally formed in 1924 as the Lithographic Technical Foundation, it is the oldest continuous organization of its kind in graphic communications and one of the world's leading centers for graphic communications research and education.

In 1999, GATF consolidated with the Printing Industries of America, subsequently changing its name to Printing Industries of America. GATF is now a division of PIA and known as the Center for Technology and Research.

Further information is obtainable on the PIA homepage.

GEAR – A toothed machine part, such as a wheel or cylinder, that meshes with another toothed part to transmit motion or to change speed or direction.

GEAR CHART – A reference chart showing the various printing lengths or repeats obtainable within the different gearing systems of rotary presses. Referred to as 'teeth per inch (TPI)' or 'teeth per centimeter (TPM)'.

GEARLESS PRESSES – Gearless presses have begun to make an appearance in the label market in recent years. They employ servo and flux vector drive components to drive each printing unit and provide a significantly enhanced quality of print, including the elimination of gear marks and the potential of much tighter registration. Gearless presses also offer the possibility of non-standard repeat lengths, as determined by gear pitch – which is seen as significant in developing global markets for printers.

GEAR MARKS/GEAR STREAKS – A flexographic or offset printing defect which usually appears as uniformly spaced, lateral variations in tone, which exactly correspond to the distance between gear teeth on the printing cylinder. Parallel streaks appearing across the printed web at the same interval as the gear teeth on the printing cylinder.

GEAR SIZE – Number of tooth of a gear.

GEL – The condition or state of an ink, coating or varnish when it has a jelly-like consistency.

GELATION TIME – The time interval between the introduction of a catalyst (or crosslinker) into a liquid material and the formation of gel, characterized by a significant viscosity increase.

GELLING – Thickening of an ink or liquid which cannot be subsequently reversed by stirring.

GERMAN LABEL ASSOCIATION – *See under VskE.*

GFF SWEDEN – GFF Sweden (part of the Sweden Graphics Media Group, SGM) was formed in 1987. In 2012 the association consisted of about 20 printing companies and an equal number of associated suppliers.

The activities of the association in addition to two well-attended association meetings per year have included the collection of statistics for materials, sales and machines, and most recently, an Industry Benchmarking & Indexing Analysis to monitor costs for self-adhesive materials.

GHOSTING – The very faint reproduction of a printed design without any actual print transfer. Appears as shadows or indistinct images in solid or reverse-printed areas and is typically caused by poor ink distribution and/or poor ink formulation.

GINITEX – The body that regulates the use of garment care label symbols throughout Europe, and whose system has been adopted by a majority of the world as an international care labeling code. Established in 1963 by the professionals of the textile-garment-care chain and with a head-office in Paris, GINEX symbols are protected trademarks in most countries and subject to an international standard ISO/EN 3758:2012 'Textiles – care labeling code using symbols' as the result of an agreement between GINEX, the owner of the trademarks and ISO. The symbols must be used in respect of the use regulations of the trademarks, together with the technical information on which they are based.

The U.S care symbols system is not identical to that used in Europe, although it is similar. The U.S. representatives to the ISO worked to harmonize the ISO standard with the ASTM (American Society for Testing and Materials) Standard D489-96c and the Federal Trade Commission's (FTC) Care Labeling Rule, which is more comprehensive and user friendly. *See also Garment labels and Care labels.*

GIPEA – Gruppo Italiano Producuttiori Etichette Autoadhesive, is the only Italian association representing self-adhesive roll label producers in Italy. The association was founded in 1989 in Milan.

The current total number of member companies is 112 of which 83 are label converters and 29 suppliers. The label converter members of GIPEA represent more than 60% of the total volume of the Italian label market.

GLASSINE – A supercalendered, smooth, dense, transparent or translucent paper manufactured primarily from chemical wood pulps, which has been well beaten during pulp preparation to secure a high degree of hydration of the stock, which gives glassine its transparency but also makes it more susceptible to atmospheric changes. Due to its curl limitations its common use as a backing paper/release liner is principally for pressure-sensitive roll-label laminates.

GLOBAL CLASSIFICATION AND LABELLING SYSTEM FOR CHEMICALS (GHS) – A global system aimed at protecting people from the mismanagement of chemicals. It classifies them according to their hazard and creates a labeling system based on universally-understandable pictograms. It was adopted in Geneva in December 2002 and was made available as a United Nations publication in 2003. The intent is that those countries that lack their own system for hazard classification and labeling will adopt the GHS as the fundamental basis for their national policies.

A harmonized system will lead to greater consistency among countries and thereby promote safer transportation and handling of hazardous chemicals. GHS was originally a voluntary system, in that it did not impose binding treaty obligations on countries, except to the extent that in countries that adopt the Code into their national regulations it will be binding on that regulated community.

Companies in the USA and Europe will be obliged to comply with the Globally Harmonized System of Classification and Labeling of Chemicals by 2015.

GLOBALIZATION – Globalization and consolidation in the label industry is not new. It is a process that has been going on for many years, but which has been dramatically speeding up since the early 1980s and even more rapidly during the 1990s and the early years of the 21st century . It has had a major impact on the way that the label user industry operates, and on the way that labels are purchased and supplied around the world.

There are number of key influences that have been driving the globalisation process:

Reduction of trade barriers and tariffs. Following each successive round of world trade agreement on tariffs and trade, reductions in trade barriers have rapidly lead to more open movement of goods, services and people around the world and, in turn, to an explosion in the growth of global trade. To facilitate the global movement of goods, banking practices have also had to become more de-regulated.

More open movement of goods, services and people is also facilitated by the ease of use and availability today of personal computers, mobile communications, e-mails and the Internet, digitization of print and digital file transfer around the world, and low-cost flights.

Moves to global manufacturing and purchasing by label buyers. In recent years there has been a growing trend by global manufacturers to move to a lower manufacturing cost basis in emerging markets, i.e. China, Eastern Europe, Mexico or India. This enables them to meet demand for their products in these markets, offers better margins and returns, and provides lower distribution and supply chain costs.

Global specification and purchasing of labels and packaging is now regularly undertaken by the major global brand owners, global consumer goods companies, global automotive and aerospace groups, global and semi-global retail groups, global drugs and cosmetics groups and by global clothing manufacturers.

The end result of sourcing by global brand owners on a worldwide basis is that they have more purchasing power which they use to:

- reduce label and packaging expenditure
- dominate purchasing negotiations
- improve supply chain management.

Impact of new technology. New technology has made a significant contribution to the ease of globalization and global trading. Communications worldwide for example, are now easier than ever and people in business can keep in touch through e-mail and mobile communications on a 24-hour basis.

Global logistics operations for the movement and tracking of goods anywhere in the world have also become increasingly sophisticated with logistics groups taking over the complete organization and movement of goods on a worldwide basis for global manufacturing operations.

In addition, the use of e-commerce, e-purchasing and e-auctions have already had a major impact on the way that label companies obtain orders, communicate and service their key accounts. Combined with sophisticated Management Information Systems (MIS)

As far as the label industry is concerned, new digital technology – digital origination, digital workflow, digital file transfer, etc, are enabling print jobs to move around the world in an instant, with digital downloading of images and files through the Internet to plants or presses globally.

Role of global label industry suppliers. It is not only label buyers that have become global in operation; key label industry suppliers, including label press manufacturers, label stock suppliers, ink manufacturers, die manufacturers, etc, now also operate on a global basis. This enables label buyers to specify the same materials, pre-press, presses, inks and controls for any label manufacturing plant in the world – and all produced to ISO, FDA or other standards.

Trading in a global world. As globalization by the world's leading label buyer companies continues to grow, so their demands on the label producer will increase, with customers expecting the label plant to offer:

- Totally digital workflow
- Digital archiving and file transfer of label jobs
- Internet communications and trading between label plant, suppliers and customers
- Supply chain management and efficiency
- On demand printing and supply of labels
- Added value solutions and products
- More global marketing initiative solutions.

GLOBAL LABEL MARKET – Up until the middle to late 1980s, the global label market was dominated by Western Europe and North America which, together with pockets of advanced label printing in Australia, New Zealand and South Africa, enabled label production to meet both indigenous demand and export growth and so achieve annual growth rates for all kinds of labels of up to 7 to 10 per cent.

As emerging label markets, such as South East Asia and the Middle East began to create their own label industries in the early 1990s and, more recently, through the development of new label markets in Eastern Europe, China, India and Latin America, so the pattern of global label production and usage has begun to undergo significant change.

Put together, South East Asia, Australia/New Zealand, China and Japan – the total Asia Pacific label market – is now bigger than either North America or Europe, and still growing rapidly in size and importance. Even the Central/Eastern European and South/Latin American label markets continue to achieve growth of close to double that of the developed markets.

In terms of label technologies, the global label market was historically dominated by wet-glue applied paper labels which, in the late 1960s, made up over 70 per cent of label usage, with gummed paper labels making up over 20 per cent. Global usage of self-adhesive labels at that time was little more than 5 – 7 per cent. Thirty years or so on, and self-adhesive labels are the dominant label usage technology in the developed markets of Western Europe and North America, while wet-glue applied label usage has declined to less than half of its original market share in these regions.

Global annual growth for self-adhesive labels is regularly estimated at between 4 and 6 per cent per annum, depending on the particular region of the world looked at, with Central/Eastern Europe, India, China, Latin America achieving the highest growth rates of all.

New label technologies developed in the 1980s and 1990s, such as sleeving, in-mold labelling, wrap-around film labels and cut-and-stack film labels continue to achieve good growth and, between them, already make up close to 20 per cent of the total global label market, and will continue to gain market share in the future, largely at the expensive of wet-glue paper labels.

In the longer term, direct digital printing onto bottles, cans, tubes and other packaging shapes is likely to gradually impinge on world label usage.

GLOSS – The characteristic of a surface which causes it to reflect light at a given angle. Relates to gloss label papers, gloss inks, gloss varnishing, etc. Gloss is measured in accordance with DIN 67530 and indicated as a percentage. It can be measured at various angles of reflection.

Gloss of paper is also measured by using a Gardner gloss meter, which measures reflected light at an angle of 75°, and is expressed in Gardner gloss units – the higher the number, the glossier the surface.

GLUE-APPLIED PAPER LABELS – The application of paper or paper-based labels to glass bottles or cans using a wet-glue was one of the earliest methods of product labeling and for many years was the major label technology. Even with the rapid rise of self-adhesive and other label technologies, glue applied labels are still, marginally, the main method of volume labeling of bottles and cans with paper labels.

Labels may be individual face, neck and back labels for application to beer, spirits, or wine bottles, or wrap-around labels used extensively on canned foods and soft drinks. depending on adhesion, application speed or drying speed.

Glue-applied paper and paper-based labels are printed on plain paper, metallized paper or paper laminate materials in either sheet-fed offset (most common) or web-fed gravure (some flexo) presses. They are often also varnished, coated or lacquered to provide surface protection during labeling, handling and distribution, before being guillotine cut to a rectangular or square size in stacks of 500 or 1,000 labels, or punched into special cut-out shapes, again in stacks, ready for placing in a hopper in the label application machine.

Glue-applied paper label substrates include:

- one-side-coated grades and uncoated white, or bleached kraft paper – 60-100 g/m2
- paper laminates with aluminum foil
- metalized paper

All of these have a similar base. For the one-side-coated grade a special printable coating is applied. Foil laminated paper is a more expensive and luxurious glue-applied label substrate and is made up of a thin aluminum foil laminated to a paper backing. Metalized paper substrates are one-side-coated papers onto which metallic aluminum has been deposited under high vacuum.

Selection of the one-side-coated, laminated or metalized paper label substrate for glue-applied labels is determined by the effects required, the performance

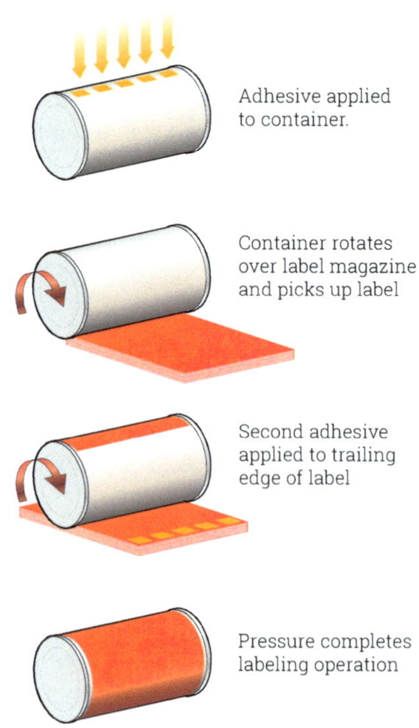

Stages in the labeling of a can.

of the finished label, the nature of the labeled product and on the label application method. Factors such as surface smoothness, opacity, stiffness, porosity, water absorbency, wet strength, grain direction and degree of curl, all need to be considered to some degree in the wet-glue label process. Corrosion and/or mold inhibitors may be required for some applications.

Label application of wet-glue labels is undertaken on a machine in which either a wet-glue or hot-melt adhesive is brushed or rolled onto the back of each label just before it is applied to the container, enabling the most suitable adhesive formulation for the specific application to be selected. The type of adhesive application can also be selected from options such as skip, pattern or stripe,

The key uses for glue-applied labels are in the high-speed, high-volume, low changeover labeling of drinks bottles and for canned foods – both human and pet food – where application speeds up to 60,000 – 80,000 or more containers per hour are available. *See also Wet-glue labeling.*

GRAB – Grab, or tack, is the ability of an adhesive to quickly adhere to a surface with a minimum of pressure. Also referred to as initial tack or instant adhesion.

GRAIN – The direction in which most fibers align when paper is formed on the wet-end of a papermaking machine. Normally synonymous with the machine direction of the paper.

GRAINING – A method of treating paper or board to give a linen, leather or other textured finish to the surface.

GRAM – Unit of weight in the metric system. The weight of one (1) cubic centimeter of water at standard conditions.

GRAMMAGE – Term used in the metric system to denote the weight of paper or board; the measurement used is the weight of a single sheet of one square meter, expressed as grams per square meter (g/m2).

GRAVURE PRINTING – Gravure, or photogravure, is a true photographic process, which is able to reproduce high quality pictures, excellent color densities and strong solid areas. It is primarily used as a long-run process for wet-glue applied labels, cartons and for flexible packaging. Gravure units are also now being incorporated into narrow- and mid-web presses for labels and flexible packaging where the ability of gravure to apply varnishes, lacquers, metallic inks, and top coatings is a particular advantage.

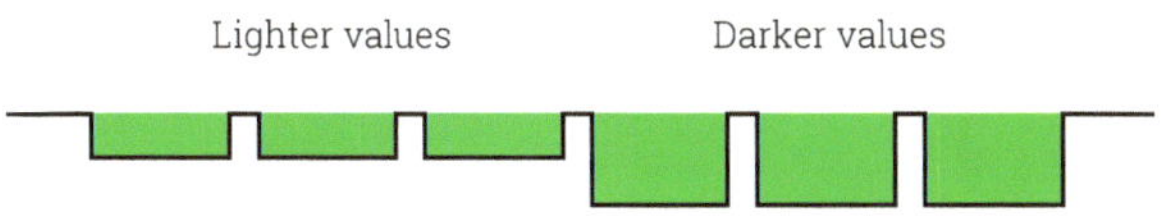

Varying cell depths used in recessed image principle of gravure.

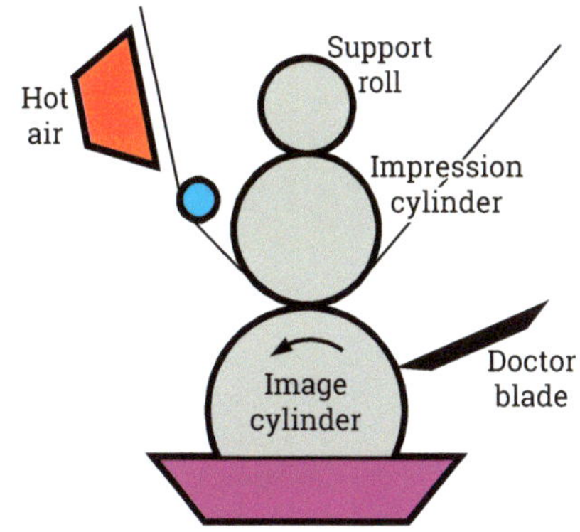

Diagram shows the basic principle of a gravure printing press.

The image carrier or printing surface for gravure primarily consist of steel cylinders with an outer shell of copper onto which images – consisting of millions of tiny cells of varying depths – are produced by chemical etching, laser etching or electromechanical engraving. The deepest cells carry more ink and print darker values; shallow cells print lighter values. Although the most expensive of all the image carriers used for printing labels, gravure cylinders have the advantage of being able to print very long runs in the millions. For additional run life the cylinders can be chromium plated.

Gravure printing presses are the simplest of all label printing presses. The printing cylinder containing the cells rotates in a thin, fluid, solvent- (or water-) based ink, which fills the recesses. Surplus ink is scraped from the cylinder surface by a doctor blade, with the ink in the cells (the image) then transferring to the label substrate using pressure against a rubber-covered impression cylinder. Hot air dryers are used between printing units to speed up drying.

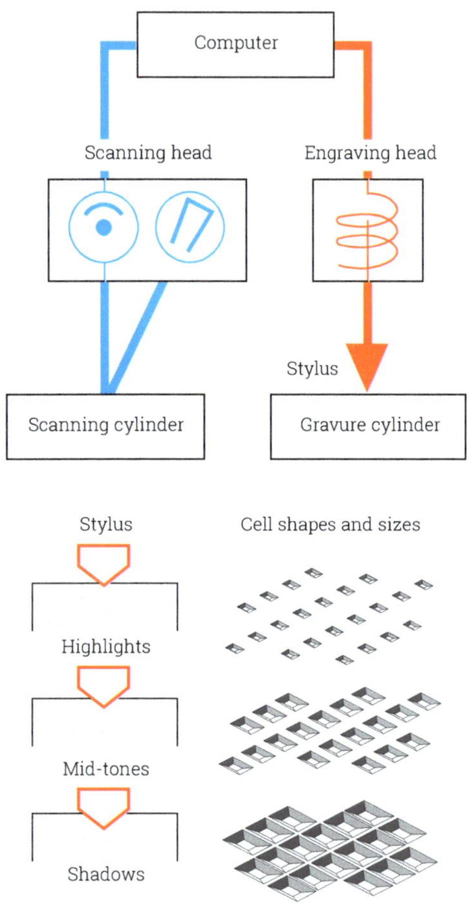

Varying cell depths used in recessed image principle of gravure.

The main methods of producing the recessed cells in gravure cylinders is through an etching process; or by using an electromechanical engraving system consisting of a rotating drum on which the original photograph or image is mounted and then scanned by a reading head. A computer processes the scanned information and converts it into electrical impulses that are transmitted to one or more engraving heads mounted over the gravure cylinder. Depending on the strength of the impulse signals, the engraving heads will engrave the cylinder to different depths, so producing different cell sizes. The shallow engraved cells will form the highlights; the deeper cells the dark tones.

Photopolymer plate systems for gravure have also been developed, as well as computer-to-cylinder systems. More recently, a direct digital laser etching process has been introduced, which is expected to eventually reduce the cost of imaging gravure cylinders.

GREEN GUIDES – The Federal Trade Commission's Green Guides outline general principles that apply to all environmental marketing claims and offer terms and examples to guide consumers in making educated purchasing decisions. Label and packaging recovery projects to reflect accurate label use within recovery infrastructures.

GREENPRINT – Avery Dennison's user-friendly way to communicate a product's environmental footprint across different stages of its life-cycle. The tool provides environmental impact data across six categories (fossil material, trees, water, energy, greenhouse gases and waste) and allows product managers, technicians and designers to quickly assess design and material options based on environmental impact.

GREYSCALE/GRAYSCALE – The depiction of gray tones between black and white. Distinct gray pixels and black and white pixels can be displayed on a grayscale monitor, but not color pixels. A strip of standard gray tones ranging from white to black, placed at the side of an original during photography to measure range and contrast.

GRINDER – A piece of machinery used to back grind flexible cutting dies using a large grinding wheel and a flat magnetic base.

GS1 BARCODES – Global data and application standards for barcodes that use the globally recognized GS1 Identification Keys to automatically identify things such as trade items, locations, logistic units, and assets. *See GS1.*

GS1 EPCGLOBAL – A new global suite of standards that combines RFID (radio frequency identification) technology, existing communications network infrastructure and the Electronic Product Code (a number for uniquely identifying an item) to enable immediate and automatic identification and tracking of an item through the whole supply chain globally, resulting in improved efficiency and visibility of the supply chain. *See GS1.*

GS1 – GS1 is an international not-for-profit association with Member Organizations in over 100 countries. It is dedicated to the design and implementation of global standards and solutions to improve the efficiency and visibility of supply and demand chains globally and across sectors. The GS1 system of standards is the most widely used supply chain standards system in the world.

With the advance of smart phones and other Internet of Things technologies, companies are increasingly forced to think beyond the point-of-sale. GS1 current research activities are focussed on adding the consumer to the currently B2B-oriented business model of GS1 and have a new initiative to explore the potential of GS1 standards in the digital world.

GSM – Grams per square meter. Also shown as g/m2. A method of measuring or indicating the substance of a paper or film substrate and adhesive or silicone coat weight on the basis of its weight, expressed internationally in grams per square meter. Some converters may still use units such as pounds per ream or pounds per thousand square inches (MSI).

GUARD BARS – Auxiliary characters or bars which appear at both ends and center of UPC and EAN symbols. They provide the reference points for scanning.

GUILLOCHE – A complex and irregular pattern design, which can be created and printed to provide protection against fraud or the counterfeiting of documents. May also be used on some security labels or packaging. Guilloches do not provide good protection against forgery by color copiers.

GUILLOTINE – A device for trimming or cutting

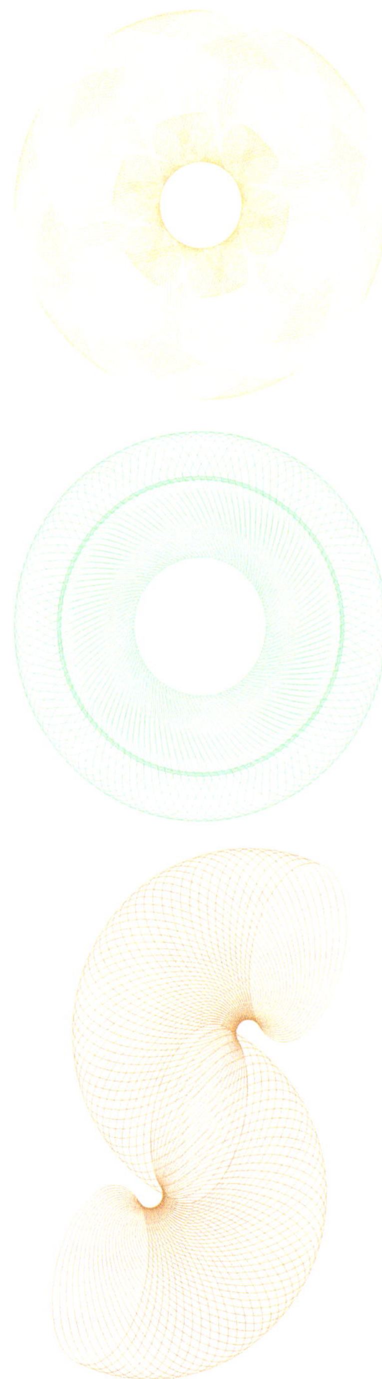

Guilloche patterns used for protecting against fraud or counterfeiting of printed products.

sheets of plain or printed label paper in stacks using a downward cutting action while the stack is 'clamped' in position. Guillotines are usually 'stand alone' items of equipment, but on some label presses they are built into the machine. Programable guillotines make up some 90 per cent of today's market and have a range of capabilities from paper-size set-up and cut-and-mark routines, through to sophisticated fault diagnosis and MIS.

Also popular are touch-screen controls that show information on screen, offer job recall, calculate cut schemes, store complicated cut programs and guide the operator through maintenance or knife changes. Effective materials handling solutions for lifting, jogging, sliding and positioning of stacks, together with sensing and safety cut-out devices can both aid productivity and reduce injuries.

GUMMED PAPER – Paper with a dextrine or PVA remoistenable adhesive coating on the underside. Needs to be moistened or wetted to activate the gum and make it stick. Normally printed as flat sheets and guillotined into cut single labels for applications that were traditionally as diverse as receipts, package, carton or case labels and some form/label combinations. Also some primary product labels that could be applied by hand. Postage stamps were one of the largest applications for gummed paper but these have been superceded in many countries by self-adhesives. Some c customs declaration slips on parcels and some window stickers are still found in gumed paper.

The wrap-around labels on rolls of plastic bags for household and garden refuse, that are applied from gummed tape dispensers and recycled paper mill reel labels that are applied automatically by a robotic arm, are rare examples of gummed paper not being applied by hand. The development of heatseal and subsequently self-adhesive applicators, plus the growth of plastic packaging helped the decline of gummed paper, which now has a market share of less than 1 per cent. Normally printed as flat sheets.

H

HALFTONE – A halftone is a film or print made up of dots (or a similar screened pattern) which allows the reproduction of tonal values using a solid color. A pattern of dots of varied sizes, the centers of which are equidistant, used to simulate a photo-realistic continuous tone photograph when printed in either color or black and white.

The halftone principle relies on the frequency of the dots being fine enough so that the viewer, at a normal reading distance, is unable to distinguish the dot pattern; the eye is fooled into seeing gray values. The darker the tone the larger the area of the dots and the less paper is visible.

Halftones can be produced with different dot sizes, called screen rulings. The finer the screen ruling, the smaller and less visible the dot. Usually, the smallest dot size possible that is compatible with the smoothness or absorbency of the substrate to be printed, is desirable. Also the ability of the press used to 'hold' a dot is an important factor.

Halftone dots can be created in a variety of different shapes, from square to octagonal or elliptical, and be both symmetrical or asymmetrical. When plates are created from film, dot gain increases the joining together of the dots and may cause sudden jumps in density rather than a smooth continuous transition.

HALFTONE SCREEN – A grid in which lines and spaces are of equal thickness and are equidistant. Used in photographing an original for plate making. Screen rulings vary from 65 lines per inch (25 lines per centimeter) up to 300 lpi (118 lpc) or above for the highest quality press work.

Magnified examples of halftone screen rulings at various lines per inch.

Lines per inch should not be confused with dots per inch (dpi) or pixels per inch, both of which are used to describe computer screen and output device resolutions.

HALO – An undesirable, peripheral outline of a printed image. An undesirable, peripheral outline of adhesive around the edge of an applied pressure-sensitive label, created by adhesive ooze or substrate shrinkage.

HANG LABEL – A label where part of the liner is peeled away, with the exposed adhesive wrapped around an object, then stuck to the back of the liner that remains. Used in the retail market place and for airline baggage tag labels.

HANG TAG – Also referred to as swing tag, coupon tag, hang ticket or swing ticket. All variations of the paperboard or plastic swinging tags that are familiar on garments and related goods to provide a brand image identity, eye-catching point-of-sale ticket,

Examples of different styles and formats of hang tags.

pricing information and/or barcode details. The tags and tickets may be rectangular, square, round, shaped or folded and are punched with a small hole so that they can be attached to garments by cotton string, elastic string, plastic tag pins, ornamental string or nylon ribbon.

They may be color pre-printed, or electronically printed/overprinted, on one or both sides to provide the necessary price, code, product information. They may also be provided as continuous tags on a roll for variable information overprinting. Perforations may be provided to separate continuous roll tags or to give a tear-off portion to the tag.

HARD DOT – A halftone dot where the fringe or halo is so slight as to be barely noticeable. The dot is very sharp.

HARDNESS – The degree of hardness of a surface or the relative resistance of a metal or other material to denting, scratching or bending. . Measured and compared by using hardness scales, such as Shore or Rockwell. The hardness of rubber printing rollers is usually described in terms of 'shore' hardness.

HARD PROOF – A proof on paper or other substrate, as distinct from a 'soft proof', which is an image on a VDT screen.

HARD TAG – Also referred to as swing tag, coupon tag, hang ticket or swing ticket. All variations of the paperboard or plastic swinging tags that are familiar on garments and related goods to provide a brand image identity, eye-catching point-of-sale ticket, pricing information and/or bar code details. The tags and tickets may be rectangular, square, round, shaped or folded and are punched with a small hole so that they can be attached to garments by cotton string, elastic string, plastic tag pins, ornamental string or nylon ribbon.

They may be color pre-printed, or electronically printed/overprinted, on one or both sides to provide the necessary price, code, product information. They may also be provided as continuous tags on a roll for variable information overprinting. Perforations may be provided to separate continuous roll tags or to give a tear-off portion to the tag.

HDPE – High Density Polyethylene. *See Films, Plastics and Polymers.*

HEAT ACTIVATED – A coating of adhesive which is dry to the touch until activated by the application of heat through the thickness of the coated substrate. All heat activated labels require the correctly heated platen, the label and the labeling surface to be in contact under pressure, at application. Variations of heat, pressure and dwell time affect adhesion performance and have to be co-ordinated.

HEAT RESISTANCE – The ability of a material to inhibit or withstand elevated and high temperatures without undergoing any physical or chemical changes. Excessive heat in a turning or milling operation can lead to the softening of a die tool. This softening allows the tool to dull or chip causing premature failure.

HEAT SEAL – The bonding together of two or more surfaces using heat. Usually for sealing plastics.

HEAT SEAL LABELS – Label papers which have a coating that melts and forms a strong bond when activated by heat. Known in the carpet industry as 'Iron-on Labels'.

HEAT-SEALING ADHESIVE – An adhesive film applied to a substrate for later reactivation by the application of heat.

HEAT-SET INK – A printing ink which dries by the application of heat after it has been printed.

HEAT TRANSFER DECORATION – Used for the decoration of containers and bottles to provide colorful, crisp, high impact graphics. The process uses reverse-printed colors from a web to decorate, glass, metal and plastic containers with a so-called 'no-label' look. End user sectors for heat transfer decoration include personal care, household products, food, industrial and beverage sectors. Originally only providing body decoration, the process now incorporates neck decoration as well. Highest impact seems to come from decoration on white, opaque containers. Strength of color on clear, transparent containers appears less powerful.

Key to the transfer decoration process is a transfer cylinder. Transfer rollers and heaters are positioned sequentially around the cylinder to supply the heat and pressure needed to transfer the color printed ink image from the web to the container. The containers are heated before and after the transfer.

The requirements for the pre- and post-treatment of the containers will be different depending on the specific container material. Web handling is similar to that of pressure-sensitive labeling, except that the transfer cylinder replaces the label peeler plate and the system can operate at higher accelerations.

Heat transfer decoration involves the mirror image printing of a design onto a special paper or film base in up to 7-color rotogravure. This can include line, vignette and halftone printing up to full color process, bright metallics, color change inks and matt or gloss finishes. Because there is no label there is no discernible label edge. Images are scratch and chemical resistant, pasteurizable, permanent and fade resistant.

Application of the images is flexible with separate transfer heads for front, back and neck labels – or a near 360 degree wraparound look. Decorated plastic containers can be squeezed and deformed without adverse effects on the quality and adherence of the bottle decoration.

Decoration success hinges on the dynamic interaction of the web, container and transfer cylinder. Where PE bottles are required, then pre- and post-flaming is required. Flexible plastic bottles are normally filled with compressed air before decoration.

Over the past few years new transfer decorations solutions for buckets and pails have evolved using digitally printed heat transfer labels.

HELIUM NEON LASER – A laser light most commonly found in barcode scanners.

HEXACHROME – A six-color process developed by Pantone, which adds a vivid orange and intensified green to the standard CMYK color printing system. A color-matching system that allows for the combination of six colors in order to create a larger gamut of reproducible color. It is claimed that the number of spot colors that can be faithfully reproduced by the Hexachrome process increases from 50% to 90%.

HICKEY – A term used to describe various printing defects, such as spots or imperfections in the printing due to dirt on the press, dried or lumpy ink, pieces of ink skin, paper dust particles, etc. A burr or defect on the printing plate or engraving. A defect or piece of foreign matter in paper.

HIGH DEFINITION COLOR PRINTING (HDCP) – A specialized color printing process that uses balanced dot frequency and screen rulings as fine as 625 lpi (246 lpc) to reproduce near-photographic quality images.

HIGH-FREQUENCY TAGS – RFID tags which typically operate at 13.56 MHz and which can be read from about 3 meters (10 feet) away. They transmit data faster, but consume more power than low-frequency tags.

HIGHLIGHT – The areas of a printed image which are nearest to white. The whitest part of a halftone when printed, represented by the area with the smallest dots or no dots at all. Some scanning and image manipulation programs require the operator to set the 'white' and 'black' points. Because the small dots of highlight areas in flexo are subject to relatively large dot gain, it is important that any highlight limitations are discussed with the printer before any separations are made.

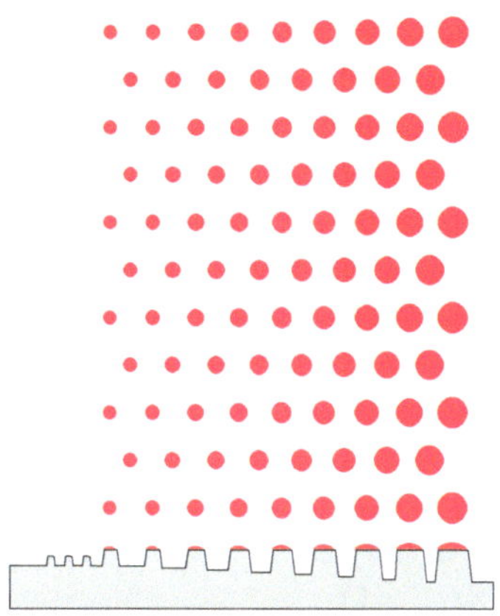

The areas of a printed image in which the smallest or "no dot" areas are the highlights

HIGH SPEED UNWIND – A device used to accelerate the unwinding of rolls of labels on a high speed automatic label dispenser.

HIGH TEMPERATURE ADHESIVE – An adhesive that enables a pressure-sensitive label to adhere or stick well when applied to a hot surface. Characteristics of high temperature adhesives include a high degree of resistance to aging or deterioration at the elevated temperatures.

HIGH TEMPERATURE BARCODE LABELS – Special computer-generated sequential barcode label products using phototypeset, thermal and laser printing methods which are designed for use in hostile environments, including operating conditions in temperatures ranging from minus 80 degrees to over 1,200 degrees Centigrade (112 to 2192 degrees Fahrenheit). Label materials, high performance adhesives, wear resistant coatings and special cutting methods have been developed for these applications. The labels are scratch and tear proof.

HISTORY OF LABELS – The history of labels, at least in a format that would be recognized by today's label producer and user, can be traced as far back as the late 1600s and early 1700s. At that time labels were mainly used to identify products being transported, such as bales of cloth and by apothecaries for identifying their medicines, or used in the front of books to donate the owner or presenters of the book.

Labeling of products for sale using printed labels had not evolved in any significant way at that time, largely because few of the population could read or write and because there was no trade of mass-produced and packaged goods.

All that began to change during the period of the Industrial Revolution and the advent of education and growth of mass literacy in the 1800s. With the evolution of the mass production of food and drink in bottles and cans, the need to put descriptive and brand information on products and to produce labels in volume quantities, now became a necessity.

By the later part of the nineteenth century the introduction of the continuous papermaking machine and the invention of lithography provided the means to economically produce long runs of the same quality (label) papers and to print them, in sheeted form, in color. It was the advent of cost-effective colorful labels produced in this way that enabled manufacturers to take advantage of individually packaging and labeling their products themselves, rather than being individually weighed and wrapped by the provision merchant. Product labeling, as is known today, had arrived.

The development of glue-applied and gummed paper labeling. Initially labels were printed on to plain paper and cut to a rectangular size by guillotine – even with scissors or punches if a special shape was required – before being applied with a wet glue or gum. Early methods of labeling used gums and adhesives, which were brushed on to the label, and then the label applied by hand. These first slow labeling methods eventually evolved into full-blown applicator machines and systems which applied a wet-glue to the back of each label and then applied the label to the bottle or can.

The first hand gumming of pre-printed paper with animal glue, sheet by sheet, was being done in 1880, but in 1886 the first continuous gummed paper tape was produced for parcel sealing (although was unsuitable for sheeting because it curled). In 1905 a patent for stabilizing, or non-curling, pre-gummed paper for labels was purchased by Samuel Jones & Company Ltd.

The non-curling process allowed air to get through to the base paper by fracturing the gummed coating

Glue-applied beer labels from around the middle of the 20th century.

into small pieces, which could then expand and contract without unduly curling the sheet, within a recommended relative humidity range. However, in very high humidity the adhesive would rapidly pick-up moisture from the air and the cracks could reconnect, leaving the sheet with a permanent curl.

For many years the principal adhesive was pure gum arabic, which continued to be used for postage stamps because it was considered more pure than dextrin. Eventually pure gum arabic became too expensive even for stamps and was replaced by starch based adhesive, obtained from potatoes or corn and converted into dextrin

Between the wars quite large volumes of luggage labels were produced from gummed paper that could stick to leather and these became quite a status symbol of the foreign holiday destinations visited, hand applied remoistenable gummed labels could not stick, saw the development of instantaneous heatseal labels, which could be automatically applied and this heralded the decline of pre-gummed paper.

Some market developments temporarily slowed the decline of gummed paper labeling. Lick and stick trading stamps grew as a retail marketing tool during the late 1950s and 1960s, but did not survive much longer. Self-adhesive trading stamps have also been used for gasoline sales subsequently, but they are not currently fashionable.

However, the development of self-adhesive labels by R. Stanton Avery in 1935 and their subsequent evolution into a highly efficient and cost-effective method of label printing and die-cutting in rolls, with automatic application in many different industries and

(left) Handmade papermaking in the 16th century.
(right) Illustration shows how handmade papermaking was still being carried out in the 20th century

particularly on cruise liners.

The non-curling gum breaking process was very successful at the time, but the development of packaging films in the 1930's and 1940s, to which applications, subsequently took most of the gummed paper market share, which has declined from over 20 per cent in the 1960s to less than 1 per cent today.

The overall dominance of the wet-glue applied

paper label for product labeling of glass bottles and cans, and gummed paper for addressing and shipping labeling of paperboard containers, continued well into the 1970s. At that time wet glue paper labeling (70 per cent of the market) and gummed paper labelling (22 per cent) were still the only key labeling technology.

Impact of self-adhesive labels. Self-adhesive labels started to make their real impact on the product decoration and labeling market in the late 1960s, initially as a means of applying price labels to products and then eventually evolving to become a high added-value method of labeling cosmetics, pharmaceuticals and shorter-run applications with both prime and secondary labels.

Since then, many factors have changed the world of labels:

- The rapid growth and emergence of computer impact address labels
- Barcode labels and variable imprinting using thermal, laser and ink-jet technologies
- A major explosion and rapid volume growth in the use of plastic containers and plastic packaging
- Relatively low growth in the use of glass bottles and cans
- The evolution of new label technologies, especially for the labeling of plastic bottles
- New demands for anti-theft or anti-counterfeit labels
- New requirements for promotional, booklet and leaflet labels
- The potential of RFID labels
- The rapid emergence of digital label printing during the early years of the 21st century.

These factors have brought many changes in the use of labels over the past thirty to forty years.

In addition, there have been many changes that have taken place in consumer lifestyles, leisure activities, global travel and legal requirements (consumer, health & safety, environmental legislation). In particular, the trend away from the purchasing of most preserved food in cans and jars to one in which the consumer makes a major use of freezer packs, convenience foods, ovenable and microwavable foods, etc, through to carry-home beers and soft drinks in cans rather than bottles, to regular eating out and a leisure life that includes overseas travel, gardening, DIY, sports and healthy lifestyle activities,

Evolution of Labels and Packaging		
Marking of goods with tar, paint, ink, chalk, etc.	*pre 15thC*	
Invention of printing	*mid 15thC*	1600
Development of packaging	*mid 16thC*	
Labeling of silks, spices & wool	*early 16thC*	
Labeling of paper	*early 17thC*	
Labeling of quack medicines	*late 17thC*	1700
Labeling of pharmacy products	*early 18thC*	
Development of heating & preserving food in can Labels needed	*late 18thC*	
Invention of paper making machine and lithographic printing. Mass production of labels	*late 18thC*	
Use of glass and metal containers by French & British armies, with paper labels	*early 19thC*	1800
Introduction of color printing and promotional labels	*mid 19thC*	
Evolution of brands and brand identity labeling	*late 19thC*	
Development of plastics packaging and aluminum foil	*early 20thC*	1900
Rise of the supermarket	*mid 20thC*	
Development & implementation of packaging and label legislation	*during 20thC*	
Introduction of VIP and digital printing	*late 20thC*	
Development of smart/intelligent labels & packaging	*early 21stC*	2000

from working out to cycling – and the consequent need for new types of labels and solutions to support these activities.

Emergence of new labeling technologies. Pressures in the early to mid-1980s to find new ways of labeling very long runs of blow molded containers led to the development of in-mold labeling, whereby the label is placed in the mold prior to the forming of the container, so becoming an integral part of the bottle.

Originally only used for decorating blown plastic containers for hair care and under-the-sink products, in-mold labeling later extended to the labeling of injection molded biscuit containers and to tubs for soft spreads and margarines. Most recently, in-mold labeling has been used for the labeling of thermoformed tubs, again for soft spreads.

Also in the 1980s came the development of shrink-sleeve labeling where a continuous web of film is printed, formed into a tube and then, for application, cut to the appropriate length, placed over the container and then shrunk to fit. Having the advantage of 360 degree decoration, sleeve labeling has created new markets and applications for labels and also added the potential to extend the shrink capability to provide a tamper-evident cap seal. An alternative and more recent technology for 360 degree bottle decoration came with the introduction of stretch-sleeve labeling in the 1990s.

Other new labeling technologies developed in the 1990s, primarily for the labeling of plastic bottles, including wrap-around film labeling, cut-and-stack film labeling, roll-on-shrink-on labeling (ROSO) and spot patch film labeling. Used for the decoration of soft drinks, carbonated beverages and some beers, these newer label technologies are achieving some of the highest growth in the label and end-user market.

The explosion of supermarkets and hypermarkets in the past 20 or more years has brought the development of 'Own Brand' labels, successfully competing with and gaining market share from the major brand owners. This, coupled with faster and faster store throughput and a significant increase in pre-packed fresh produce, has brought about the need for shorter and shorter run lengths, a major requirement for price-weigh labels, ever reducing lead times, Just-In-Time (JIT) manufacturing practices, improved supply chain management and demands for improved quality standards.

Most recently, digital printing has facilitated shorter run lengths, the possibility to personalize labels or easily produce multi-variations or versions, as well as incorporate sequential number or codes. Even newer is the growth and interest in interactive labels used with smart phones to access more information or become involved in social networking.

All these trends and pressures have had an influence on label evolution and on the types of labels used today, as well as the printing process required, on pre-press technology, on label application and on label usage.

Anyone interested in understanding more about the evolution and history of labels and label technology would be advised to obtain a copy of *The History of Labels book.*

HOLE PUNCHING – Hole punching (pinfeed) is normally used for making EDP holes but other shapes are also possible. The holes are cut by using either a male/female system or using a shaft with movable EDP rings. The latter is used in anvil position to cut onto the face material and the waste is being removed by the waste matrix when stripping the face. *See EDP.*

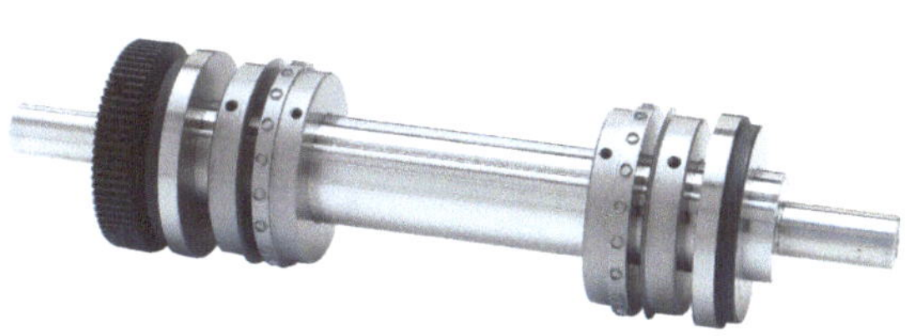

Shaft with movable EDP rings. *Source: Rotometrics.*

HOLLOW DIES – Hollow cutters. The edges of the dies are usually bevelled about 20° and sharpened. The material is punched through into a wood or soft metal block in order to not dull the edges. These die can be pressed into the material with or a mechanical press.

HOT-MELT ADHESIVE – Thermoplastic material with 100 per cent solids that is heated to temperature above their melting point and applied to substrates in the molten state. Unlike water-based or solvent-based adhesives, hot-melt adhesives do not require drying; they set as quickly as they can cool down to their solidification point, which makes them ideal for picking up labels in high-speed labelling lines.

Hot-melt adhesives are used for many adhesive applications where their immediate speed of setting/bonding is ideal for fast production. Self-adhesive hot-melt coated material is usually passed over a chilled roller after laminating to its backing liner, which cools and sets the adhesive coating leaving it with a very tacky pressure-sensitive surface.

They are often used for more difficult labeling applications or low energy/apolar plastics and moist chilled environments where emulsion adhesives have difficulties. At higher adhesive coat weights adhesion to rough surfaces is improved, but hot-melt self-adhesive formulations are generally unsuitable for labeling plasticized surfaces, where the plasticizer will migrate into the adhesive and degrade it so causing adhesive stringing and strike through of label papers, which may result in bond failure.

Self-adhesive hot-melt formulations can be obtained in permanent, peelable, chill-permanent, high-tack and permanent and deep-freeze label laminates.

Hot-melt adhesives are popular because they form a virtually invisible line on glass, metal, PET, and other plastic containers. In the self-adhesive sector, hot-melt adhesives can have very high tack, which makes them ideal for some very difficult applications.

Key criteria in the use of hot-melt adhesives are temperature – which controls viscosity – and adhesive film thickness, which affects speed of setting, tack and open time. The thicker the adhesive is applied, the longer it will take to set.

HOLOGRAM – A microscopically fine diffractive structure by which three-dimensional images are generated. The top metallized layer in the structure of a hologram provides high reflectance to exhibit defined color shifts, both in normal incidence and oblique angle viewing, depending on the design of the hologram. In general, aluminum metallization is used for highly reflecting images, while high refractive index coating – such as zinc sulphide – is used for see-through security documents such as passports and identity cards. The intrinsic security of holograms is captured within the origination process as they are difficult to manufacture and lasers are required for the actual recording of the hologram images.

Holograms provide powerful visual appeal as well as prestige and anti-counterfeiting/brand protection properties. In addition they:

- Are virtually impossible to replicate
- Cannot be effectively reproduced by photocopiers or scanners
- Will expose any attempts at alteration or removal from a document or label
- Are easy to authenticate with the naked eye
- Require a great deal of specialist know-how and expensive equipment to originate and manufacture
- Cannot be replicated by reproduction using any conventional printing process.

Holograms have become powerful tools in the prevention of counterfeiting of documents, labels, seals, tags, cards and packaging and a part of a rapidly growing global security industry's armoury. Both overt and covert machine readable features, variable data and unique serial numbering, etc, can also be incorporated. Holograms offer a wide variety of different features which can be matched to different levels of security requirements, from those used in relatively low-cost commercial applications such as packaging and labels, to a more sophisticated security method of protecting currency.

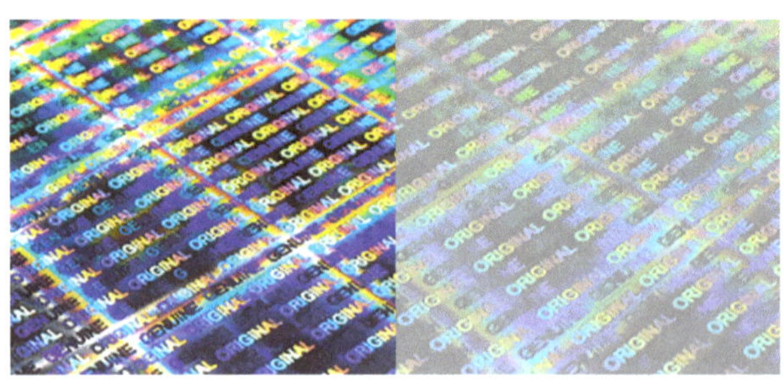

Holographic metallic cold foil used to provide anti-counterfeiting and product enhancement

A variety of hologram (OVID/DOVID) devices have been devised and patented, including rainbow holograms, gyrograms, movigrams, holographic stereograms, dot matrix holograms, pixelgrams, excelgrams and kinegrams. Amongst these, dot-matrix and kinegram elements are the ones most widely used for their ability to provide a variety of iridescent and cinematic effects, and relatively low sensitivity to light conditions.

HOT-FOIL PRINTING (HOT-FOILING) – A dry printing process which uses very thin aluminium foil in a variety of metallic colours – such as gold, silver, red or blue – rather than inks from which to print. Hot- foil printing is achieved by transferring the coloured metallic pigment coating from a ribbon or plastic material known as the 'carrier' onto the surface of the label material to be printing using a hard printing plate which bears the image to be hot-foiled.

The transfer is achieved through the application of heat, pressure, and the length of time the heated coating area is in contact with the substrate – known as the dwell time. The balance and control of these elements is critical and must be individually calculated for the surface to be printed, and the type of ribbon or foil being used.

The printing plate used for hotfoil blocking needs to be of a hard material and have a raised image similar to that used by the letterpress process. The fact that image transfer relies upon both heat and pressure restricts plate materials to either a very hard thermoformed plastic plate for very short runs or plates produced from brass, steel, or zinc, for the longer runs.

While rotary foil blocking has gained some ground for long-run labelling applications where other print modules may print the detail, the majority of hot-foil blocking for label printing is short-run text and solid color, produced on small flat-bed machines.

Units designed for the hot-foil printing or decoration of labels come in a variety of configurations and widths. Stand-alone hot-foil blocking machines are generally narrow-web, but the standard self-adhesive production widths are used where combinations print processes include foil blocking. Larger sheet-feed presses are used for foil blocking or large sheets of glue-applied or in-mould labels.

The hot-foil process can print on a wide range of surfaces and produces bright effect from metallics, or prints high opacity colours relatively simple equipment. It is used to provide a luxury (metallic) look on many cosmetics, toiletries, health and beauty labels, on wines and spirit labels and in other higher added-value label applications.

A more recent development of hot-foil blocking is the cold foil process, in which

A print unit is used to print a special adhesive on the label web where the metallic effect is required. When the metallic foil is brought into contact with the adhesive it adheres to it to produce the printed foil design on the label. *See Cold foiling.*

HOT-MELT COATING – A method of coating with a 100% solids adhesive compound of wax, resins of polymer, or mixtures thereof, heated to a fluid state and applied to a substrate by, for example, a roll, gravure or extrusion process, and then cooled with a subsequent chilling device.

HOT STAMPING – A printing process that uses heat and pressure to transfer the image from a metal printing plate to a label substrate using a metallic or colored foil to produce the desired result. *See also Hot foil printing.*

HOT STAMP ROTARY DIE – Special flexible die in which copper and steel are bonded in combination to provide both etchability and magnetic attraction (for mounting on a heated magnetic base) which reduces the cost of rotary hot-stamping and decreases turnaround time. The dies are 30-40 per cent of the cost of solid rotary hot stamping cylinders.

HUE – Color is measured in terms of hue, saturation and luminance. Hue is the predominant wavelength of a color which distinguishes it from other colors, such as red, blue or green; saturation is the color intensity (from white to pure color); and luminance is the brightness of the color (from light to dark).

HYBRID PRESS – Press lines which combine a CMYK inkjet print engine with at least one conventional press station on the same transport, including flexo printing, in-line converting, embellishment and finishing units. The category includes presses that are designed and built as hybrids as well as ones that have resulted from the retrofitting of a conventional press with an inkjet module.

Hybrid digital/analogue presses have been around for some years but had failed to make a major breakthrough in the label market. New generations of hybrid inkjet/flexo presses are now offering viable solutions.

HYDROPHILIC – Readily wet by water (water-loving). Used to describe the non-printing areas on a lithographic printing plate.

HYDROPHOBIC – Water repellent (water-hating). Used to describe the water repellent image areas on a lithographic printing plate.

HYDROSCOPIC – The ability of some materials to absorb moisture. Paper fibers are hydroscopic and will take up atmospheric moisture and expand. May cause wavy, cockled or creased paper, poor print registration and other print problems. Paper in storage or transit should be protected against humidity and moisture changes.

I

ICC COUNTERFEITING INTELLIGENCE BUREAU – The International Chamber of Commerce formed the ICC Counterfeiting Intelligence Bureau (CIB) in 1985 as a focal point for industries and other affected interests worldwide to fight the growing problem of counterfeiting. It is a not for profit membership organization comprising members from numerous countries and sectors, include large multinational companies, trade associations, law firms and technology producers.

The CIB, which is a specialized bureau within the ICC Commercial Crime Services, protects industry from the damage caused by counterfeiting by gathering intelligence, carrying out covert investigations and providing expert advice and training.

IDENTIFIER – Printed indication on packaging, either as a label or by direct printing, to communicate the fact that the product is protected by an electronic article surveillance (EAS) tag.

IDLER ROLLERS – Rollers used on web-fed paper or film converting lines to support, smooth or direct the web in its course through the machine. The rollers are not driven.

IGT – The successor to the world-renowned TNO-Institute for the Graphics Techniques, founded in1939. IGT Testing Systems specializes in developing, manufacturing and marketing of printability testing equipment and tack testing equipment for the worldwide paper, printing and graphic arts industries.

IGT test equipment is used by label stock and label ink manufacturers, press manufacturers and label printing companies for testing tack, pick, gloss, density, etc, of flexo, gravure, offset and letterpress materials, inks and prints.

IHMA – THE International Hologram Manufacturers Association is made up of more than 60 of the world's leading hologram manufacturers and is dedicated to promoting the interests of members worldwide, and to helping users to achieve security, packaging, graphic and other objectives through the efficient use of holography.

IHMA members adhere to a strict code of practice, promising to supply the best holograms for all uses. They can also protect their security holograms and customers' products and documents against counterfeiting of holograms on the Hologram Image Register, a secure registry of holographic images published by the IHMA to safeguard hologram copyright and underpin the use of holograms in authentication and security printing.

IMAGE CARRIER – That part of a printing press which is responsible for transferring ink from the ink distribution system to the substrate. The printing plate or plate sleeve on the press that carries the image.

IMDG CODE – The International Maritime Dangerous Goods Code is accepted as an international guide for the transport of dangerous goods by sea and is recommended to governments worldwide for adoption or for use as a basis for national maritime regulations. It provides guidance for not only the shipping industry but also for those in maritime service-related sectors, including packaging, labeling, marking and handling. In practice, from a legal point of view, the Code is mandatory, although there are some provisions that are recommendatory in nature.

IML – In-mold label. A pre-printed paper, synthetic or film label which is placed inside the mold before a plastic is blown, injected or thermo-formed to produce a plastic bottle or tub. The label is inseparably combined with the container and becomes an integral part of it without any label edge visible. There is no subsequent requirement for label application equipment. The in-mold labelling process has found increasing application, meeting the demand for precision labeling of high-quality, low weight molded containers using the same type of material as that of the container (e.g. PP on PP) for ease of recycling. The process is economical only if extremely large production quantities are involved.

In Europe, it is estimated that some 80% of in-mold labeling is with injection molded or thermoformed tubs for dairy foods, such as soft spreads, margarines, cheeses, sauces and ice-cream, with 20% used with blow-molded containers for under-the-sink products, household chemicals and industrial, laundry products, personal and hair-care products and some juices. In the States, it is in-mold labeling of blow-molded containers which dominates the market.

In-mold labels are mainly printed by sheet-fed offset or by web-fed gravure. Narrow- to- mid-web flexography, letterpress and offset are also used by some printers – with off-line die-cutting to shape. Inks are critical to the process with labels protected on the surface with a UV or EB curable top coat. Die-cutting is also critical, particularly when labels printed in sheets are stacked and rammed through a tunnel, emerging cut to size. Edge welding may be a consequence of this process.

In-mold label substrates. Initially, the process involved placing paper labels coated with a heat-seal back layer into the mold before blow molding. This layer must fuse to the bottle during the blow molding process. More recently, for recycling and performance purposes, synthetic paper materials such as Polyart or Synteape have become more common as labels on under-the-sink labeled products, as well as specially developed OPP films (which fuse directly to the container and eliminate any tendency to an orange-peel effect) which are used for the labeling of injection molded tubs for soft spreads. All in-mold label substrates must have good lay-flat properties for trouble-free feeding from the molding machine hoppers.

In-mold label application. Because of the high cost of the basic molding equipment and molds, plus the need to modify these to be able to insert and position labels accurately into the mold, in-mold labeling has had a somewhat limited acceptance in the market place. It has little more than a 2-3 per cent market share. Relatively long runs have traditionally been needed to make the process viable, which is particularly difficult when short-run, 'just-in-time' label manufacturing is being increasingly demanded by the label user.

Against this, the requirement for both recyclable and/or pre-labeled containers, the potential of higher packaging line speeds and in-case filling, improved appearance and better squeeze resistance, spurred on by environmental and economic issues and more economic methods of producing and inserting in-mold labels, are all expected to aid the growth of the process in the coming years.

A recent development of in-mold labelling is with

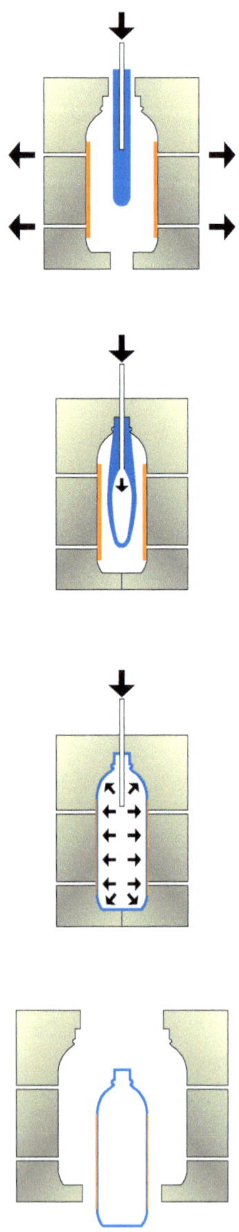

1 Labels are placed against the internal walls of the open mold.

2 While a pre-form plastic tube passes into the mold, suction is used to keep the labels firmly fixed in place until the mold is fully closed.

3 Air is then forced into the heated plastic pre-form forcing it to fill the void inside the closed mold.

4 The heated plastic expands rapidly, pressing against the inner walls of the mold and fixing the labels in place on the bottle.

5 The completed bottle is then released from the mold.

Diagram showing the operations involved in in-mold labeling.

the use of single portion thermoformed pots for yoghurts and cream-based deserts where the label is applied in the mold – yet the labelstock is on a reel with a conventional pressure-sensitive adhesive. The machine used for this application forms the pot, and fills and seals 12 or more pots per cycle.

IMPACT PRINTERS – Printers which transfer an inked coating from a ribbon to the substrate by mechanical means, e.g. by a daisywheel or dot-matrix system. Impact printers are easy to service and can be used for the production of multiple carbon or carbonless copies. Their disadvantage lies in their loudness and their limited graphics ability.

IMPRESSION – The image transferred from the printing plate or cylinder to the label substrate.

IMPRESSION CYLINDER – The cylinder on a printing press over which the label substrate passes and is supported while the inked image is being transferred from the printing surface to the substrate.

IMPRINTING OR OVERPRINTING – The printing of changeable or variable copy onto blank or pre-printed labels, tickets or tags using a secondary printing device such as a computer printer, dot-matrix printer, etc.

INACTIVE TAG – An electronic article surveillance (EAS) tag that is not active and cannot be detected. Either the tag can be re-activated or not.

INCH TO MILLIMETER CONVERSION – As both inches and millimetres are used for measurement in different regions/countries within the global label industry, conversion may need to be undertaken. This can be done by taking the inch figure and multiplying by 25.4 to convert to millimeters, e.g. 4" x 25.4 = 101.6 mm. Some of the more common approximate inch and millimeter equivalent sizes used in the label industry are shown under the entry for 'Converting metric to imperial'.

INCHING – Intermittently operating a press for very short increments of web travel. Presses are designed not to operate at speed while in an 'inching' mode.

INDESIGN – A major printing/publishing industry standard software package from Adobe that enables pages to be designed and made-up to incorporate images and text.

Size in inches	Size in millimeters
1 inch	25mm
6 inches	150mm
7.5 inches	200mm
10 inches	250mm
12 inches	305mm
14 inches	360mm
16 inches	400mm
18 inches	450mm
20 inches	500mm

Equivalent sizes in inches and millimeters

INDIAN LABEL ASSOCIATION – *See under LMAI.*

INDIGO – The name given to commercial and industrial digital offset presses manufactured by HP Indigo. The printing process on Indigo presses is in three stages: a latent image is created on a photo imaging plate (PIP) foil; the image is developed by a binary ink developer (BID) and transferred first to an offset blanket and then from the blanket to the substrate; the whole operation then repeats itself for each color.

Image generation in the electrographic process used by Indigo presses uses light to create latent images on the PIP foil. This is achieved by first negatively charging the foil, then using a laser beam from the writing head to discharge the image areas, with the discharged areas composing the latent image.

The latent image is converted to a developed (through the BID) physical image on the PIP foil by applying Electroink. The developed image can then transferred to the substrate in two stages – firstly to the blanket (ITM Drum) and then from the blanket to the substrate. After printing the blanket and PIP are cleaned ready for the next cycle.

In commercial press models of the Indigo, sheet paper is taken by grippers and placed around the impression cylinder. Each color is passed to the

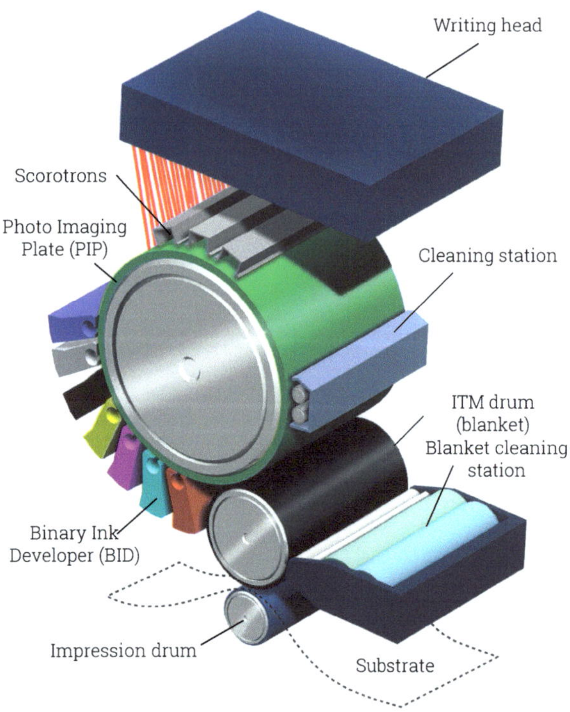

Schematic diagram of the Indigo digital offset press.
Source: HP Indigo

blanket and from the blanket to the substrate in the same revolution.

Web-fed industrial Indigo presses used by the label industry – are 'one shot' presses in which all the layers of Electroink are first accumulated on the blanket before being transferred to the substrate in 'one shot'. Each ink layer is added onto the blanket in perfect register with the previous one. The machines are 4, 5, 6 or 7 colors, including CMYK, Indichrome – Indigo's version of 6 process color printing – and spot colors.

The HP Indigo presses used in the label industry offer a streamlined digital workflow and color printing, which includes Pantone matched colors, white and fluorescent colors, on most substrates. UV invisible inks for security applications are also available. The offset printing process and the liquid Electroinks produce offset-litho like quality labels.

A variety of in-line or off-line finishing solutions, ranging from UV over-varnishing, lamination, die-cutting and matrix removal to cold-foil stamping, are available. Variable data printing capabilities enable text and images to be changed for each label, enabling serial numbering and barcoding to be undertaken. *See also Digital printing.*

INDUSTRIAL LABELS – Labels which are required to withstand a range of conditions and extremes of climates which may include oil, water, chemical or grease resistance, high or low temperatures, abrasion or scuffing.

IN-FEED NIP – The mechanism which controls the forward travel of the web into a printing press.

INFRA-RED DRYING – A means of drying inks or coatings using heat generated by wavelengths below red in the spectrum.

INFRA-RED INKS – Inks which show particular properties when viewed under infra-red light. There are also special inks that are invisible when illuminated by IR light. These are clearly visible in normal conditions but disappear when viewed by CCTV cameras specially adapted for use under IR conditions.

INFRA-RED LIGHT – Wave lengths below red in the spectrum. These longer waves, or rays, are used as a source of heat in infra-red drying systems.

INHIBITOR – Any substance which slows or prevents a chemical reaction under a given set of conditions. A compound that retards or stops a chemical reaction such as oxidation, corrosion or polymerization. Inhibitors are intentionally used to maintain formulated silicone release compounds in a coatable condition for extended periods of time, preventing premature increases in viscosity or gelling.

IN-HOUSE COATING AND LAMINATING – The cost of the self-adhesive laminate is a major part of the label printer/converters costs in producing a finished label for a customer. Depending on the size of the laminate order by the printer to the laminate supplier, the volume of the annual business transacted between them, the importance of the ultimate end user and the overall size/importance of the printer in the market place, so he may be able to negotiate some level of price discount with his laminate supplier.

By investing in a coater/laminator and installing it in his own plant, however, the converter can effectively produce his own laminate in-house at a lower cost than buying-in from a laminator. The overall saving in materials costs can be in the order of 25-30% – excluding overheads. Even then, there is still a significant cost saving by manufacturing the laminate in-house, providing waste costs are well controlled. Some label makers who have installed laminating machines have eventually found the price of their problem material and subsequent waste too much and reverted to buying-in their laminate

The label printer would need to have a sufficient annual volume laminate requirement to be able to keep a coating/laminating machine in straight-forward production without standing idle for much of the time. In practice therefore, it is only the top handful or so of label converters who would be in a position to viably undertake in-house coating.

INITIAL RELEASE – A test for release performance run immediately after coating and laminating of pressure-sensitive label stock.

INITIAL TACK – The degree of stickiness or grab of an adhesive label surface when it is first applied to a product or surface. The adhesive strength displayed by a label applied to a surface with minimal contact pressure.

The FINAT Test Method 9 'Quick-Stick' Tack Measurement, where a loop of adhesive coated material lightly contacts a glass plate and the separation force is measured, simulates initial tack.

INK – The black and colored inks used in the various label printing processes. Flexo and gravure inks are very thin and fluid, while letterpress and offset inks are thicker, paste type, inks. The colored inks used for process printing are usually cyan, magenta, yellow and black, and are defined by international standards. These may be supplemented by additional colors when printing by the Hexachrome or Opaltone process.

In the label and package printing industries there are frequently many special brand owner 'house' or 'brand' colors which may need to be matched to meet a particular hue or color saturation.

INK BLEED – The penetration of an ink into a facestock after printing in such a way as to cause discoloration of the facestock around the printed image, or impinge on another ink or color that has been printed adjacent to the ink that is bleeding.

INK FOUNTAIN – That part of an offset press inking system which stores and supplies ink to the inking rollers in a controlled manner during the printing operation.

INKJET METALLICS – With a need to incorporate larger and more abrasive metallic particles, UV inkjet has struggled to develop a successful metallic ink. Alternative routes to producing metallic effects with inkjet include digital cold foiling in which a patterned adhesive is inkjeted onto the web (replacing a flexo adhesive plate). A cold foil is then bonded to the web and can be rewound in the normal way.

Metallic effects can also now be produced using the Process Metallic Color System developed by Color-Logic. In this system, software cuts white masks and permits graphic designers to specify the precise metallic colors required.

INKJET PRINTING – A non-impact, plateless, process that prints from digital data and uses jets of very fine ink droplets fired at the substrate to form the same or variable information images on to paper or other substrate without a press. There are two main types of ink jet printing devices: continuous and drop-on-demand. Ink jet printing can be applied to virtually any surface. It offers a number of benefits for black-and-white, spot-color or full-color digital printing, namely:

- It is a simpler method of printing than any other process
- It is the least complex of all the digital printing technologies
- It can produce good quality results, is silent, and has relatively low-cost consumables
- It does not depend on lasers, electrons or light to produce the image.

Most commercial and the industrial inkjet printers found in the label and packaging sectors use a piezoelectric material in an ink-filled chamber behind each nozzle instead of a heating element. Applying a voltage makes the piezoelectric material change shape, so generating a pressure pulse in the fluid and forcing a droplet of ink from the nozzle. Piezoelectric (also called Piezo) inkjet technology provides a wider variety of inks than thermal inkjet as there is no requirement for a volatile component, and no issue with build-up of ink residue. However, the print heads are more expensive to manufacture due to the use of piezoelectric material.

The DOD process uses software that directs the heads to apply between zero to eight droplets of ink per dot, but only where needed.

Piezo inkjet technology has long been used on production lines to mark products. For instance, the 'use-before' date is often applied to products with this technique; in this application the head is stationary and the product moves past. Requirements of this application are high speed, a long service life, a relatively large gap between the print head and the substrate, and low operating cost.

Most recently, it has been the introduction of high quality peizo color inkjet printing in CMYK, plus white and, in a number of press systems, with orange, violet or green as well that has started to gain ground in the narrow-web label industry for the printing of industrial, household products, pharmaceutical, consumer goods, etc, labels.

UV color inkjet provides a good complement to UV flexo, enabling converters to produce short runs of high quality color labels, have a flexible production operation, and develop new levels of service. *See Continuous inkjet, and Drop-on-demand inkjet.*

INKJET PRINT RESOLUTION – *See Digital print resolution.*

INKJET SCALABILITY – Inkjet printing is a more scalable process than other digital printing technologies. It can be made narrower or wider; made faster or slower, and it can be built in a modular fashion, particularly if converters want to put inkjet print stations inside or on a conventional analogue press.

INK MANAGEMENT SOFTWARE – A user-friendly production management tool that enables the operator to see details of existing recipes, store new color specifications and find any return inks that are available and suitable in real-time. This significantly reduces waste and storage space, as well as providing transparent per-job ink costs to manage profit margins. *See also Automated ink management.*

INK MIST/INK FLY – Minute flying ink filaments or threads in the form of a mist created on a press during printing through the use of long inks with low yield values.

INKOMETER – An instrument designed for measuring the tack of a printing ink.

INK PUMP – An electric or air driven mechanical pump on a printing press that circulates the ink or coating from a holding tank to the ink fountain.

INK RUB – A printing defect, often associated with matt-coated papers, in which parts of a dried ink film are removed through pressure or friction from another surface.

INK TAG – A variant of the removable anti-theft hard tag principle, which contains a permanent dye that will stain and render the product unusable if the tag is tampered with or removed without a special tool. This tool is only carried by store staff.

INK TEST – A test for measuring the surface tension of a plastic by coating it with a special ink. Ink keying is not possible unless the inner tension of the ink corresponds to the surface tension of the ink. The sooner an ink gathers into globules on the surface of the specimen, the higher is its surface tension and poorer its wet-ability.

IN-LINE PRESS – A printing press which is coupled to or incorporates other operations in-line and driven

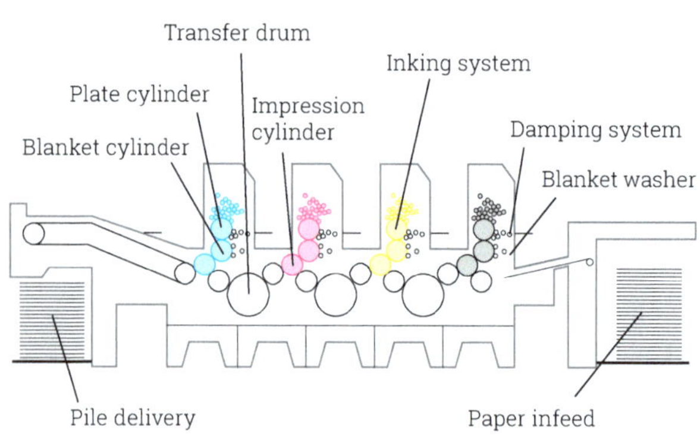

Diagram of an in-line sheet-fed 4-color press.

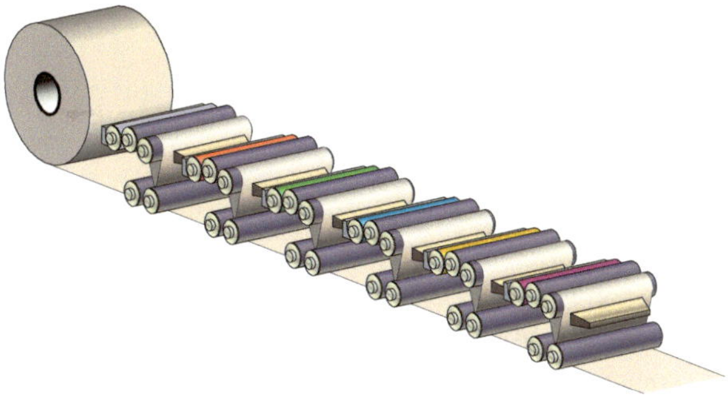

Principle of an in-line roll-fed press.

by the same drive shaft, or an electronic drive shaft in the case of a servo-driven press. The advantage of the in-line configuration is to allow finishing units to be added on, so turning a press into a complete processing and converting line which can undertake in one in-line pass operations such as printing, sheeting, die-cutting, perforating, punching, over-laminating, folding, creasing, embossing, etc.

Also refers to a multi-color printing press in which the printing stations are all mounted horizontally in a line.

IN-LINE WEB INSPECTION – In-line web inspection on a label or packaging press is a first line of defence for the press operator to detect a range of streaking, color variation, mis-register, varnish, hickey, etc, faults – any one of which could get a job rejected. Inspection enables operators to inspect the quality of print in real-time, zoom-in on images and detect defects far more effectively than with the naked eye.

Web inspection began with stroboscopic flash systems and rotating mirror systems, but the general trend today is to use video web inspection systems which enable the press operator to view a web moving at many hundreds of meters/feet per minute on a video display monitor as if the web was stationary. It is also possible to compare real-time images with stored, recorded master images using a split screen.

Video web systems consist of a CCD (charge-couple device) video camera with a zoom lens to control magnification, as well as a strobe light, video picture grabber, computer software and electronics, plus a color monitor. In operation, a full video image, or frame, is created by firing a high-intensity xenon strobe light to illuminate the web and act as a shutter. The very short flash intervals are synchronized with the press speed. The system's electronics and controlling software take digitized images from the camera's electronic chip and save them to a high-volume memory.

Systems may have options for selecting and storing random positions on the web in different zoom ratios, while software-controlled cameras with a motorized traverse enable movement in both lateral and linear planes. Some models can inspect both sides of the web simultaneously.

With many different video inspection systems on the market – at varying price levels – care should be taken to choose the most appropriate in-line web inspection system for the particular application required.

IN-MOLD LABELS – *See IML.*

INTAGLIO PRINTING – A method of printing in which the image to be printed is etched or engraved below the surface of a print cylinder or plate. The printing process is carried out under pressure. The ink is consequently drawn up from the recesses onto the substrate and dries leaving a slightly raised image, which can be detected by feel. Used in security and banknote printing. *See also Gravure.*

INTEGRATED CIRCUIT (IC) – A microelectronic semiconductor device comprising many interconnected transistors and other components. Most RFID tags have ICs.

INTEGRATED TAG – An electronic article surveillance (EAS) tag that has been incorporated into a product or packaging at source, or a manufacturing stage, either by the packaging converter or by the product manufacturer.

INTELLECTUAL PROPERTY RIGHTS – The means by which organizations or individuals can protect their creative and innovative ideas using patents, copyright, trademarks or designs, in much the same way that physical property can be protected. Although some intellectual property rights (IPR or IP) are protected automatically, most cannot be effectively protected unless they have been applied for, registered or granted. To ensure that intellectual property is correctly protected it is necessary to have some understanding of the main kinds of IP protection that are available:

Patents - A patent is issued to protect the invention or discovery of new and improved products or processes, machines, articles of manufacture, or any new useful improvements thereof, that are capable of industrial application.

Copyright - A copyright protects an original artistic or literary work in the fields of publishing, music, films, sound recordings and broadcasts, including software and multimedia.

Trademarks - Trademarks are used for brand identity of goods and services so as to allow distinctions to be made between different manufacturers and traders.

Designs - Used to protect the whole or part of a product's appearance in terms of shapes, colors, textures, lines, contours and/or the materials of a product and/or its decoration or ornamentation for an article of manufacture.

The seal of Cotton and Woolmark registered trademark symbols

Intellectual property rights can be extended much wider than these four basic types of IP to cover such things as Certification Marks (a BSI 'Kite Mark' or Good Housekeeping 'Seal of Approval'), Regional Marks (Champagne, Cheddar, Stilton, etc), Collective Marks ('Wool Mark' or 'IATA') and Service Marks – which promote services and events rather than products (the Olympic Games and the interlocking Olympic circles).

Once protected, the owner of intellectual property rights can control the nature and way it is used and be rewarded for its misuse. Action can be taken through the courts to sue for infringement or, in certain circumstances, such as counterfeiting, initiate criminal proceedings.

INTELLIGENT/SMART LABELS – Intelligent and/or smart labels and packaging solutions go beyond conventional label or pack applications so as to provide a whole range of clever solutions in the manufacturing, distribution, sales, end-use or after-use chain and provide functions other than the standard tasks of protection, display or information carrying.

Many intelligent/smart solutions may be automatically read – some remotely, some can be activated in response to a trigger event, some can be switched on and off in response to certain conditions and some have the capability to have information added to them.

One way of assessing intelligent/smart label and pack solution is to categorize them by their action or function, as can be seen in the diagram. Even this chart does not include developments in contact or near contact smart reading technologies incorporating data matrix codes, 2D codes, optical or magnetic codes, nor the reactive smart authentication technologies such as biocodes.

Intelligent and smart labels and packaging now covers a vast array of technologies that enhance pack functionality, product branding, and benefit the supply chain and consumer. These include labels and packs which can indicate time or temperature, provide evidence of correct sterilization or autoclaving, indicate freshness, detect microbial growth, absorb gases, oxygen or moisture, provide evidence of damage or shock, automatically route or track goods, deter counterfeiting, confirm authentication, reduce retail theft or provide evidence of tampering. The continuing evolution of nano-coatings, nano-taggants and nano-codes are expected to rapidly expand the range of intelligent/smart solutions available to the label converter and label user.

Adopting a smart solution is a complex balance between the technologies that are available, the performance and usage benefits of using them and the cost factors involved. New enabling technologies and consumer demands are now driving 'smart/intelligent' technology even further forward. Advances are enhancing products in new and exciting ways, even becoming part of the product itself. Smart/intelligent technologies are a key to this evolution. Their potential to increase brand and product value is enormous.

For the future, labels and packaging that incorporates a disposable device that interacts with an individual, a process, a personal computer, microwave or electronic shopping list may not be as far away as

Smart / Intelligent Labels

Smart labels	Smart active labels	Intelligent labels
Labels store information and can communicate with a reader. More sophisticated systems can be write and read. Do not require line of sight.	Labels become active in response to a trigger event (i.e. filling, release of pressure or gasses, exposure to UV or moisture).	Label function is able to switch on and/or off in response to external/ internal conditions. Ability to sense and inform.
RFID labels EMID Chip labels Chipless labels	Oxygen scavenging labels Anti-microbial labels Ethylene scavenging labels Odour and flow absorbing labels Moisture absorbing labels Heating/cooling labels	Microbial growth indicators Physical shock indicators Leakage, microbial spoilage indicators Light protection Time/temperature indicators Freshness indicators

The role and function of smart, smart-active and intelligent labels.

one might think. Such devices may well revolutionize the world of packaging and the role that it performs in a fast-changing product and consumer environment. *See also Smart active labels, RFID, Food freshness indicators, Temperature indicators, etc.*

INTELLIGENT TAG – A miniaturized complex electronic circuit capable of storing and communicating discrete information to a receiver. Unlike a regular electronic article surveillance (EAS) tag, this is not a binary on/off device, but often contains a 64-bit integrated circuit capable of storing millions of variable pieces of information, thus providing unique identification. Available only in Read Only or Read/Write formats.

INTERACTIVE LABELS – Labels which interact through graphical or design elements with smart phones or other mobile devices to make available additional content of various kinds. Additional content can be supplied through a website; by enabling access to a downloadable or streaming file – including video; or providing access to an Augmented Reality (AR) experience viewed directly through the handset screen.
See also Augmented Reality, Mass customization and Mobile Action Codes.

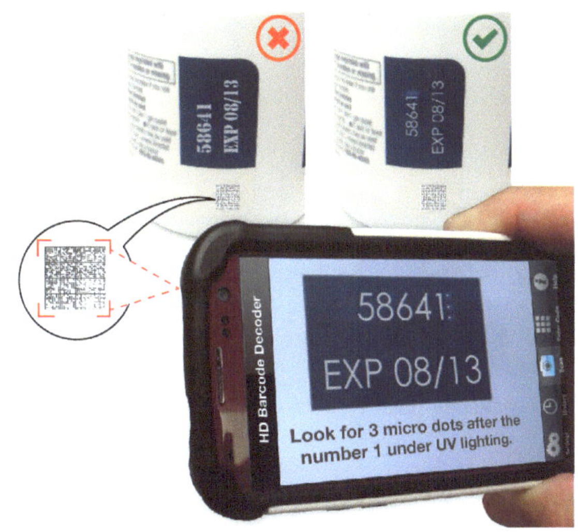

HD Barcode from Complete Inspections Systems

INTER-DECK DRYING – modern narrow-web presses are fitted with inter-deck drying heads allowing each individually printed colour to be cured/dried before printing the next colour. This is called printing wet-on-dry and eliminates the problems which can be encountered when printing wet ink onto a wet ink.

The drying or curing head is positioned so that the printed substrate is exposed to the drying head for the correct dwell time to ensure that the ink is fully cured/dried.

There are three types of drying systems used in the self-adhesive label and narrow-web sectors, Infra-Red, Hot Air and Ultra Violet known as UV.

INTERFACE – The area of contact between two adjacent surfaces.

INTERLEAVED 2 OF 5 – A numeric-only, bi-directional, self-checking barcode symbology consisting of five bars, two of which are wide. In this code both spaces and bars carry information, and there must be an even number of characters in the code.

5 0 1 2 3 4 5 6

Interleaved 2 of 5 medium density barcode

Interleaved 2 of 5 (ITF) is the most popular numeric only barcode in non-retail distribution and is the recommend symbology for corrugated shipping containers. It has also been selected for use on other shipping containers in the grocery industry. Its main advantage is its density. The standard ITF contains 14 digits, with an optional 6-digit suffix.

Mainly used for warehouse inventory, the code is also accepted in airline ticketing and baggage and cargo handling.

INTERMITTENT FEED PRESS – A roll-label press, which prints from a reel by stopping the web during each print cycle in a stop-go action.

INTERNATIONAL HOLOGRAM MANUFACTURERS ASSOCIATION – *See under IHMA.*

INTERNATIONAL LABOR ORGANIZATION (ILO) – Founded in 1919, the ILO is a UN specialized agency which seeks the promotion of social justice and internationally recognized human and labor rights. It formulates labor standards in the form of Conventions and Recommendations which set minimum standards of basic labor rights: freedom of association, the right to organize, collective bargaining, equality of opportunity, and other standards regulating conditions across a wide range of work-related issues. It provides technical assistance in fields such as: vocational training, occupational health and safety, working conditions, labor statistics, industrial relations, etc.

INTERNET OF THINGS – System to connect multiple physical objects via the internet using unique identification protocols, including systems to monitor and control objects remotely

INTERROGATOR – *See RFID reader.*

INVISIBLE BARCODES – These make use of near infra-red fluorophones which are invisible to the naked eye. They produce visible fluorescence when subjected to laser light from a scanner. The inks used for invisible barcode reading are available for both ink jet and flexographic printing and are targeted at applications in the packaging of cosmetics, pharmaceuticals and luxury goods.

ION DEPOSITION PRINTING – Printing using a non-impact electronic image process. A latent image to be printed is first created by ion deposition on a sensitized drum. The image is coated or developed with a dry toner powder before being transferred to the label or tag substrate by cold pressure. The drum is then cleaned before the next image is created. Each image can be different.

Ion deposition printing is capable of printing barcodes and other information onto substrates such as paper, boards, vinyl, polyester and other filmic materials at relatively high speeds. The image may need to be over-varnished or laminated to improve durability, rub or scuff resistance.

IR-SCANNABLE – An image or barcode capable of being read by an infra-red scanner.

IRIDESCENT INKS – Inks which change color when viewed from different angles. The ink is colored with newly-developed liquid-crystal polymers instead of conventional dies and pigments and is a mixture of two different liquid crystal components that can be cured with UV light. The inks must be printed on top of another color for the effect to be obtained; the best results being achieved when printing on a black background. Maximum effect is obtained when a heavy deposit of ink is used. Applications are in the packaging of cosmetics, beverages and pharmaceuticals.

IRRADIATION – To treat with ultra-violet light or other high-energy ray.

ISO – International Organization for Standards is the world's largest developer of voluntary International Standards. It provides world class specifications for products, services and good practice, to ensure quality, safety and efficiency – all developed through global consensus. Indeed, since 1947, ISO has published over 19,500 International Standards covering almost every industry.

An international network of national standards institutes of 161 countries on the basis of one member per country, with a Central Secretariat base in Geneva, Switzerland that co-ordinates the system.

ISO is the source of ISO 9000 – Quality management, ISO 14000 – Environmental management; ISO 22000 – Food safety management, and many more standards used by the label and packaging industries.

ISO 9000 – Achieving certification to the internationally recognized quality management standard BS EN ISO 9000 has become the most common way for the label industry to introduce and implement a quality management system and demonstrate its commitment to quality. For many label industry suppliers and converters – as well as label buyers – it has become the expected standard.

Implementation of ISO 9000 involves making a large number of small improvements across a wide range of business processes. Activities can be managed as a process, with each activity adding value to the last, all ultimately leading to customer satisfaction. Simple, user friendly, documentation can be created around a company's existing business practices.

The benefits of obtaining and maintaining ISO 9000 can be viewed on various web sites or through the International Standards Organization. *See also Quality management.*

ITALIAN LABEL ASSOCIATION – See under GIPEA.

J

JAM – A printing press or other equipment stoppage (crash) due to problems with sheet or web feed, double feeding on sheet-fed presses, misfeeding, mechanical failure, incorrect press setting, static and/or poorly conditioned paper.

JAPAN FEDERATION OF LABEL PRINTING INDUSTRIES – *See under JFLP.*

JDF – Job Definition Format (JDF) is a comprehensive XML-based file format industry standard for end-to-end job ticket specifications combined with a message description standard and message interchange protocol. It is designed to streamline information exchange between different applications and systems, and to enable the entire industry – including media, design, graphic arts, on-demand and e-commerce companies – to implement and work with individual workflow solutions.

In simple terms, a JDF file can be regarded as the digital equivalent of a label printer's job bag. It contains all the information needed for a job to be produced and can have notes and instructions added to it, either by operators or automatically by the equipment being used at each stage. Being digital, a JDF file offers huge advantages over a physical job bag. Anyone who needs to – at any point – can access details and amend them.

A common file such as this, with all the details of the job, enables customer service personnel to see where a job is at any one time and advise the client accordingly or, if the system allows it, enable the customer to check the status of their job(s) themselves. The production director should be also be able to see all stages of production and react to or pre-empt potential problems; the finance director can access quotes or production costs and review profitability.

One of the key factors about JDF is that it makes the biggest impact where it is implemented first and most rapidly; in pre-press and MIS – both of which are reliant on software and which can be easily updated to handle JDF. As jobs become shorter, pre-press and administration account for an increasing proportion of jobs costs. Consequently, the savings achieved can make a big contribution to the bottom line.

JFLP – The Japan Federation of Label Printing Industries (JFLP) was established in 1961 and currently has a membership of over 700 companies that produce identification labels and seals, sealing, sales promotion and adhesives. The Federation is based in Tokyo.

Most of the companies that make up JFLP are small and medium sized; 80 per cent have less than 20 employees.

JIG – A device used to assist the correct placement of a hand or semi-automatically applied pressure-sensitive label. The jig is usually made to fit the shape of the product being labeled. A jig may also be used to aid manual converting operations.

JIT – Just in Time manufacturing in which the product is produced only after receipt of the order. JIT provides a major application for Print-on-Demand (POD) and is a key driver for the increase of digital printing in label applications.

JMF – Job Messaging Format (JMF) is the little brother of JDF, created by equipment to provide feedback on its status or on an operation carried out on a job to the central controller and repository of the JDF.

JOG – To vibrate a stack of paper or printed sheets of labels into a compact pile for guillotining, packing or handling. To run a label press intermittently in very short increments. *See Inch.*

JPEG – The JPEG File Interchange Format is a lossy compression algorithm which allows images to be exchanged between a wide variety of platforms and applications. Compression ratios down to 25:1 can be achieved with very little perceived loss of quality on continuous tone files.

Lossy compression reduces a file by permanently eliminating redundant information in the file yet, when the file is uncompressed, all of the original necessary data in the file is completely recovered and restored. Commonly used for photographs and complex images.

JUMBO ROLL – A full size roll of pressure-sensitive material as it comes off the laminating machine.

K

KB – Kilobyte, or 1024 bytes. The most common measurement of computer file length.

KELVIN – A unit of measurement of light temperature.

KEY – The adhesion of an ink, coating or adhesive to a material. May also be a secret value using in encrypting algorithm and known by one or both of the communicating parties (similar to a combination number for a vault).

KEY – In mechanical engineering , a key is a machine element used to connect a rotating machine element to a shaft. Through this connection the key prevents relative rotation between the two parts and allows torque to be transmitted through. For a key to function the shaft and rotating machine element must have a keyway, which is a slot or pocket for the key to fit in. The whole system is called a keyed point. A keyed point still allows relative axial movement between the parts.

KEYLINE – An outline drawing of finished label art, which indicates the exact shape, position, and size for elements such as text, line diagrams, halftones, etc.

KINEGRAM™ – A proprietary line generated optically variable device (OVD), which can provide a very high level of security using optically variable ultra-high resolution lines of infinite variety. They resemble a hologram but, on closer examination, there are a number of differences and, in effect, the image appears to come alive.

Kinegrams can incorporate traditional security print design features such as guilloches, rosettes and micro-lines, all types of micro lettering and other complex computer enhanced design elements. *See also OVDs.*

KISS CUT – A die-cutting operation in which self-adhesive face material is cut through to the release liner backing, but the liner itself is not cut.

KISS IMPRESSION – The lightest pressure that will transfer a film of ink from the inking rollers to the printing plate and from the printing plate to the material being printed.

KNOCKING UP – Making the edges of a stack or ream of paper straight and regular. Knocking up can be done manually, but is more commonly achieved mechanically with pushers or by vibration. *See Jog.*

KRAFT – A long-fibered strong paper containing unbleached wood pulp (brown in color) made by the sulfate paper-making process.

L

LABEL – A label may be a piece of paper, paperboard, laminate, metallic foil, plastic film, fabric or synthetic material that can be affixed to almost anything to indicate its contents, nature, direction, ownership or destination. Labeling is the art of applying or attaching the label to a particular surface, item, or product.

Originally, labels were used as a means of identifying a product or to supply information about the properties, nature, or purpose of the item being labeled. Today, their use is often required for legislative purposes, and they are seen as sales and marketing aids in the total design of a package or product. In some cases, labels have become products in themselves.

LABELEXPO EXHIBITIONS – Labelexpo shows are the world's leading series of exhibitions and conferences for the label and related industries. Founded as an annual national coding, marking and label exhibition event in England in 1980, but by 1985 had become a pan-European show in Brussels supported by the key international label press manufactures and materials suppliers. Within two years, rapid Labelexpo growth saw the exhibition move from down-town Brussels to the much larger Parc des Expositions – still its European home today – and begin its transition into a major biennial international show for the world of labels. Labelexpo Europe is now the world's largest label and conference event with over 500 exhibitors and in excess of 25,000 visitors.

Following demands from the main exhibitors for a Labelexpo show to be also established in North America, Labelexpo USA was launched in 1989. It has since developed into a major biennial show attracting visitors from both North and South America, Europe and Asia. Now re-named as Labelexpo Americas, this major international label exhibition and conference event, like Europe, attracts some 400 exhibitors and falls in alternate years with the European show.

By the late 1980s, new and emerging markets for labels were emerging around the world, particularly in South-East Asia. To continue to support and stimulate the growth of labels and the major global suppliers, Labelexpo Asia was launched, originally in Singapore, but is now in China.

At the end of the 1990s, Labelexpo shows covered around 80-90% of the more developed global market through the shows in Western Europe, North America and Asia. However, more recently, rapid label industry growth has been experienced in Central and Eastern Europe, in Latin America, Africa and India. Not surprisingly, the leading label industry suppliers were soon seeking a means of targeting sales into these new regions of the world.

From this demand came a new label exhibition in India and conference/summit events in Mexico, Brazil, Thailand, Indonesia and South Africa.

LABELING MACHINE – A machine that dispenses a self-adhesive label and applies it to a container, pack or product. The machine incorporates a means of removing the backing material (release liner) by pulling it over an acute metal beak, so leaving the label free to be applied to the product. *See also Applicator, self-adhesive.*

LABEL PRINTING AND PRODUCTION – The printing and production of labels is undertaken on a wide variety of presses – from sheet-fed to web-fed – and using almost every different printing process available. The type of press or process used is determined by:

- the specific label or printing requirement,
- the availability and quantity of labels,
- the nature and quality of the printing, the number of colors,
- whether subsequent converting operations (die-cutting, embossing, metallic foiling, laminating, waste stripping, etc.) are carried out in-line on the press or are separate stand-alone operations,
- whether the finished labels are required in cut sheet labels, punched labels or for automatic application in a continuous web.
- how the labels are to be shipped to the packaging or bottling line (in slit reels, cut to size or punched to shape in stacks).

Apart from the mechanical printing processes used to produce pre-printed labels – rotary and semi-rotary letterpress, flexo and UV-flexo, screen process, offset litho, gravure, hot or cold foiling – there are a range of

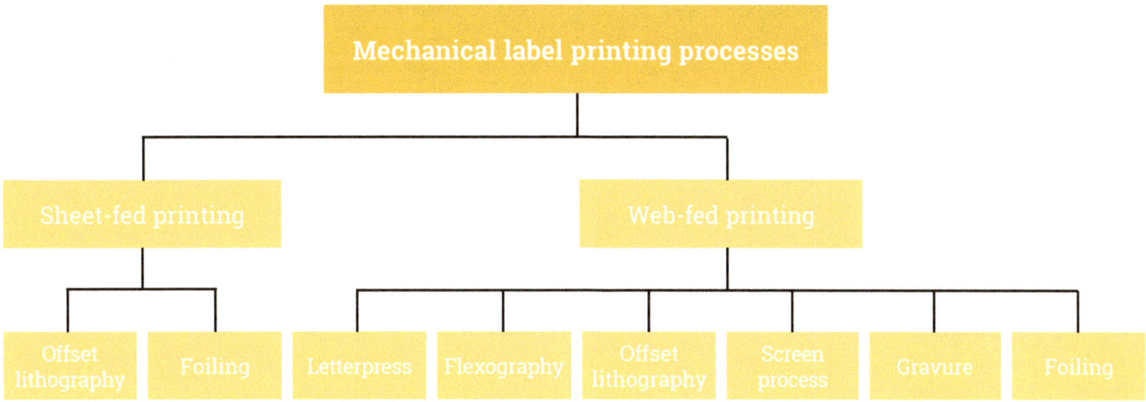

Mechanical label printing and production processes.

on-demand, variable information printing solutions used by the label industry. These include thermal printing, laser and ink-jet printing and, more recently, complete digital color printing presses.

The mechanical label printing processes can be categorized by the way in which the area of the printing plate or cylinder that carries the printing ink (the image carrier) is defined when compared to the non-printing area. Printing plates and cylinders that print from a raised, inked surface are known as relief printing processes. These include letterpress and flexographic printing. Printing processes that carry the printing ink in a recessed pattern of etched or engraved cells are known as intaglio printing. These include gravure and rotogravure, while printing processes that print from a flat, chemically-defined, printing plate are known as planographic printing processes. The most commonly used planographic printing process used for labels is offset litho.

Each of these mechanical printing processes has undergone developments in recent years, which have led to better quality, color and speeds. As such there is no 'best process'. Each of them is capable of high quality printing at economic speeds and each can impart a quality or characteristic to the overall design and image of the finished label. These characteristics include such things as evenness or brightness of colors, thickness of the ink film, fineness of reproduction, quality of halftones, the texture and feel of the label, etc. Some can also impart a range of

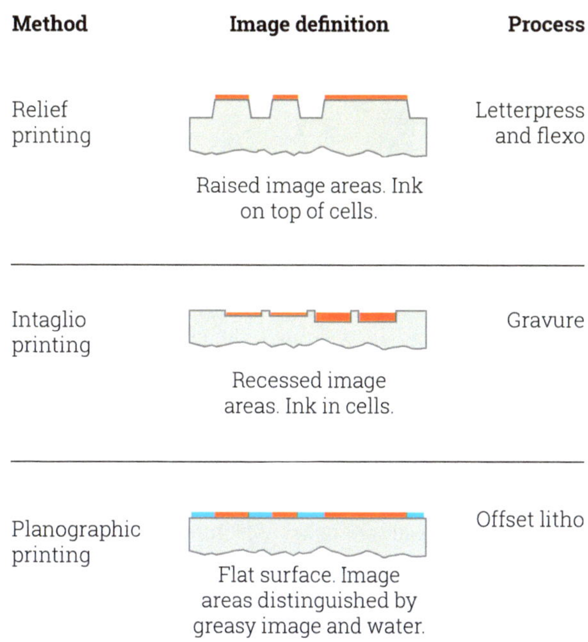

Diagram shows the key differences between the main label printing process.

special effects like scent-impregnated inks, rub-off or scratch-off inks, color or temperature-change inks or raised tactile images for warning purposes.

The one aspect that none of these mechanical label printing processes can handle economically is that of printing variable images that involve barcodes, batch and date codes, sequential numbers, price and/or weight information, or very short runs. To meet these requirements a number of methods of electronically printing variable images or short runs have been developed. Some can be run in line with fixed-image mechanical presses, some are stand alone machines, and some are meant for adding variable information off-line.

Most recently, digital printing technology – in which the image is defined or created by computer – has rapidly found a place in the label printing industry for full-color and spot-color short-run printing of labels. By the end of 2013 there were over 2,000 web-fed digital presses in label plants worldwide.

LABELS ACROSS – The number of labels in the across web direction.

LABELS AROUND – The number of labels around (per revolution) of a printing cylinder, cutting cylinder and/or flexible die.

LABELS & LABELING – The world's leading trade magazine for the label industry, with a readership in over 140 countries. Available in both printed and internet formats, in English and selected foreign languages. Founded in 1978, Labels & Labeling is the main source of news worldwide on technology, materials and product developments for label printers. The sponsoring magazine for all Labelexpo shows and Summits worldwide. *See also Labelexpo.*

LABEL STOCK – Sheets or continuous webs of paper backed with adhesive, which may be pressure-sensitive, heat-sensitive or remoistenable gummed. The label face material or laminate from which labels are printed and/or converted. The term is most commonly applied to pressure-sensitive roll laminate materials.

LABEL TESTING, PRESSURE-SENSITIVE ADHESIVES – Testing in the early days of the pressure-sensitive industry was often limited to seeing if the material was sticky and would adhere to a surface. What testing there was tended to be largely subjective, with tack being measured by the 'feel' of the adhesive mass and with adhesion measured by peeling the product by hand from a surface and recording the adhesion level as a rating of 1 – 5, along with an indication of the label fiber tear or the rate of failure.

With the difficulty of measuring a label's adhesion to a container or pack these methods are still used, but as the industry grew, standard test methods started to be developed for the more 'standard' pressure-sensitive properties such as peel adhesion, tack and shear resistance, with trade associations such as FINAT, PSTC, ASTM and TLMI working to standardize procedures and equipment in order to improve the consistency of results. In some instances new test equipment needed to be developed.

Lower-cost, durable test equipment is available for the shop floor – at the converter and end-user – to undertake quick checks on material quality and processability, while more sophisticated equipment is available for QC and testing laboratories and for research purposes. Today, the key pressure-sensitive adhesive test requirements can be summarized as:

Peel Adhesion. This is defined as a measure of the adhesion of an adhesive laminated to a surface using a defined amount of laminating pressure and after a defined period of time. This is regarded as the 'bond strength' of the adhesive. As pressure-sensitive adhesives attain their bond strength by flowing and 'wetting' a surface, laminating pressure and residence time before peeling are important variables that need to be controlled. To overcome these potential variables, test methods, equipment and procedures have been developed by industry trade associations and other bodies, and should be used as appropriate.

Tack. A measure of the instantaneous adhesion of a pressure-sensitive adhesive with little or no external laminating pressure. This is related to the 'stickiness' of the adhesive. Quantifying pressure-sensitive adhesive tack is again carried out by using the relevant test methods and equipment that have been devised for measuring this property.

Shear resistance. Is defined as a measure of the cohesive strength of the adhesive mass and is often related to the load bearing characteristics and cold flow of the adhesive – most commonly measured as the dead load shear. This method involves bonding a given area of pressure-sensitive product to a test surface and allowing it to reside long enough to build up a good adhesion to the surface then, using the prescribed test equipment and procedures, measuring the time required for the adhesive layer to separate cohesively. This gives a measure of the shear resistance.

A list of all the current FINAT Methods and Reference handbooks can be found under the entry for FINAT Test Methods.

LABEL TYPES – Historically, the range and variety of label types was fairly limited, with over 90 per cent of label usage in the early 1970s coming from wet-glue and gummed paper label solutions. Since the 1970s however, there has been a major shift in the range and variety of label technologies used by the packaging industry and end-user, moving away from dominance by wet-glue and gummed paper label technologies to dominance by self-adhesives from the mid-1990s onwards – at least in the mature markets of Western Europe and North America.

Today, self-adhesive labels make up well over 40 cent of all label usage in these markets, although wet-glue is still well in excess of 30 per cent. Gummed paper labels, so widely used in the 1960s and 1970s, are now less than one percent of usage, much of their traditional usage largely superseded by self-adhesives.

New technologies emerged during the late 1970s, the 1980s and 1990s, such as sleeving, wrap-around film, in-mold, cut-and-stack film – primarily developed for the labeling of plastics bottles and containers – and these make up around 17 per cent of the total label market in the developed regions of the world.

The other key shift in label usage has been a steady and continuous move to the usage of non-paper labels to meet the growing demand for new ways of labeling plastic bottles and containers. Driving forces behind this growth of filmic labels have included:

- high annual growth of plastic bottles (compared with glass bottles and cans)
- the requirement for compatibility of labels and

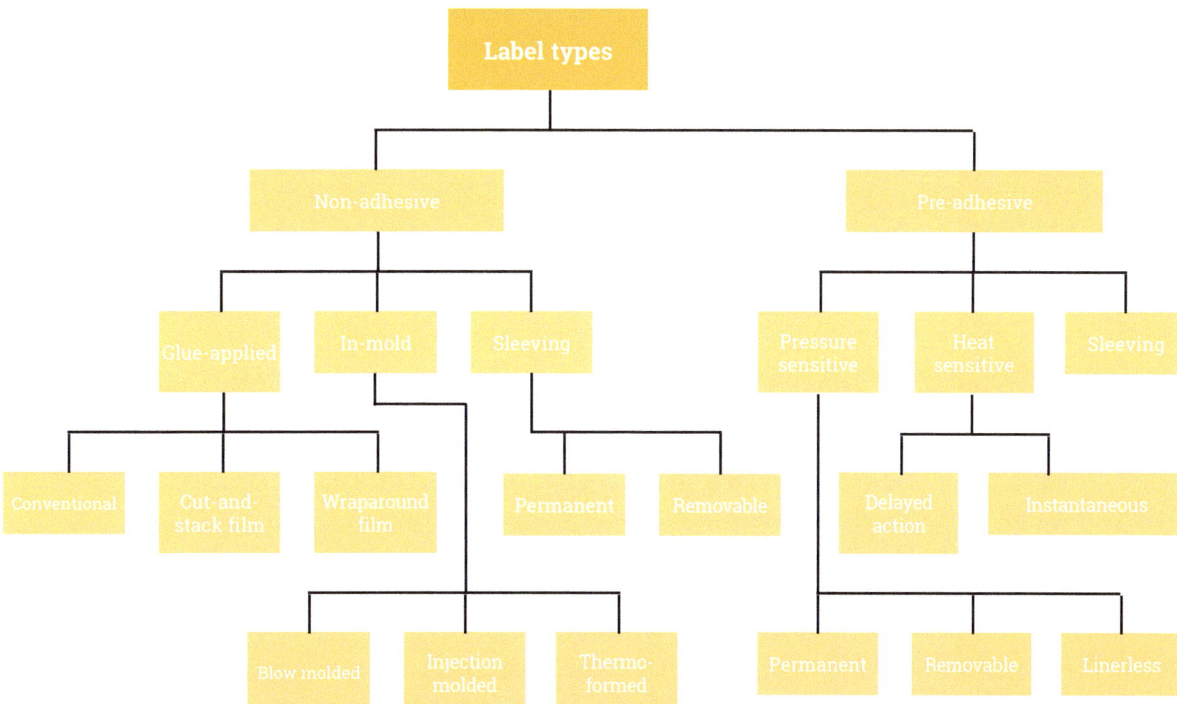

Labels come in many different styles and formats depending on label technology, materials used and method of application.

bottles due to waste and recycling issues
- demand for clear, no-label look packaging
- the ability to provide white, colored, silky, iridescent or pearlescent surfaces
- the need to provide moisture, chemical or performance labels used in demanding applications.

From almost total dominance by paper labels in the 1980s, the market has evolved rapidly to a point where non-paper film labels now make up over 25 per cent of all label requirements, and are growing at several times faster than paper labels.

When looking at label types today, they can all be regarded as falling into one of two key categories: those that are printed and converted in the form of paper or synthetic substrates (film, metallic foil, metallized, etc) and to which an adhesive, glue or gum is applied at the point of application; and those substrates which already have the adhesive or gum on them before they are printed. This adhesive or gum is then activated by pressure, moisture or heat at the point of application.

Even then, the range and variety of label types can be split further into specific sub categories. Sleeving into shrink or stretch labels; in-mold into blown, injected or thermoformed label categories; self-adhesive to include linerless products. All these main label types can then be put together into a 'label tree' to provide a quick reference source for label users.

Factors in the selection of labels. Each label technology or label type has its own advantages or disadvantages in the provision of label solutions for different applications. Factors to be considered when selecting a label type include:

- cost of label material (non-adhesive or pre-adhesive, opaque or clear, metallized, etc)
- cost of printing and converting (number of colors, in-line options, sheet or web printed)
- visual appearance and image (paper, film, design capabilities, aesthetic appeal, premium quality, tactile 'feel', 'no-label' look)
- durability (resistance to scuffing, scratching or image deterioration)
- production flexibility (ease and speed of label line change-over)
- environmental considerations (recyclability, returnability, waste disposal)
- volumes required (short, medium or long-runs) speed of application (high-speeds required?)
- capital investment of label line equipment (wet-glue high; self-adhesive lower)
- running cost/hourly rate of label line (operator and machine)
- down-time (set-up, clean-up, removal of misapplied labels).
- performance needs of labels (chemical or water resistance, labels for autoclaving or sterilizing, wet-strength, high or low temperature usage or application)
- information needs on the label (reverse-side printed, leaflet or booklet label)
- security features (tamper-evidence, anti-theft, brand protection, hologram)

Rather than look at any one or more criteria in isolation when selecting a label type, it is more usual today to look at the chosen solution or possible solutions in terms of total applied label cost. That is, what is the total cost of labeling a bottle, jar, can or other product with the required quality and performance label, at the desired application speed and the optimum cost – bearing in mind materials costs, printing and converting costs, investment and operation costs of application technology and end-usage requirements. In many cases it will be found that there is more than one labeling solution for any one particular application.

LACQUER – A clear resin/solvent coating, which may be glossy or matt, applied to a sheet or web of printed labels to provide protection or resistance against scuffing, rubbing, chemicals, moisture, etc. Carried out as an off-press operation, it is performed on a machine equipped with a roller coater after the inks have been dried. Since lacquering is a separate operation using expensive materials it increases the unit cost of the labels.

LAMINATE – The term used to describe a self-adhesive label construction and the process that brings and adheres two surfaces together.

LAMINATE – A clear plastic film applied to a sheet or web of labels by heat or adhesive to provide and enhanced, glossy or matt, appearance or for protection. Film laminates are more expensive than liquid coatings, but generally give a brighter gloss and better protection.

LAMINATING ADHESIVE – An adhesive used for the laminating of transparent films onto printed surfaces. There are both conventional and UV-curable laminating films for in-line and off-line processing.

LAMINATION – A clear plastic film (polyester, polypropylene) that is bonded to a printed web through the use of adhesive, heat and/or pressure to provide enhanced protection, performance or image. Laminates are available which are non-printable or for overprinting with thermal transfer printers.

LASER – An acronym for Light Amplification by Stimulated Emission of Radiation. A laser is a source of emitting continuous or intermittent pulses in the form of waves of monochromatic, extremely stable and coherent light with a small spot size, which makes them ideal for scanning, recording, etching and creating images on light-sensitive surfaces. Used in digital imaging devices to produce images by electronic impulses from computers or facsimile transmission. There are four main types of laser: crystal, gas, liquid and semiconductor.

LASER DIE-CUTTING – The cutting out of a label through the use of a computer-controlled laser. The laser cutter uses programs developed from the step-and-repeat function of label origination to guide the cutting head around each separate label profile. The laser cuts through the face material, either paper or filmic, and stops at the adhesive layer.

Since the early days of digital laser die-cutting, label finishing technology has changed significantly. Laser die-cutting technology for labels today – and now even for folding cartons – is far superior to that of even 10/15 years ago. It's faster, more efficient, has a higher cut quality and, if used for the right kinds of applications and markets can certainly both reduce a converter's costs and create value-added opportunities.

Key benefits of laser cutting technology are seen as:

- Firstly, the technology offers 100% savings in cutter tooling. No dies are required, whether flat, rotary or flexible. If a converter has enough die and size changes a day then laser cutting very soon starts to become a very viable option.
- Tool-less non-contact laser production offers multiple depth cutting possibilities and can include kiss-cutting, thru-cutting and perforation in one pass.
- Converters using laser cutters estimate that the savings in set-up waste can be as high as 60%. Because laser cutting is an entirely digital process, pattern changes are simple, requiring only edits to drawing files without any downtime for set-up or die creation
- Existing laser converters estimate that there can be as much as a 40% – 60% saving in time and labor costs – a significant element of any converters costing.
- Lasers today can be used to cut all shapes and sizes. There are generally no size limitations, except that the more labels across the web and the more complex shapes that need to cut, the lower the web cutting speed will be.
- With few exceptions, lasers can cut most types of substrates. PVC is not possible due to the toxic gases given off when cutting. Aluminium foil is also not possible as its wavelength is too close to that of a C02 laser.
- Apart from cutting, laser systems can be used to etch OCR fonts, one and two dimensional barcodes, serial/sequential numbers and codes. The type of information that can be laser-etched is limitless – and all in one pass with the cutting operation.
- Laser cutters can be used off-line with re-registering of webs, or linked in-line to digital or conventional label presses.
- Laser cutting can also be combined with other laser processes such as perforating, scoring, kiss-cutting, etching and ablating
- When using digital printing from inkjet or Xeikon presses there can be an unlimited print and cut length with laser cutting to provide extended label/banner opportunities.
- Again with inkjet/Xeikon technology jobs can be batched across or along the web to maximise short-run production, enabling multiple label size and shape changes in one run.
- Laser cutting can be used to convert difficult materials, such as abrasives and adhesives, with ease.

Laser die-cutters are able to take any vector-based digital image – perhaps generated on an Esko system – and import this into the cutter's operating software so as to generate job set-up within a few minutes. There are a number of companies that provide laser die-cutting machines that can be linked to roll-label presses or used as stand-alone devices.

LASERGRAVURE – A process used for engraving plastic coated gravure cylinders using a high-powered laser device.

LASER HARDENED – A process employed to harden flexible cutting dies, increasing the hardness of the cutting line from 50 HRC to approximately 64 HRC. The amount of hardness can be adjusted depending on the application.

LASER PAPER – Paper that is suitable for laser printing. A label stock used with a laser printer to print labels in sheets (A4 labels). The label stock must be produced using paper with suitable electrical surface resistivity and low moisture content, which must be maintained throughout self-adhesive lamination, slitting, transportation to the label maker, label production and transportation through a distributor to the end-user. Moisture-proof packaging is essential.

Failure to do this will risk the toner powder not attracting to the labels evenly and may cause laser printer jams due to paper curl. The label stock must be able to withstand heat produced in the laser printer and able to conform to bending around rollers in the machine.

LASER PRINTER – A form of electrostatic printing in which an image is created by reflection from an original, but in which a laser light is switched on and off according to digital information from a computer. A method of printing labels which utilizes a laser beam to put images onto a substrate.

In operation, desk-top laser printers used for label printing employ laser light to create a laser image on a light-sensitive drum. This image is then developed with a dry powder toner, which subsequently fuses to the paper label substrate (sheet-fed or web-fed) using heat and pressure. The light-sensitive drum is then cleaned ready to accept the next image.

Care needs to be taken when selecting label materials for laser printing. The heat generated will distort some face materials and cause adhesives to ooze at the edges during printing. Laminate prepared specifically with laser printing in mind, is readily available from most label stock suppliers.

LASER SCANNER – An optical scanning (reading) device, which uses a low energy laser light beam as its source of illumination. A device used for reading barcodes using low energy beams as a light source. Also used in four-color process reproduction.

LASER SCORING OF FLEXIBLE PACKAGING – Laser scoring of flexible packaging involves the use of a focused laser beam to remove a specific portion of a material's surface, so weakening the substrate and allowing it to easily fold or tear. While a number of applications for laser scoring exist, the most prominent in the packaging industry is for flexible packaging.

These applications are able to laser score a tear line into a film substrate to create an easy-open package with a controlled tear across the packaging's top. The package film is scored to an exact depth to weaken a specific layer of the film without affecting the package's barrier layers or performance.

Laser processing applications include straight line cross web scoring, straight line down web (or machine direction) scoring, contoured machine direction scoring, and cross web patterns. These processing capabilities add any number of easy-open features to flexible packaging, like tear strips, pour spouts, recloseable bags, resealable pouches, microwaveable packages, and peel away windows.

LAST IN, FIRST OUT – Stock which is used as soon as possible after it has been received.

LAST IN, LAST OUT – Stock which is used in rotation in the order that it was received.

LATMA – LATMA, the Label and Tag Manufacturers Association, Australia, represents the Australian label industry and has nearly 80 voting members across the various Australian states. Started over 30 years ago, LATMA has a rich history and was originally made up of separate State associations, operating under the umbrella of LATMA National, the national body founded in 1989.

In 2009 the state associations were consolidated into a single national body and today LATMA represents all members in Australia.

LAY FLAT – A label material which has good non-curling characteristics to make it suitable for any form of printing or processing where a flat sheet is required. The ability of a label construction to remain flat within a prescribed relative humidity range, providing that it has been temperature conditioned in its original packaging, in the press room, before printing.

LCD – Liquid Crystal Display. A method of creating alphanumeric displays by reflecting light on a special crystalline substance. A type of digital output display that is frequently used in such things as calculators.

LEAFLET LABEL – *See Booklet label.*

LED – Light Emitting Diode. A semi-conductor that produces light at a wavelength determined by chemical composition as a result of electrical stimulation. Commonly used as a light source in wand type readers and in some non-impact, electrographic printers where an LED array writes the image rather than a laser beam. Also now being used as part of LED UV ink curing systems. *See LED UV curing.*

LED UV CURING – A new alternative to conventional UV curing lamps, It is utilized for drying UV sensitive materials through polymerization, by enabling thinner, heat-sensitive substrates, while reducing harmful by-products such as ozone, thus improving workplace safety. Additionally, UV LEDs can be turned on and off instantly and utilize input power efficiently enabling lower operation costs over the application lifetime.

UV LED technology makes use of high intensity light sources that are created from arrays of Light Emitting Diodes (LEDs), rather than using conventional lamp technologies. It is used to cure UV adhesives, UV digital, screen and flexo inks, as well as UV curable coatings.

LEGGING – The stringing out or threading (filamentation) of a pressure-sensitive adhesive when a label or tape is drawn away from a substrate, release liner or its matrix.

LETTERPRESS PLATES – As a relief printing process, plates for letterpress printing need to have the image areas raised above that of the non-image areas. Originally, plates were photoengraved by a powderless etching process on to zinc or magnesium or, for copper plates, by exposing a pre-coated plate to a negative, then developing the plate and etching it in a special machine using powderless etching chemicals.

Narrow-web letterpress printing plates are normally photopolymer plates. These are usually pre-coated and can be used as original and wraparound plates. Nylon-based plates, on a steel base and then mounted on magnetic cylinder, are also popular in narrow-web printing.

LETTERPRESS PRINTING – Letterpress printing of sheets was the earliest method used to print labels back in the 18th and 19th centuries and is still a key technology in the printing of quality prime and secondary roll-label self-adhesive labels today.

Modern letterpress printing uses photo-sensitive polymer plates onto which an image (ink carrying area) is produced by photographic or direct imaging plate-making techniques. After exposure, the plate can be treated in a special wash-out solution (chemical or water-based) to develop the image and non-image areas of the printing surface. The relief image area plate can then be affixed around a printing cylinder and placed in the printing press. During the printing cycle the raised surface is rolled with a sticky

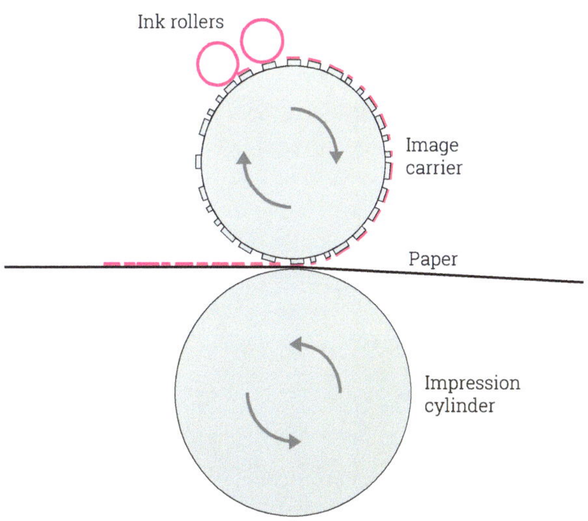

Diagram shows the principle of rotary letterpress.

(high-tack) ink and the inked image then transferred to the label substrate under pressure.

Letterpress printing roll-label presses, primarily used for self-adhesive label production, are narrow-web – 200mm (7.5"), 250mm (10"), 360mm (14"), 400mm (16") or 450mm 18") machines – and are either web-fed semi-rotary, intermittent feed, machines or web-fed full rotary printing presses, both of which can include up to six, eight or more printing units, together with all the necessary reel unwind, die-cutting, waste-stripping and rewinding capabilities. Almost all utilize UV-curing inks and drying systems.

In intermittent feed roll-label letterpress presses, the web of label stock is progressively and intermittently advanced to the printing position, printed and then moved on for die-cutting and waste stripping operations. The presses are slower than full rotary presses, but do offer advantages for short (10,000 or so labels) to medium (25,000 to 50,000) label runs where they can provide press flexibility and quick changeover capabilities and, due to the variable feed length that can be selected on an intermittent feed system, there are no costly changes involved in changing cylinders, gears, etc.

Rotary letterpress in-line, common impression drum or stack roll-label presses started to be installed in label printing and converting plants in the late 1970s and by the mid-1980s made up almost 70 per cent of all new roll-label presses being installed in Europe. Even today, with the rapid growth of UV flexographic label printing, rotary letterpress still has a substantial portion of the global total installed press base.

Compared with intermittent feed letterpress machines, rotary letterpress offers significant market advantage in that they can produce up to three times or more output – and all to a high quality, consistent result with, probably, less dependence upon operator skills than is required by other printing processes.

Rotary roll-label presses may have an in-line printing head configuration, or the print heads may be placed around a common-impression drum or arranged in a satellite or stack pattern.

LEVELING – The ability of a coating to flow sufficiently so that its surface will be relatively flat and free from irregularities resulting from the coating process.

LIDDING FOILS – Printed aluminum lidding foils are produced in either reels for top closure lidding of pots and tubs containing yoghurts, creams, spreads, ice cream, processed cheese; or in the form of pre-die-cut round, square or rectangular easy peel lids, again for dairy product pots and tubs.

The lids are usually coated with a heatseal primer or protective coating, printed in up to 8-colors (gravure for very long runs, UV flexo for narrower-web printing, or digital for shorter, customized or multi-varieties), either dotted or fully embossed with different textures, and laminated. Different lid thickness and structures are used depending on the end application.

Metallized lids provide a barrier to light, oxygen, water and aromas. Paper-based lidding material are also used for PS-based yoghurt and dessert pots. *See also Narrow-web packaging.*

LIFE – TLMI's Label Initiative For the Environment (LIFE) scheme is an EMS specifically developed for the narrow web printing industry. It provides a standard that addresses clean production technologies, minimizes waste, targets improvements in water and air pollutants, and reduces greenhouse gases.

Obtaining certification enables companies to address issues such as recycle-compatibility of adhesives, the source and destination of liner materials, label stock material constructions and the feasibility of adopting thinner substrates that generat less waste.

LIFE-CYCLE ASSESSMENT – A technique to assess the environmental aspects and potential impacts associated with a product, process or services. The international Standards Organization (ISO) developed LCA standards in the 14000 series from 1997 through to 2002, such as 14040 and 14044

LIFE-CYCLE INITIATIVE – UNEP's life-cycle initiative promoted in conjunction with the Society for Environmental Toxicology and Chemistry (SETAC)

LIGHT-FAST INKS – Labels and stickers used for outdoor and other applications where the printed image is exposed to daylight or fluorescent light, which is rich in UV radiation, may fade. Where inks may be subjected to such conditions the ink maker should be advised so that the best ink solution can be provided. Testing of printed samples for fading and 'accelerated aging' is routinely carried out by the ink manufacturer.

LIGHT FASTNESS – *See Light resistance.*

LIGHT PEN – A hand-held wand scanner which is used as a contact barcode reading device.

LIGHT RESISTANCE – The ability of a label material or ink to resist fading after exposure to sunlight, ultra-violet light or weathering. Nearly all plastics, including pressure-sensitive face materials, tend to darken under such conditions. May also be called light fastness when referring to printing inks.

LINE COPY – Any copy suitable for reproduction without using a halftone screen.

LINER, RELEASE – Component of pressure-sensitive label stock, which functions as a carrier for the label. Prior to label application, it protects the adhesive and readily separates from the label at the time of application.

LINERLESS LABELS – A roll of self-wound, most commonly a direct thermal coated and top coated label material, onto which the surface to be printed is coated with a release coating and the reverse side with a pressure-sensitive adhesive. When the roll is wound up the face stock functions as the release surface. The labels are then butt cut (or may have limited shapes cut to the top or bottom) and applied on a print apply label applicator, or using one of the proprietary applicator systems.

There are a number of proprietary linerless technologies on the market today, including Catchpoint, Ravenwood Packaging, NULabel, ETI Converting, Ritrama CORE and Pagomat. *See also Backingless labels.*

LINE WIDTH REDUCTION – The degree of gain that is allowed for when making up barcode films.

LINEWORK FRACTALS – A type of security design element used in the creation of complex security backgrounds. They are very difficult to reconstruct and offer a good protection against color copiers.

LITHOGRAPHY – Lithography, mostly referred to as offset lithography or simply as offset, is the most commonly used printing process for the printing of glue-applied paper labels using a wide variety of sheet-fed presses. More recently, offset is also being used in roll-fed press formats for high quality self-adhesive label production and in both sheet-fed and roll-fed versions for in-mold label production and for some new developments in cut-and-stack and patch film labels.

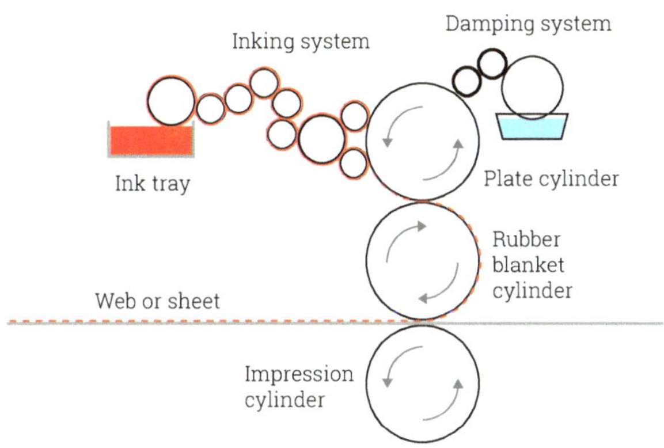

Diagram shows the principle of offset lithography

Offset lithography is a high quality planographic process in which the image and non-image areas of the printing plates are on the same plane (flat) surface, but are differentiated chemically in a way in which the image areas are made ink receptive and water repellent, while the non-image areas are water receptive and ink repellent.

Platemaking is relatively simple. Thin sheets of a grained and light sensitive coated (aluminum) plate are exposed through a negative or positive film – or by computer imaging – to create the image and non-image areas and then treated chemically to aid ink or water receptivity as required.

Once mounted on the plate cylinder in the press (whether sheet-fed or roll-fed), the plate is repeatedly contacted after each print with damping and inking rollers; damping, which is mostly water, to the non-image areas and the ink to the image areas.

For self-adhesive label production, web-fed in-line offset presses will also incorporate multiple print heads, die-cutting, waste stripping and rewinding in one pass. More commonly in self-adhesive label production, offset litho print heads are used in combination with other process, i.e. screen, hot stamping or UV flexo varnish for high added-value applications in, say, the cosmetics or wine labeling sectors

In the last few years a new generation of mid-web (400-600mm wide) roll-label offset presses have been developed to target more efficiently the higher added-value in-line printing and converting of traditional sheet-fed glue-applied labels for the wines and spirits markets. *See also Offset-lithography.*

LMAI – The Labels Manufacturers Association of India is registered with the Government of India and represents label printers from all over India.

LOGISTICS LABELS – While primary and secondary labels are found on unit packs – the individual bottles, cans or packs – there is a need for labels on transit packaging that are used to help move goods around, track and trace their movement in the supply chain and ensure they reach the correct delivery address. These logistics labels are frequently printed with computer-generated graphics, variable text and unique numerical or barcode symbols using thermal, laser or inkjet imprintable labels.

Logistics, barcode and variable information printed labels are predominantly printed in one color (black) only, often during the transit packaging, palletising or warehouse/distribution chain stages, and applied to cartons, trays, boxes, cases and pallet loads prior to shipment. Print and apply label equipment is commonly used for these applications, with the variable information entered through a keyboard or keypad.

More recently, major developments have been taking place in what are called smart, intelligent or chip labels, which are finding applications in the logistics chain for identification, traceability, track-and-trace and smart logistics applications. Many of these new label solutions are based on smart label RFID (Radio Frequency Identification) technology where the label/tag costs less than one dollar.

Most smart labels are in the form of very thin labels or laminates to get the price down and to make them suitable for targeted applications that are, in the main, new ones where the label must usually be disposable.

Smart (RFID) labels in the logistics chain are used to:

- implement a step change into the supply chain operation
- provide full visibility of the stock in the supply chain
- automate the processes associated with the management of stock

See also Barcode, RFID, Variable information printing.

LOGO – Logotype. The name of a company or product in a special design used as a trademark in advertising, promotion and brand/product image identity on labels and packaging.

LONG INKS – Ink that flows well on the inking rollers of a press. However, if the ink is too long it may break up into filaments on the press and cause 'ink fly' or 'ink misting'. *See Misting.*

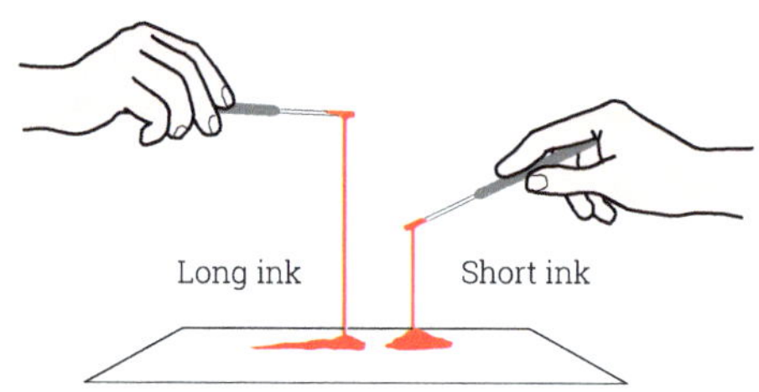

Long and short inks. Long inks may cause ink fly on the press

LOOK-ALIKE PRODUCTS – Copy or look-alike products are produced to perform the same function as the genuine article and may be deliberately designed to resemble authentic products (color, shape, wording). However, they are produced by an independent manufacturer under its own name, even using a brand name that may be very similar to the original. Such look-alike products may have been created with a minimum of R & D, minimal advertising and promotion and selected as being a product area with a fast turnover and/or high profitability.

It is important that manufacturers and brand owners of genuine goods provide some protection against this practice. This may be carried out by registering Trade Marks, registering shapes, packaging, smells or sounds and by multi-class actions or following procedures for the international registration of marks.

LOW COST RFID – Typically refers to tags costing less than one dollar each and with a range of up to one meter, or under $5 dollars for ranges above one meter. They differ from conventional RFID tags in that they can be applied in different, new applications and have therefore become of interest to many different end users and suppliers.

Low cost RFID tags provide an alternative to a barcode, magnetic stripe or printed label in that they offer advantages in terms of orientation or obscurity tolerances. They are usually cheap enough to be disposable and thin enough (or flat enough) to go into new applications, even to be incorporated inside sheets of paper or label laminates. Flat, low cost RFID versions are usually called smart labels. Typical end users are currently in airlines, libraries, toys, passports, and laundry. *See also Smart labels.*

LOW DETECTION – Where detection of an electronic article surveillance (EAS) label is reduced to a level below its specification by environmental or other causes.

LOW FREQUENCY TAGS – RFID tags which typically operate at 125 KHz. A key disadvantage of low-frequency tags is that they have to be read from within three feet/one meter and the rate of data transfer is slow. However, they are less expensive and less subject to interference than high frequency tags.

LPI/LPC – Lines per inch. Lines per centimeter. A measure of the screen resolution of a halftone film, plate or printed image. 150 lpi = 60 lpc. Today, lpi and lpc more accurately stands for the number of screen cells per inch or per centimeter, which is also referred to as the screen ruling. *See Screen ruling.*

LUMINANCE – Color is measured in terms of hue, saturation and luminance. Hue is the predominant wavelength of a color which distinguishes it from other colors, such as red, blue or green; saturation is the color intensity (from white to pure color); and luminance is the brightness of the color (from light to dark).

LUMINESCENT INKS – Inks which can be invisible but become luminous when exposed to UV light. Mainly used in security printing applications.

LUMINESCENT PIGMENTS – Pigments that produce striking luminescent effects in the dark. There are two types of luminescent pigments: one that is activated by ultra-violet radiation (usually black light) to produce a very strong luminescence; the second uses phosphorescent pigments, which do not require any separate source of radiation.

M

M – Roman numeral for 1,000

3M FRAGILE MATERIALS – Tamper indicating materials based on a range of films and papers that provide tamper evidence due to their low structural integrity. Available as destructible paper labeling materials, destructible films and acetate tapes, they are used in applications such as the production of safety warning labels, warranty seals, package seals, sample seals and licensing labels.

MACHINE DIRECTION – The direction in which the majority (over 50 per cent) of the cellulose paper fibers lie due to the motion of the paper-making machine. This is with their lengths parallel to the web's forward movement. The machine direction of the paper is important in many converting, label printing and label application stages. *See also Grain direction and Cross direction.*

MACHINE-GLAZED PAPER (MG) – Paper which has had one side made smooth and glossy by drying in contact with a large, heated, polished metal cylinder, which forms part of the drying section of the machine. The other side of the paper remains relatively rough.

MACHINE READABLE INKS – Magnetic inks containing magnetic iron oxide and which can be read by machine have been used on checks and similar documents for many years. Used with MICR characters (*See MICR*) they enable automatic sorting operations to be carried out.

Machine readable MICR characters

Non-magnetic characters can also be read by machine using OCR (Optical Character Readers) which can recognize characters in a suitable black or dark color in a specific place on a document, form or label. Guidelines, which are invisible to the reading instrument but can be seen by eye, are used to ensure the information to be read is in the correct position on the document.

MACHINE SHARPENING – Flexible dies are all sharpened on a flatbed CNC milling machine. As the tolerances on a flexible die are extremely tight, this milling process takes place under conditioned circumstances. In principal, no hand sharpening is necessary for the 'finishing touch'.

Rotary cutting dies are either milled or spark-eroded in creating the cutting lines/contour. The actual sharpening of the cutting lines is done on a CNC multi axis milling machine and should also take place under conditioned circumstances.

Manual sharpening of the cutting lines is still sometimes necessary to finish a rotary cutting die and is mainly used for repairing rotary dies.

MAGENTA – Hue of the subtractive primary color and a 4-color process ink. It reflects or transmits blue and red light and absorbs green light.

MAGENTIC BASE – A stainless steel-based foundation for mounting flatbed flexible dies, especially used for fine cutting contours and if more accuracy is needed

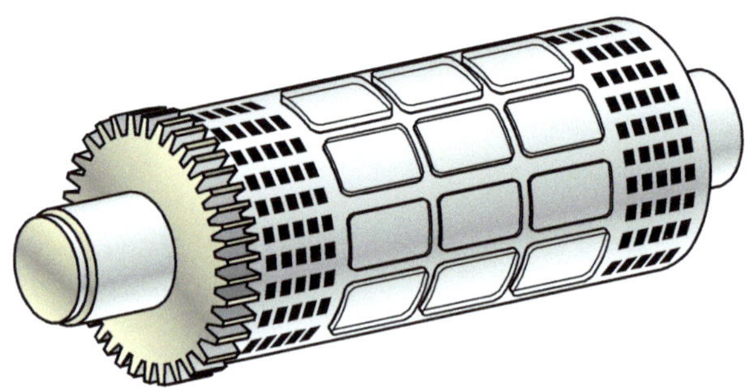

Permanent magnets around the periphery of cylinder hold the flexible die in place

MAGNETIC CYLINDER – A stainless steel-based cylinder having a series of permanent magnets inserted around its periphery and used in die-cutting to hold flexible steel dies in place. Magnetic cylinders will fit on to any press that takes rotary dies, and there are no size limitations outside of those relating to the press dimensions. They are installed in exactly the same way that conventional rotary dies are installed. This can be done by the regular press operators or maintenance staff.

When using magnetic cylinders, press operators will need to become aware of the high standards of cleanliness required, together with an appreciation of the careful handling required to avoid creasing or damaging the plate. The skills required for mounting are similar to those required for mounting printing plates.

Magnetic cylinders are also used for the mounting of metal-backed printing plates. *See also Flexible dies.*

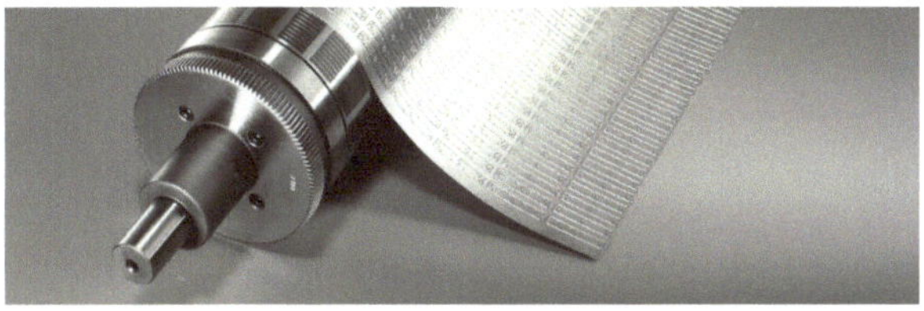

Magnetic cylinder and flexible die. *Source: Kocher & Beck*

MAGNETIC DIES – A thin, etched, flexible steel 'foil' that carries on its surface the cutting edges used for die-cutting. The flexible magnetic die is mounted on to a base cylinder magnetically. *See also Flexible dies.*

MAGNETIC INKS – Developed to increase speed and efficiency in the handling of checks and other security documents. The special inks can be magnetized after printing so that the printed characters can be read by electronic reading equipment. *See MICR.*

MAGNETIC STORAGE – Any disc, film, tape, drum or core that is used for the storage of digital information.

MAGNETIC STRIPE – Magnetic stripes are a well-established read/write auto-ID technology that may be applied to labels, tickets, tags and cards for applications as diverse as travel ticket documents, access control, storage, credit cards, debit and pre-payment cards. The stripe contains particles of magnetic material that can be programmed to contain and store data in magnetic form. The magnetic stripe may be applied to the material by adhesive using magnetic tape or by printing a magnetic stripe slurry.

ISO standard magnetic stripe cards have three tracks of data, the first being assigned to the airline industry (accepts alphanumeric entries), the second to financial transactions (numeric content only), and the third being a re-writeable track designed for off-line magnetic stripe-based transactions. Non-standard and multiple stripes are used in large numbers in access control and storing value.

MAGNETOGRAPHY – A digital printing technology in which a magnetic charge is produced on a magnetic drum by creating a digitally-generated image. To this image is applied a magnetic toner, which is then printed on the label or document substrate by heat and pressure. After printing, the cylinder is cleaned before the generation of the next image.

Magnetography has mainly been used for the printing of barcodes, labels, tags (including postal mailing tags), lottery tickets, addressing and sequential numbering.

MANAGEMENT INFORMATION SYSTEMS (MIS) – *See MIS.*

MANAGEMENT WORKFLOW SYSTEMS – Designed to provide a complete production management workflow the latest systems allow users to store, back-up and monitor all aspects of print quality from one (remote) access point regardless of site structure and location, enabling managers to monitor, analyze and control the quality of the production process from their desktop computers.

MANDREL – A shaft upon which cylinders or other devices (perforation blades, numbering boxes) are mounted. Also refers to an unwind or rewind shaft on which rolls of label stock or finished labels are mounted.

MANOMETER – a pressure gauge used to measure force applied to a cylinder

MARKETGATE – Walmart's Sustainable Packaging Scorecard modeling software that is provided by ECRM. Only third-party audited entries are made visible to its buyers.

MASS CUSTOMIZATION – The rapid growth of digital printing on labels and packaging has introduced a new marketing tool for brand owners by offering short or ultra-short runs, personalization, multiple versions or product variations, as well as new types of on-pack competitions, advertising and promotions. Mass customization also allows unique codes to be printed on packaging and labels which, when used with a smart phone, offer consumers the possibility to authenticate a product, its packaging or content.

MASTER ROLL – A full width roll off a paper making machine, extruder or laminator that has completed primary manufacturing processes and is as yet untrimmed or unslit.

MATCHPRINT – The name for a range of analog, inkjet, digital and virtual proofing systems originally developed by the 3M Corporation and now distributed by Kodak Polychrome Graphics (KPG).

Matchprint proofs made from film output are an exceptionally accurate rendition of what will eventually appear on a printing plate, whether by conventional or computer-to-plate methods. Matchprint inkjet proofs are digitally prepared on high-end color printing devices normally composed of CMYK, plus two additional colors. They have no specific screen pattern and more closely resemble stochastic screening. Matchprint virtual proofing enables on-line viewing of the many rounds of proofing that occur before a file is approved for printing, so speeding up the approval process.

MATRIX CODES – 2D matrix codes are symbologies based on the position of black elements of a regular profile (round, rectangular, etc) within a two-dimensional matrix. Each element is the same size and it is positioning of the elements that defines the data contained in the code. Matrix codes have many times the data storage capacity of conventional barcodes and may include photographs.

Matrix codes may be printed on labels or packs with invisible UV-sensitive ink that only becomes visible when illuminated with light outside the visible spectrum ('black light'). High-resolution printing is not required to achieve reliable recognition. Printing is feasible by all types of marking devices – computer printers, dot matrix, ink jet, laser, thermal transfer, and lithography – in any color combination that provides about 205 contrasts.

MATRIX WASTE – The skeleton of face material and adhesive waste surrounding self-adhesive labels after die-cutting. The matrix waste is normally removed on the press converting line by stripping it from the web of die-cut labels and re-winding it on matrix re-wind roller. Waste may also be removed by shredding into small pieces and/or by extraction systems that convey the waste away for bagging or baling. Also known as the waste skeleton.

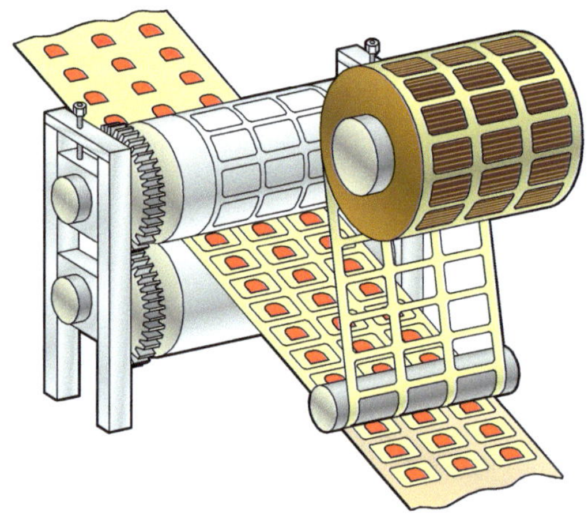

Matrix waste being removed and re-wound after die-cutting.

MATT (MATTE) FINISH – Paper coatings, inks, varnishes, lacquers and laminates may all be obtained with a matt finish, which diffuses the reflection of light to provide a matt, non-glossy, surface. Paper finish with a level, smooth-coated surface and little or no gloss.

MEGABYTE (MB) – 1,024 kilobytes (Kb). One million bytes of information. One million character codes on the computer.

MEMBRANE SWITCH – Self-contained sealed film lamination that has been printed with conductive inks that form conductive paths for electrical currents. They are primarily composed of pressure-sensitive polycarbonate and/or polyester films, which have been designed to replace traditional glass-type switches.

MEMORY – A means of electronically storing data in an organised way prior to subsequent retrieval. Memory can be non-volatile, Read-Only Memory (ROM) or volatile Random Access Memory (RAM). The action of a computer depends on the use of both ROM and RAM types of memory. An important factor is the speed with which memory can be read, because this determines how fast the computer can operate in practical programming terms.

MEMORY – Apart from a computer term, memory is also used to describe the property within a material that causes it to try and return to its original dimensions and shape.

MEMORY CARD – A chip card without a processor, i.e. not a true smart card. Also used for data storage devices of the type being standardised by the American PCMCIA and the Japanese JEIDA organisations.

MESH – The weave dimension and angle of the fabric of stencil materials used in the screen-printing process. Mesh describes the number of screen openings per unit length through which the liquid material is screened to apply it to a substrate.

METAL FOIL – A thin, flexible layer of metal – most commonly aluminum – which is used as a label face material. Some thinner gages are often laminated to paper for improved strength.

METALLIC INKS – Inks which consist of powder-like metallic flakes, such as aluminum and copper alloys, which are mixed with a suitable varnish or pigment carrier. The varnish or carrier dries rapidly and binds the flakes to the surface. Coated papers give the best results while gravure, due to the ink film thickness, produces the best optical effects with gold and silver metallic inks. On rough surface papers, a base ink is usually printed first, allowed to dry and then overprinted with gold.

Metallic gold and aluminum inks can be printed by the offset, gravure or letterpress printing processes, although care should be taken to use a neutral fountain solution when printing offset to avoid tarnishing of the bronze powder.

METALLIZED PAPER OR FILM – Plastic film or paper materials that have been 'coated' on one side with a very thin layer of metal (about 1 micron thick), usually aluminum. Metallizing is produced by melting and vaporizing the aluminum in a vacuum while passing a web or paper or film around a chilled roller and over the point of vaporization. The vaporized molecules then collect on the cool web, so providing the paper or film with a metallic finish.

Metallizing may be carried out by direct metallizing onto the material surface, or by transfer metallizing where the vaporized metal particles are attracted in the vacuum chamber to a very smooth plastic carrier web and then transferred to the chosen substrate under pressure. This gives a higher finish than direct vacuum metallizing.

METAMERIC INKS – pairs of inks that are visually similar under daylight conditions, but change when viewed in different light sources. For example, if any attempt is made to copy a document on a color copier, the colors will change, e.g. orange can change to yellow, flesh tones to red, or greens to blues.

METRIC SYSTEM – The decimal system adopted by most countries other than the USA for solid, liquid and distance measurements. The metric system term for paper basis weight is grammage. Expressed in grams per square meter (g/m2), such measurements are independent of paper size.

Length measurements in the metric system are in kilometres, metres, centimetres or millimetres. Temperature is measured in degrees Centigrade.

In the metric system there is a basic paper size, known as the 'A' series. The series consists of sheet sizes in which the area from one size to the next varies by a factor of 2 or 1/2. The A4 paper size is that used in most ink jet, laser and copier label printing applications. *See A4.*

MICR – Magnetic Ink Character Recognition. A printing method that uses magnetic inks or toners to produce special machine readable font characters – such as E13B or CMC7 – to print documents that can be read and processed automatically by magnetic sensing machines. *See also Magnetic inks.*

MICR PAPER – Usually a high-quality bond paper with good surface characteristics and dimensional stability for printing with magnetic inks.

MICROCOMPUTER – A PC is a microcomputer, in

which the Central Processing Unit (CPU) is contained on a single chip.

MICRO-DOT SECURITY™ – A patented system for permanently microscopically marking and identifying a wide range of items with coded micro-dot identifiers the size of a pinhead. Sometimes used in woven labels as an extra security thread. The thread can only be viewed with high magnification and acts as an authenticity device to prove manufacturer.

Micro-dots can be custom encoded with up to five lines of company information containing up to 15 characters per line and be consecutively numbered. They can easily be hidden on a product and concealed with an UV dye, which is only detectable with a scanner.

Actual size of custom coded MICR-DOT identifiers.

Information on Micro-dots as seen through a 30x microscope.

Microscopically marking with micro dots the size of a pin head.

MICRO-EMBOSSING – Embossing where the effect is achieved with minimal depth and intricate and complex designs. Used in security applications such as event ticketing, anti-counterfeiting and legal documents.

MICRONS – Units in the metric system used to measure the thickness of paper, film, labelstock, printing plates, etc. One millionth of a meter (approximately .00004 inch). In some countries the measurement of the thickness of thin materials may be undertaken using gage or mils, where 25 microns = 1 mil = 0.001" = 100 gage.

The following table gives a comparison of some of the more common micron thicknesses used by the label industry for measuring film thickness, etc, compared with their approximate equivalent thickness in mils.

MICRONS	MIL
12	0.50
25	1.00
30	1.20
36	1.50
50	2.00
57	2.30
100	4.00
125	5.00
250	10.00

Conversion of common micron thicknesses used in the label industry.

MICROPERF – Microperforation. Very small perforations or minute pinholes in paper that enable a section or part of the paper to be easily separated. Microperforation leaves a smooth edge without the normal more jagged edge found with standard methods of perforation.

MICROPROCESSOR – The heart of modern computing systems, it appears as a single integrated circuit supported by many other elements so as to provide the primary intelligence of a computer.

MICROTAGGANTS™ – Originally developed by the 3M Company in the late 1970s to trace the route of explosives from manufacturer to detonation, but now added to numerous products to enable subsequent identification of the genuine product and/or as a safeguard against counterfeiting and gray market diversion, as well as for liability/warranty claims. They can be incorporated into a clear varnish to be printed onto labels, cartons or documents.

Microtaggants are comprised of microscopically tiny particles of plastic hardly discernible to the human eye made up of a series of layers of different colors. The selection of colors and their sequence constitutes a unique code for each use. The codes can be easily read with a pocket microscope.

MICRO TEXTS – Very fine micro printed texts that are difficult to determine with the naked eye. What may appear to be a normal straight-line image may contain

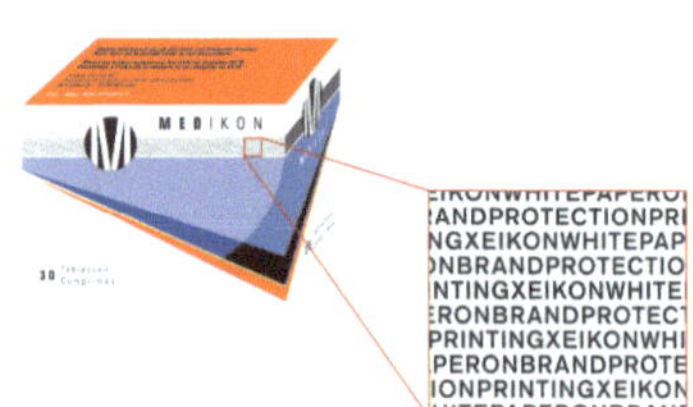

Micro text is almost impossible to see with the naked eye but can be viewed and read under magnification. *Source: Xeikon.*

hidden information or a logo. Can be used as a hidden security/authentication feature in labels and packaging.

MICROWAVE TAGS – Radio frequency tags that operate at 5.8 GHz. They have very high transfer rates and can be read from as far away as 9 meters (30 feet). They are relatively expensive and use a lot of power.

MIDTONE – The tonal value of dots located approximately between the highlight and shadow values of a halftone image.

MIGRATION – The transmission of elements of an adhesive or ink into the face material or other substrate onto which it has been applied.

Migration – The property whereby a UV photo-initiator can migrate through semi-porous packaging layers into the contained product. In the case of food labels, components of the face material, ink or adhesive may migrate through the primary packaging into the foodstuff. Glass, metal and appropriately specified laminates are barriers to migration. *See also Food contact.*

MIL – Thickness measurement of thin materials used in some countries. 1 mil = 0.001 inch = 100 gage = 25 microns. A comparison of some of the more common thicknesses in mils used in the label industry with their micron equivalent is shown under the heading of Microns.

MIL	MICRON
0.50	12
1.00	25
1.20	30
1.50	36
2.00	50
2.30	57
4.00	100
5.00	125
10.00	250

Approximate conversion of common mil thicknesses used in the label industry into microns.

MILLIMETERS TO INCHES CONVERSION – With both millimetres and inches used for measurement in different regions of the global label industry a quick conversion can be undertaken by taking the millimeter figure and dividing by 25.4 to convert to inches, e.g. 150 mm ÷ 5.4 = 5.9". Approximate conversion values for the more common millimeter to inch conversions used in the label industry are as follows:

SIZE IN MILIMETERS	SIZE IN INCHES
25 mm	1 inch
150 mm	6 inch
200 mm	7.5 inch
250 mm	10 inch
305 mm	12 inch
360 mm	14 inch
400 mm	16 inch
450 mm	18 inch
500 mm	20 inch

Approximate equivalent sizes in millimetres and inches
See also under Converting metric to imperial.

MINUS CORRECTION – *see under Corrected anvil roller*

MIS – The development of integrated Management Information Systems (MIS) has made a considerable impact on a label industry faced with ever smaller label print runs, more diverse technology and products, more complex printing and converting operations, plus the need for special color mixes. Also, for supporting customers that require up-to-the-minute proofing or delivery information almost on demand and direct links for order purchasing and invoicing.

It is against this kind of requirement that management information systems have evolved. The more complex and flexible the production mix, the more the need for a highly flexible and versatile management system that communicates all job information and changes, to all staff instantly.

Today, MIS systems enable everything from materials purchasing, to production, distribution, invoicing and accounts information to be held in one computerized system, integrating with other applications, including desktop productivity tools, press management control systems, accounting and administration systems and supply chain integration.

Systems are fully integrated so data only has to be

entered once and changes ripple through to other modules automatically. For example, if a customer changes the quantity required on an order, the computerized system automatically adjusts the works order, re-times the job, adjusts the production schedule and stock control – and then re-calculates the quantity of raw materials allocated.

At the hub of an MIS system is a database that holds all the details required to manufacture, stock and sell the label products, including press requirements, tooling, colors/inks, plates and finishing. Materials order/stock modules include barcode-driven traceability and details of substrates. An origination module provides full control of design, proofing, platemaking, etc; the specification of the label drives the automated production planning process.

Although based on standard components, bespoke programming is often involved in tailoring individual MIS systems to an individual company's needs. Modern MIS systems operate in real time, allowing the operator to move from window to window and enabling electronic commerce through flexible and adaptable architectures.

MISREAD – Occurs when the data output of a reader does not correspond with the encoded data presented.

MISTING – Minute particles of adhesive, coating or ink that are thrown off as rollers rotate at high speeds. Apart from unsatisfactory build up on the machine or press and wastage of materials, misting may constitute a health hazard if operators breathe in the particles.

MMSI – A million square inches of material.

MOBILE ACTION CODES – Printed 2D/QR codes, matrix codes, digital watermarks, Microsoft MS tags, embedded NFC, or other forms of printed mobile action codes, placed on labels, packaging, hang tags, posters, displays, etc., to enable individuals to instantly access a brand owner's website or on-line services with their smart phones, in turn triggering an interactive experience that can bring life to a printed label or packaging substrate.

Mobile action codes now appear widely on product packaging, in shopping centers, malls and retail outlets and take consumers to different types of locations, such as loyalty programs, social media, location based discounts, special offers, opportunities to win prizes, additional information about products, and the downloading of user instructions.

MODEM – A device used to transmit data between two computers using telephone lines by converting computer signals (data) into high-frequency voice communications signals and vice versa.

MODULAR – Machine design in which the individual components can be placed in any order within the web path. Modular printing presses enable the customer to determine in which order they require printing processes, i.e. 1 rotary screen unit + 4 rotary letterpress + 1 UV flexo, and where and in which order finishing stations are to be positioned.

In label application, some of the latest labeling lines are modular systems and enable the customer to choose from glue applied application, self-adhesive application, wraparound film application or sleeving, in any combination or sequence.

Modular machines and equipment can also mean that modules can be added after manufacture on the customer's site to increase or change the equipment's capabilities.

MODULAR LABEL APPLICATORS – One of the more recent developments in application technology is the introduction of modular labeling systems in which wet-glue, self-adhesive, shrink or wrap-around film labeling heads can all incorporated into the one labeling line. This offers bottlers complete flexibility in their bottling plant for body, neck, front and back labeling using any one – or combination of – label types and application method.

Modular systems are seen as cost-effective to purchase, flexible in application, and offer brand owners a wide label and bottle image choice. Modular systems are one of the fastest growing label application methods and are expected to continue finding a key role in bottle decoration and branding.

MODULE – The width of the smallest individual element that can be used in a barcode. Codes are made up of exact module widths. As they are magnified or reduced in size, the module relationship between bars remains constant. *See Barcode.*

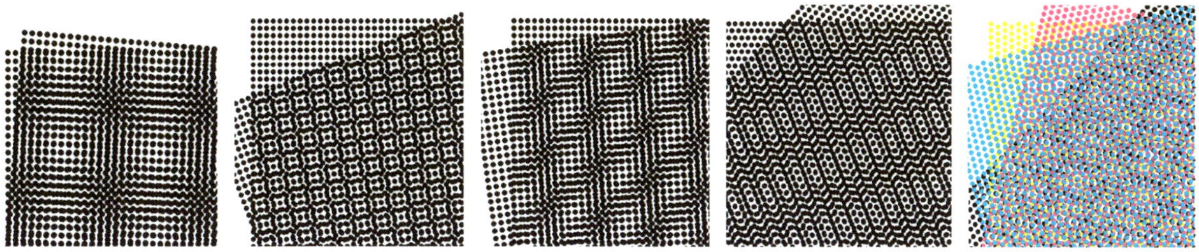

Examples of moiré patterns caused by incorrect halftone screen angles. Right-hand illustration shows correct screen angles.

MOIRÉ PATTERN – Found in black and white or color halftone printing, it describes the irregular unwanted interference pattern of screen dots caused by combining one, regular, halftone pattern with another similar one, so causing disturbing patterns or patches – either over the whole image or in certain color combinations. Results from incorrect screen angles being used when overprinting colors or when reproducing from an already printed subject.

MOISTURE CONTENT – All paper and paperboard materials contain some moisture within the paper fibers and the sheet or web is never normally completely dry. The amount of water or moisture in the material is determined by specific test methods and is normally quoted as a percentage of the total weight.

The moisture content of paper is important in storage, transit, handling, printing, finishing and labeling. Uneven or changing moisture content (exposure to moisture of stacks of un-printed or printed label sheets, or to the outside edges of webs) may cause print register problems, wavy or cockled edges, creasing, jams, curl, poor fusing of toners and inking problems. Low moisture content (dryness) causes static problems leading to jams and misfeeds.

Changes in moisture content may be caused by changes in Relative Humidity (RH), storage in cold conditions, pressrooms at significant temperature variance to warehouse, etc. When paper is the correct moisture balance with its surroundings, i.e. neither losing or absorbing moisture, it is said to be in equilibrium.

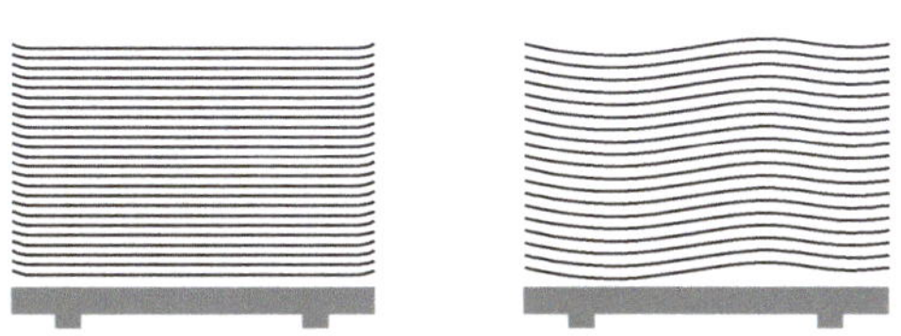

Uneven or changing moisture content may cause print register problems, wavy or cockled paper, creasing, jams, curl or inking/toner adhesion problems.

MOISTURE PROOF – A packaging material or label substrate that is impervious to moisture or water vapor.

MOISTURE RESISTANCE – Material which resists the uptake or passage of moisture. The moisture resistance of paper labels can be improved by varnishing or overlaminating.

MONOCHROME – A printed image showing all the necessary tonal shades – from black to white – which has been reproduced in a single color. Although synonymous with black and white printing, the term can also be used correctly to describe any single ink printing, regardless of the color.

MONOMERS – The primary chemical units which are able to link together in long chains or networks to create polymers – and which can be classified into thermoplastic or thermosetting plastic materials. These are used for applications such as film extrusion or blow molding of bottles. There are also photo- or UV-sensitive polymers used in platemaking. *See Films, Polymers, Blow molding, Extrusion.*

MONO WEB – The brand name for a linerless self-wound printed pressure-sensitive label web. A release coating is applied over the printed surface of the web so that the adhesive applied to the back of the web will not stick to the face when re-wound. Die cutting is carried out on the label application line. Developed by Waddingtons in the 1980s.

MOTORCYCLE HELMET LABELING – Head injury is the leading cause of death in motorcycle crashes and un-helmeted motorcyclists involved in crashes are more than three times as likely to suffer injuries, than those wearing helmets. Compared with car occupants, motorcyclists are around 16 times as likely to die in a traffic crash. Consequently many Countries, States, National or International bodies, have made the wearing of motorcycle crash helmets mandatory or have issued standards or regulations establishing minimum performance requirements for helmets – which may include various permanent labeling requirements, such as size and location of labels.

Motorcycle helmet labeling

Helmets meeting mandatory standards are required to be labeled with the manufacturer's name or identification, the helmet model description, size, date of manufacture and a symbol or mark that the helmet conforms to the applicable standards. The color, size and position of the label may be defined, as well as the nature of any instructions to the purchaser regarding helmet construction and other safety-related information.

As helmets may be damaged by some common substances without visible damage, only approved or recommended paints, cleaning agents or adhesives (including self-adhesive labels) may be used, as appropriate.

In all cases, it is advisable to check the appropriate regulations before producing helmet labels.

MOTTLE – Occurs when a printed image – usually a solid color – has a spotty or uneven appearance.

MOUNTING – The process of affixing printing plates on to a printing base or cylinder in the correct positions for color-to-color register or print-to-die register. Special plate mounting devices have been developed to ensure that plates are mounted in the correct position.

MOUNTING TAPE – A thin, double-side adhesive coated tape used for the mounting of printing plates on the printing cylinder. Mounting tapes may be compressible or non-compressible substrates and are available in a variety of types and thicknesses. *See also Tapes and Plate mounting tapes.*

MSI – A thousand square inches of material.

MULTI-PROCESS PRESS – A printing press, usually narrow-web, which combines two or more mechanical, electronic or digital printing processes in-line on the one machine. May also be called combination press. *See also Combination press, In-line printing and Modular.*

MYLAR™ – A DuPont registered trademark for a clear, tough polyester film, which has a good resistance to moisture, solvents, oils and other chemicals.

N

NACDS – The National Association of Chain Drug Stores, North America's largest pharmacy organization with a membership of more than 40,000 retail community pharmacies, and NACDS 125 chain member companies include regional chains and national companies, as well as an international membership of 40 members from 13 countries. The Association has served as the voice for chain pharmacies since 1933. Further information from the NACDS website.

NAMEPLATE – A rigid or flexible material that carries detailed information about a product, e.g. Serial Number, Voltage, Type, etc, and which is fixed or mounted in position by screws, pins or adhesive. Materials used may be metal plates, metallic foils, metallized plastics, rigid plastics or other materials designed to withstand exposure and adverse conditions.

Pressure-sensitive film labels often replace traditional metal nameplates containing information about a product, and therefore need to be as durable as the product itself as they carry information which is essential for the use of the product throughout its working lifecycle. In such cases, permanent adhesives are used in conjunction with PET films.

NANO-BARCODES – Nano-barcodes and taggants use track and trace applications that have been developed using nanoscale fibres that can be placed inside paper. These are able to serve as a 'fingerprint' that can prove that the product is genuine. At about 150 nanometers in diameter, the fibres are smaller than living cells and invisible to the naked eye. They contain even smaller nanoparticles that carry an electricalmagnetic or optical signature. The product simply needs to be scanned using a special scanner to provide authentication.

NANOCOATING – A coating process for deposition of discrete layers of nano-particles into a moving web. The technology enables the coating of a wide range of nano materials (e.g. conductive particles, drug molecules, electrochromic materials) on a series of substrates. Potential market applications for the technology include transparent gas barrier, anti-microbial coatings, anti-corrosion coatings and transparent conductive coatings.

NANOCODING – Interesting developments are taking place in the application of nanocoatings to labels, nameplates, documents and tags in which codes are carried on the coating layers in a similar manner to a barcode, and which can then be read by a laser scanner in much the same way as a supermarket scanner.

The scanner captures the nano surface structure of the coded information, looks it up in the coding system – using the original code as a reference – and compares the characteristics of the nanocode with the stored code memory. Both fixed and variable data can be incorporated and, with it being impossible to replicate the coating and code structure, secure authentication can be assured.

Research and development to refine and implement nanocoding technology and create new label/tag applications is continuing.

NANOMETER – A unit used for expressing wavelengths of light. A nanometer is one thousand-millionth part of a meter.

Major developments are now taking place in the implementation and use of nano-technology – the use of nano-particle coatings and taggants – in a wide variety of applications of potential interest to the label and tag industry.

NANOSENSERS – The possibility of tagging food and food packages from farm to fork – during processing, while in transit, in restaurants or on supermarket shelves, eventually through to the consumer – is now becoming possible through the use of nano- 'Senser' (Silicon Enhanced Nanoparticles for Surfaced Enhanced Raman Scattering) tags. Similar to nanobarcodes, nanosenser tags can be incorporated into paper, packaging or labels and be read by an automated reader up to a metre away. This will enable labels or packs to be read at the checkout – like RFID tags – or to be read covertly at various ports.

Using nanosensers in this way, it is expected that it will be possible to monitor packs and labels for pathogens, temperature change, leakages, etc, not just for food related applications, but also in medical and hospital environments.

NANOTAGGANTS – Powder-like substance composed of millions of individual micro particles so small that they are slimmer than a human hair. From nanometer, the units in which wavelengths of light and other radiant energy are expressed. The taggants can be made from inert polymers – gelatine, starch, sugar or other bio-polymers that are used to coat tablets and other medication (so already approved by the FDA) – and can be used to protect virtually anything from counterfeiters by incorporating them into coatings that can be applied to paper; they can even be added when the paper is still in a wet pulp state.

The particles can range in size from a few hundred nanometers to 200 microns (8 mil), with multiple taggants being able to reside on the end of a human hair, and be made up to bear text, graphics, symbols, shapes, barcodes or digital information that can be used to protect products, packaging, documents or labels forensically, or read with special equipment or scanners. They can also be used to carry unique customer information.

NANOTECHNOLOGY – Nanotechnology is defined as research and technology development at the atomic, molecular, or macromolecular levels using a length scale of approximately one to one hundred nanometers in any dimension; the creation and use of structures, devices and systems that have novel properties and functions because of their small size; and the ability to control or manipulate matter on an atomic scale.

In the metric system of measurement, 'Nano' equals a billionth, with a nanometer being one-billionth of a meter – about one hundred thousand times smaller than the diameter of a human hair. References to nanomaterials, nanoparticles, nanoelectronics, nanodevices and nanopowders simply mean the material or activity can be measured in nanometers. To further appreciate the size, a human red blood cell is over 2,000 nanometers long, virtually outside the nano-scale range.

In the form of micro-particles it has become feasible to coat a wide range of nano-materials (e.g. conductive particles, electrochromic materials, etc) onto a surface to provide anti-microbial coatings, anti-corrosion coatings, transparent conductive coatings, gas barriers, anti-counterfeiting, or other specific features.

In other areas, nanotaggants – powder like substances composed of millions of nano or micro-particles – can be made from inert polymers (gelatine, starch, sugar, or other bio-polymers, etc) that can be used to protect virtually anything from counterfeiters by incorporating them into coatings that can be applied to the surface of paper, or even added when the paper is still in a wet pulp state.

These nanotaggant particles can range in size from a few hundred nanometers up to 200 microns (8 mil), and may be made up to bear text, graphics, barcodes or digital information that can be used to protect documents, labels or packaging forensically, or to be read with special equipment or scanners.

Nanotechnology offers the potential to manipulate these various types of nano-materials for use in particular applications through chemical and/or physical processes so as to create new paper or board materials with specific enhanced properties or characteristics.

NAPCOR – National Association for PET Container Resources. NAPCOR was founded in 1987 is the leading trade association for PET plastic packaging industry in the United States and Canada. It promotes the introduction and use of PET packaging, facilitates its recycling and communicates the attributes of PET containers as an environmentally sustainable package.

NARROW WEB – Printing presses in which the web width may be as narrow as 150 mm (6 inches) and range up to as high as 500mm (20 inches), 600 mm (24 inches) or even 700 mm (28 inches), according to which press manufacturer is consulted. In practice, there appears to be no actual agreement as to what constitutes the exact cut-off width between narrow and wide-web presses. Common practice is beginning to increasingly define presses with web widths between 600 mm (24 inches) and 800 mm (32 inches) as mid-web.

NARROW-WEB PACKAGING – Narrow-web presses have traditionally been used for the production of self-adhesive labels and for some shrink-sleeve and in-mold applications. More recently, narrow-to-mid-web presses have begun to find increasing usage for the printing of a whole range of shorter-run, custom personalized, versions and variations of flexible packaging printed on unsupported and barrier materials made from plastic, metallic foil, mylar or paper – depending on the end-use application.

Today, the range of packaging produced on narrow-web presses includes:

- Sachets for food, medical, pharmaceutical and consumer goods
- Sachet envelopes
- Stand-up pouches
- Wet wipes
- Blister foils
- Disposable bags
- Soup packets
- Flow wrapping
- Lidding films and foils
- Snap packs
- Barrier films

While volume production of packaging such as sachets and pouches is widely produced on high-tech UV and water-based flexo presses – some printing in up to 8 or 9 colors – as well as UV offset lithography and specialist print finishing machinery, the potential of using narrow-web conventional and digital presses has become increasingly attractive as run lengths have decreased and interest in personalization and multi-versions has grown.

Lower pre-press, plate and tooling costs for narrow-web package printing and the potential to customize the print have made narrow-web packaging more economic and flexible for shorter-run and personalized solutions.

NATURAL AGING – The time in which a material may change some physical or chemical property under normal environmental or usage conditions.

NATURAL RUBBER-BASED ADHESIVE – An adhesive whose main component is natural rubber, normally containing a solvent. It is used for a wide range of applications and especially for the labeling of rough surfaces (e.g. carpets, fabrics), convex surfaces (e.g. ampoules, vials or syringes) and other difficult substrates. Not as resistant to sunlight, oxygen or ozone as most synthetic rubber adhesives.

NCR – No carbon required. A term originally introduced by the National Cash Register Company – which formerly owned the patents for their NCR Paper system. This utilized encapsulated dyes coated onto the underside of one surface which when pressed against the silton clay coating on the following form part caused the dye capsules to break and create an image in the receptive coating underneath. Unlike carbon paper, NCR Paper would only copy onto silton clay coated surfaces, so was cleaner in use and reduced the number of plies in a multipart form set.

The term 'carbonless paper' has now superseded this term.

NEAR FIELD COMMUNICATION (NFC) – A short-range high frequency wireless communications technology that enables the exchange of data between devices over a 5cm (2 inches) distance. NFC tags contain data and are typically read-only although some are re-writable. They can be custom encoded by their manufacturers and embedded in or attached to a variety of form factors that do not require batteries, including to labels and packaging, stickers, posters, or other promotional material.

NFC devices are able to communicate with existing ISO/IEC 14443 smartcards and readers, and with other NFC devices. They are mainly targeted at mobile phone usage, with their primary application being for use with NFC handsets and smart phones to expedite payments and instant contactless transactions and authentication procedures.

NFC is still largely regarded as a next generation technology for the label and package printing sectors, but with printed electronics likely to bring cost savings to NFC tag fabrication, and with NFC enabled phones growing in usage, then NFC is seen has as having interesting application in product authentication, traceability and brand protection.

NEW ZEALAND LABEL ASSOCIATION – *See under SALMA.*

NICHE MARKET – Used to describe those sectors of the label market in which specialized knowledge, technology or solutions are required and where price competition is less critical than in more developed markets or sectors. The term applies both to an existing market which is not being fully serviced and a potential market that could be serviced with a little creativity on the part of the label printer.

The pressure-sensitive label sector is ideal for

creating and building niche applications. The very concept of the in-line label press is the key to versatility and versatility is often the answer to filling new requirements and creating niche markets. A well-designed press will permit the adding of additional printing heads, different printing processes, variable imaging, digital technology, de-laminating, re-laminating, embossing, reverse side printing, over-laminating, and a variety of post-print operations.

The product coming from the press can look very different from the original simple label and will be accepted by the buyer with less emphasis on price – and more emphasis upon the job it performs. In fact it may no longer even be considered a label, more a product in its own right.

More specialized markets where there are still niche opportunities include creative marketing and promotional labels, brand protection and authentication, on-pack promotions, leaflet and booklet label solutions, digital printing and in the future, major opportunities with intelligent, smart, smart active, smart re-active and performance labels.

NIP – The contact point between two driven rollers. Usually a metal driven roller and an un-driven rubber covered roller. May also be the contact point between two press cylinders, i.e. the plate and impression cylinders. Sometimes referred to as the draw or pull rolls in a web press.

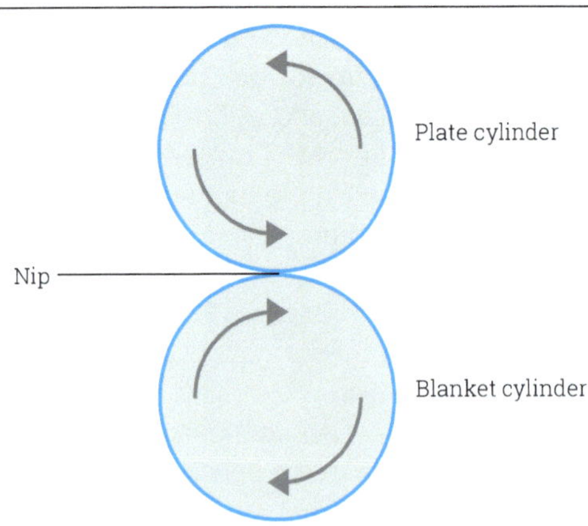

Nips occur at the contact point between two driven rollers

NIP – Non Impact Printing. *See Non-impact printer.*

NIP GUARD – A guard put in place to eliminate the risk of hands or clothing being drawn into the nip existing between a pair of driven rollers or between two press cylinders. *See Nip.*

NITRO CELLULOSE OR VINYL LACQUERS – Lacquers are used on both paper-backed foils and some synthetics to make the material printable by giving the ink something to key on. On foil materials it may also be used to add color to the surface of silver aluminum foil.

NODE – A device connected to a network cable, usually a workstation or server, but may also refer to a repeater or passive network junction box.

NOISE/INTERFERENCE – All electronic article surveillance (EAS) detector systems experience noise/interference. The greater the level of interference, the worse the performance of the detectors. This noise/interference can be caused by metal structures and electrical wiring positioned near the detectors and from the atmosphere (ambient noise).

Interference with an electrical signal which may come from other electrical devices such as motors or fluorescent lighting. Coaxial cable is more immune to noise than twisted pairs, while optical fibres have complete immunity.

'NO-LABEL' LOOK – The 'no-label' look using clear or translucent, thin film labels, has become a popular trend in the beverage, cosmetics and food sectors, as well as for personal care products and over-the-counter pharmaceuticals. In addition to allowing the consumer to see how much remains in the bottle, it also allows the color of the product, or container, to shine through the label and creating the impression that the text has been printed directly to the bottle and in this way the container or product contributes to the overall appearance of the labeled pack. The 'no-label' look can be achieved with wrap-around film, cut-and-stack film, patch film, in-mold or pressure-sensitive labels.

The highly flexible thin film labels used for the pressure-sensitive 'no-label' look adhere well to straight, curved or recessed glass and plastic better than paper labels, and can also tolerate freezing, pasteurization and moisture without losing adhesion, strength or readability.

The key to the success of the 'no-label' look is the marrying of plastic face films with an exceptionally smooth film liner and transparent (clear) adhesive to produce a uniform and very smooth adhesive layer, and a label that consequently appears invisible on the packaging.

The use of glass-clear labels on crystal-clear bottles has been rapidly taken up by brand owners and converters in recent years, for the successful image decoration of high-volume bathroom products, as well as for the labeling and decoration of opaque and dark-colored containers.

NOMINAL RANGE – The read range at which an RFID tag can be read reliably.

NOMINAL SIZE – The standard size for a barcode symbol. Barcode symbols may be printed over a range of magnifications, most commonly from 0.80 to 1.20. *See also Barcode and Barcode symbol.*

NON-BLOCKING – Used to describe a coating or a remoistenable or heat-sealable adhesive label paper, which resists adhering to itself within a reel or stack of sheets, under specified storage conditions.

NON-IMPACT PRINTER – An electronic printing device that prints on a substrate without any direct or indirect mechanical impact or contact of characters with the surface. Non-impact printers include laser printers, thermal transfer, copiers, thermal printers and ink jet printers. They can reproduce graphics and barcodes in good to very good quality, depending on their resolution. *See Variable data printing, Digital printing and Ink jet printing.*

NON-ORIENTATED FILM – A film material in which the polymer chains have not been subjected to stress and alignment to improve certain key properties. *See also Biaxially orientated.*

NON-PRIME LABEL – A label that only supplies secondary or supporting information to a product, rather than the main label used to identify or display that product.

NON-READ – Refers to a lack of data output when a barcode is scanned. May be due to a defective code, incorrect scanner orientation, scanner failure, speed of scan or operator error. Also called No Read or No Scan.

NON-READ INK – Inks that prohibit detection by an optical scanner due to their high reflectance. Used as visual guides that will not interfere with data readability.

NON-TOXICITY – The state of being physiologically inert, e.g. harmless to health. Used with reference for example, to adhesives and printing inks. Indicates that the material, product, ink or coating is free from any poisonous qualities. Important in applications where the label may be in food contact, or for toy labelling, etc.

NON-VOLATILE – That portion of an adhesive, coating or sealer that does not vaporize or evaporate at relatively low temperatures.

NON-WOVEN – A manufactured synthetic material of directionally or randomly orientated fibers bonded by friction and/or cohesion and/or adhesion. May be given a woven appearance through a finishing procedure. *See Synthetic paper.*

NUMBERING BOX – A mechanical numbering device that can print alphanumeric characters on a sheet or web during the printing process. Traditionally used for consecutively numbering business forms. With the growth of electronic printing technology in which variable numbering sequences are incorporated into the printing program, numbering boxes are not required.

NUMERIC – A character set (in a barcode) or a machine vocabulary that only uses numbers, rather than including letters of the alphabet. *See also Alphanumeric.*

NYLON – A DuPont trade name for a strong plastic film that has a high oil and gas resistance.

NYLOPRINT – The trade name used for BASF's photopolymer plate-making material.

O

OBJECT-ORIENTATED – Bitmapped images are defined by values attributed to pixels, while object-orientated graphics are described by mathematical co-ordinates. Graphic elements described as a set of mathematical values that describe an outline and fills and which can be imported into other packages and re-sized without needing to revert to the original package in which they were created.

OCR – Optical Character Recognition. An electronic method of reading (scanning) printed text copy with software capable of recognizing printed or typed text matter and then converting the scanned images into an electronic equivalent – digitized files that can be saved on disk and edited as a text file.

OEM – An abbreviation of Original Equipment Manufacturer. A company that produces finished assembled products, such as photocopiers, laser and ink-jet printers and other machines, as distinct from an added value reseller or distributor who may re-badge a machine with their own name.

OFF-CUT – Material that is not used when a roll or sheet of paper, film or labels is trimmed to the correct size or width. Trim width that is not utilized but usually calculated within the costs.

OFF-LINE – A device or operation that is independent of a large processing unit or of an in-line printing or converting operation. Also refers to computer usage where, for example, a computer printer may be disabled by taking it off-line.

OFF-LINE SILICONE COATING – The manufacture of release liners in a separate operation from the adhesive coating process. The self-adhesive coating and laminating process by which purchased silicone coated papers or films are made.

OFFSET – The unwanted transfer of an ink or varnish from one printed surface to another surface. The transfer of ink before it is fully dry from the face of one sheet to the back of the sheet above in the delivery stack of a sheet-fed press, or from the printed surface to the backing liner in a roll. Also called set-off.

OFFSET BLANKET – A covering for the cylinder on an offset litho press which accepts the inked image from the printing plate and then transfers the ink to the surface of the material being printed.

OFFSET, DIGITAL – A digital printing press patented by Israeli company Indigo (now HP Indigo) which involves the transferring of 'electro' inks onto a blanket by varying the potential of a charged cylinder. Once the image is completed – with up to seven inks – it is transferred to the substrate in one 'shot'. *See also Indigo.*

OFFSET LITHOGRAPHY – The lithographic printing process, mostly referred today as offset lithography or simply as offset, is the most commonly used printing process for commercial printing and for the printing of glue-applied paper labels using a wide variety of sheet-fed presses. More recently, offset lithography has also begun to find a growing usage in roll-fed rotary or semi-rotary press formats for high quality self-adhesive label production (especially for wine label production) and in both sheet-fed and roll-fed versions for in-mold label production. Also for some new developments in cut-and-stack and patch film labels.

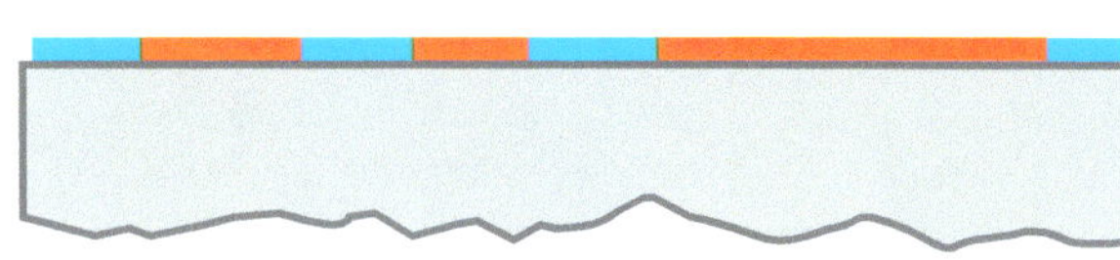

The image and non-image areas in offset lithography are differentiated chemically on a flat surface

Unlike other label printing processes, offset lithography is a high quality planographic process in which the image and non-image areas of the printing plates are on the same plane (flat) surface, but are differentiated chemically in a way in which the image areas are made ink receptive and water repellent, while the non-image areas are water receptive and ink repellent. There are no raised or recessed image areas like those found in the letterpress, flexo or gravure processes.

The manufacture of printing plates for the offset process is a relatively simple operation. Thin sheets of a grained and light-sensitive coated (aluminum) plate are exposed through a negative or positive film – or by computer imaging – to create the image and non-image areas and then the plate surface is treated chemically to aid ink or water receptivity as required.

Once mounted around the plate cylinder in the press (whether sheet-fed or roll-fed) the plate is repeatedly contacted after each print with dampening and inking rollers; damping, which is mostly water in a fountain solution, is applied to the non-image areas and the ink to the image areas. The plate then prints onto a rubber (offset) blanket and from the blanket onto the substrate to be printed.

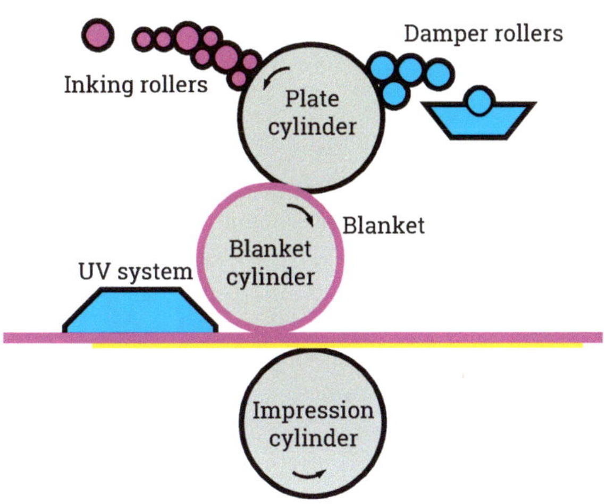

Diagram shows the elements of an offset litho press

Waterless offset presses which use special waterless offset plates with a silicone rubber coating in the non-image areas are also used in self-adhesive label production. These require no dampening solution or roller on the press.

For self-adhesive label production, roll-fed in-line offset presses will also incorporate multiple print heads, die-cutting, waste stripping and rewinding in one pass. More commonly in self-adhesive label production, offset litho print heads are used in combination with other processes, i.e. screen, hot stamping or UV flexo varnish for high added-value applications, for example in the cosmetics or wine labeling sectors. More recently, there have also been developments in offset/gravure combination presses.

In the last few years a new generation of mid-web (up to 600-800mm wide) roll-label offset presses have been developed to target the more efficient, higher added-value in-line printing and converting of traditional sheet-fed glue-applied labels or in-mold labels. *See also Label printing and production.*

OFFSET POWDER – Used in sheet-fed printing. A fine powder mist sprayed onto the still wet ink on each sheet of stock in the press delivery tray to prevent the ink from 'setting-off' or transferring onto the back of the next sheet coming into the tray.

OIL CANISTER LABELS – Oil canisters that are blow molded are filled with oil immediately after the blowing process. Residual volatile matter in the canister polymer necessitates the use of a label featuring special properties.

The adhesive must be strong and permanent and the face material must be gas permeable in order to prevent escaping volatile matter from forming blisters in the label.

OLIGOMER – A reactive chemical compound whose molecules consist of a group of linked monomers. These are compounds intermediate in size between a single monomer unit and a large polymer molecule.

OMNIDIRECTIONAL – A barcode symbol that can be read in any direction or orientation whenever the surface on which the symbol is printed is in a plane parallel or near parallel to the output part of the scanner.

ON-DEMAND PRINTING – The ability to print text or graphics in a variable data/text format from computer-generated instructions using electronic or digital printing devices as required (on-demand), rather than being printed ahead of time. On-demand printing of labels reduces stock and control requirements and eliminates wastage.

ON-LINE – In computing, an operation in which peripheral devices are connected directly to the central processing unit of the computer. The transfer of data from one device to another via a direct link. On-line computer connections require that the linked devices understand their data formats and structures. They must also be able to keep pace with the speed of data transfer.

In printing and converting, a term used to describe whether operations or products are enhanced 'on-line' rather than 'off-line'.

ON-PACK RECYCLING LABEL (OPRL) – Launched by the British Retail Consortium to tell the consumer which part of a package can be recycled

OOZE – This occurs when the adhesive squeezes out from under the backing in a pressure-sensitive laminate. This results in the edges of rolls becoming sticky or tacky.

OPACIMETER – An instrument used to measure the degree of opacity of a material, such as paper. The amount of light that gets through the sheet or web depends on its opacity or ability to block the passage of light and is measured in an opacimeter.

In an opacimeter, sheets of paper or other material are placed over an aperture in a pile thick enough to be opaque. A photo-electric cell measurement is set to give a deflection of 100 divisions. If the pile is then reduced to one sheet of material, and the measurement is taken again, it will fall to a lower value because less light is scattered back. If the deflection is now 94, then the opacity of the material would be 94 per cent.

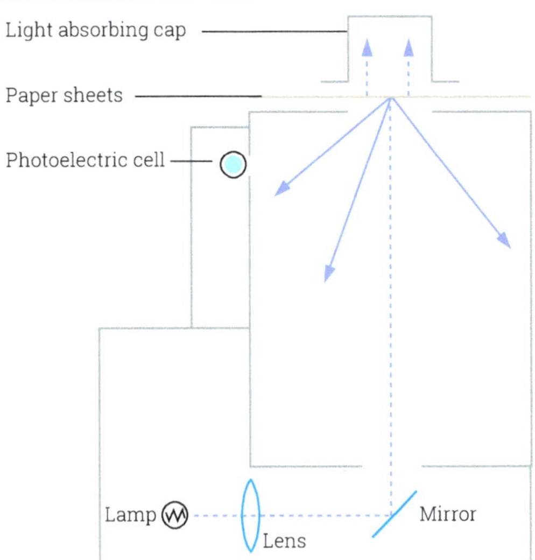

Diagram shows the principle of an opacimeter

OPACITY – The extent to which a paper or other material is capable of obscuring matter printed on the reverse or on an underlying print, or obscuring any other image or surface. Ability of any substance to prevent light from passing through. Opacity may be stated numerically as the percentage of light falling on the material which is scattered back by the fibers and particles of the loading. It can be measured with an opacimeter. *See Opacimeter.*

OPALTONE – Opaltone is a 7-color printing process which utilizes CMYK plus an R/G/B (Red, Green and Blue) ink set to deliver an expanded color range beyond CMYK. The system, licensed by Opaltone Inc, incorporates the OMS™ (Opaltone Matching System), digital color separation software, OMS swatch books and a standard Seven Color Process ink set.

OPAQUE – A material or surface that is not transparent or translucent. Does not transmit light.

OPEN SYSTEMS – Systems in which general interconnection and inter-working between data processing systems of different computer equipment manufacturers are based on common standards for communication.

OPEN TIME – Refers to the period of time during which a delayed action heatseal coating stays sufficiently tacky for application after removal of the heat source.

OPERATING SIDE – The side of a label press or other equipment from which the operator makes adjustments to the printing, converting or finishing operations. The operating side of a machine, usually the opposite side to the machine drive or gears.

OPP – Orientated polypropylene. Film substrates used by the label industry may be orientated in one direction or biaxially orientated in both directions. By biaxially orientating, the film's molecules are aligned in the cross (transverse) machine direction as well as the machine direction. Once biaxial orientation has taken place, the film exhibits a marked improvement in optics, strength, moisture barrier and low temperature durability. The increased tensile strength of the film improves the on-press stability, with less stretch and subsequently improved print performance. *See also Biaxially orientated and Polypropylene face materials.*

OPTICAL CHARACTER READER – A device that is able to 'read' and process machine-readable or hand-written characters.

Machine readable MICR characters

OPTICAL DISC – Optical discs (disks) use light to store and read information.

OPTICALLY VARIABLE INKS – Inks which appear to change color when viewed from different angles. The most effective of these are produced using a very thick ink film which can only be achieved using intaglio or screen printing. There are a range of gravure-printed optically variable inks which give reasonable effects. These are used for visual verification.

ORANGE PEEL EFFECT – Surface coating or printing irregularities, which resemble the skin of an orange in texture. The effect is caused when an ink or coating does not flow-out or level properly. Also refers to the appearance of a film after lamination and to the appearance of some films after being used for in-mold labeling.

In flexo printing, an orange peel effect may result when the flexo printing ink has been over-thinned, making it impossible to achieve good even color coverage. The printed color in what should be solid areas appears blotchy with an orange peel like appearance.

ORIENTATION, POLYMER – The production of a highly uniform structure in polymer materials by aligning their crystalline nature by stretching or cold drawing during their fabrication. *See also BOPP.*

ORIENTATION, BARCODE – Barcode symbols may be positioned so that they can be scanned horizontally (the bars are vertical in a 'picket fence' orientation) or vertically (the bars are horizontal in a 'ladder' orientation).

ORIGINATION – A term used in repro to describe any or all of the reproduction processes that may occur between design and printing.

Picket fence

Ladder orientation

Picket fence and ladder orientation of barcodes.

OSHA – Occupational Safety & Health Administration of the U.S. Department of Labor, whose mission is to ensure safe and healthy workplaces in America. It has three key areas of strategy:

Strong, fair and effective enforcement. OSHA seeks to assist the majority of employers who wish to do the right thing, while focusing its enforcement resources on sites in high hazard industries or those with high injury and illness rates.

Education and enforcement assistance. OSHA is able to play an important role in preventing on-the-job injuries and illnesses through outreach education and compliance assistance using an extensive website, training and information materials.

Co-operative program. OSHA encourages trade and professional organizations, businesses, educational institutions, government agencies and labor organizations that share an interest in workplace safety and health to collaborate with them in preventing illness and injury in the workplace.

The OSHA Act covers most private sector employers and their workers, in addition to some public sector employers and workers in 50 states and certain territories and jurisdictions under federal authority. *See OSHA website for more information.*

OUTPUT RESOLUTION – The output resolution of a given device describes the number of separating image points, spots, that the device is physically able to write. The output resolution of printers or imagesetters is often referred to in dots per inch (dpi) or dots per centimeter (dpc).

OUTSOURCING – Outsourcing or contract sourcing of packaging and labeling – even product manufacturing – has become a multi-billion dollar global business, with an estimated three-quarters of all items found in kitchens, bathrooms or garages not packed and labeled by the company whose name is on the label. The basis for this outsourcing of packaging and labeling services is that brand owners focus on what they are good at, leaving the specialist services companies providing their non-core activities – cost-effectively, efficiently and flexibly.

Outsourcing service providers include contract packers and labelers, re-packers, co-packers, fulfilment providers, contract manufacturers, pack and label designers and consultants. The packaging, labeling and product manufacturing sectors they provide services for cover almost every kind of product and manufactured item, from cosmetics, toiletries and healthcare, through pharmaceutical and medical products to foods, chemicals, drinks, electronics, toys, hazardous materials to the automotive aftermarket and military equipment sectors.

Outsourcing can offer many benefits to brand owners, product manufacturers, down to even the smallest of companies, looking to outsource their packaging and labeling requirements. Services offered by the different packaging and labeling service providers include:

Creativity – Package engineering, R&D, pack concepts, access to formulations, formulation development. Product launch concepts

Consultancy – Performance testing, legal services, regulatory guidance, shelf-life testing, development expertise, distribution testing. Project management

Design – Packaging, product and label design, point-of-purchase display design, graphic design

Packaging – Packaging in almost every form of container or pack, from bottles and cans, to bags and sacks, cartons and flexibles, corrugated containers, blister packs, clamshells, pouches, crates or cargo containers. Shrink wrapping and overwrapping. Retail packaging and repackaging. Hand assembly

Filling – Liquid or hot filling; clean room, aseptic, modified or controlled atmosphere filling; capsule, powder or tablet filling; coupon and literature inserting

Labeling – Self-adhesive, glue-applied, shrink, stretch, wrap-around and cut-and-stack labeling solutions. Tagging. Coding and marking.

Compliance – Regulatory approval. FDA, EPA, DEA, USDA or kosher registration, certification or compliance; quality control services

Distribution – Warehousing, drop-shipping, product shipping, fulfilment, computerized inventory control and EDI transactions.

OVDS – Optically Variable Devices (OVDs) are devices whose visual appearance changes depending upon the angle of viewing and/or type of illumination. They fall into two main categories:

Diffractive Optically Variable Devices (DOVIDs) which is a term used for all types of diffractive devices. Iridescence is based on the diffraction of light by microscopic gratings, and this group includes holograms.

Other types of OVDs, but which are not diffractive. This includes light interference type devices, optically variable inks, coatings and films.

OVDs incorporate specific characteristics which, on their own, can provide security features but, when combined with other security features, can provide even more formidable protection against counterfeiting and/or forgery. Because OVDs are variable, depending on the direction of viewing or lighting, they cannot be photocopied or scanned; they will also leave evidence of attempts at alteration or removal, and are difficult to replicate. They can be easily authenticated by the naked eye and do not require any special operator training or equipment.

The use of OVDs as security and anti-counterfeiting devices on paper or plastic labels, hang tags, swing tickets, textile labels (threads), plastic cards and security documents continues to grow. *See also Holograms, Security Holograms, and DOVIDs.*

OVER-LAMINATING – The application of a clear self-adhesive film to a label stock to provide increased protection or to enhance the image or quality of the label. Over-lamination is usually carried out as an in-line process on narrow-web presses, or a separate operation for sheet-printed products. Heat-seal over-laminating materials are also available, although these are not normally used in label applications.

OVERLAP – An overlap occurs when a label is wrapped around a container or bottle and the two ends overlap one over the other so that the label adheres to itself.

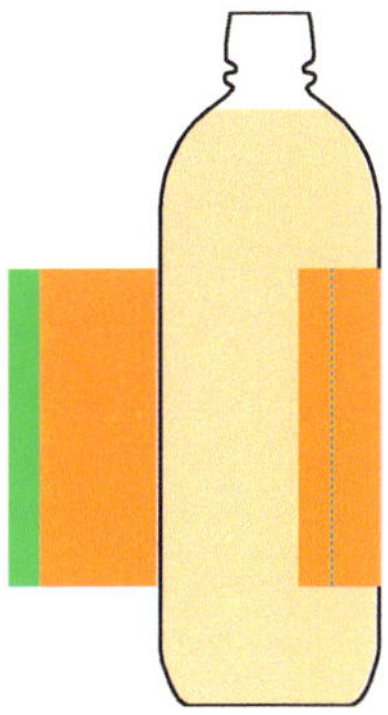

Label is glued at trailing edge and overlapped on itself

OVERPRINTING – The printing of additional images or text onto a pre-printed label, either over an existing image or within a space left during an earlier printing operation. The material being overprinted may be fixed or variable information. *See Variable data printing.*

OVERS – To allow for wastage during press set-up, printing and finishing operations it is usual for the printer to allow for an additional percentage of label stock, inks, laminates, etc, so that the agreed specified order quantity can be delivered to the customer. Depending on the length of print run, so the percentage for wastage will vary from a few percent for very long runs to 10 percent or more for shorter runs. Good quality overs may be included in the quantity delivered to the customer although, for some market sectors like pharmaceuticals, exact quantities only may be specified and delivered.

Also refers to any material, stock or products left over and unused after printing.

OVERT FEATURES – Overt features are authentication devices built in to labels, documents and packaging which are visible and show dynamic visual effects. Their main advantage is the fast and easy, on the spot, authentication where no additional devices are needed. Overt features are expected to fulfil three main criteria:

- Communicate to the verifier
- Be easy to identify
- Be hard to copy and imitate.

Physical secure solutions offering overt features include holograms, optically variable devices (OVDs), watermarks, thermochromic inks, threads, foils and laminates and security papers. Overt devices can be made more secure by combining them with covert features so as to give the benefits of both systems.

In such cases the person verifying the overt device can raise the level of authenticity evaluation by reading the covert information.

OXIDATION – The process in which an ink, coating, adhesive or other material takes in oxygen to form an oxide. Traditional, non UV-curable letterpress and lithographic inks dry by the oxidation process. Adhesive films may deteriorate due to atmospheric exposure. Hot-melt adhesives may break down due to prolonged heating and oxide formation.

OXIDATION (TOOLING) – The result of a reaction between the tool face and oxygen exposure. In order to prevent tooling rusting (oxidation) tools shoul be clean and oil carefully before storage. Oxidation can have a detrimental effect on the cutting results of both rotary dies and well as flexible dies.

P

PACKAGING HYGIENE GUIDELINES – Published by the Chilled Food Association (CFA) in the UK. The Guidelines set out the requirements of chilled food manufacturers regarding the production, handling and storage of packaging materials supplied to the industry for ready to eat products.

Key points to be taken into account when developing and producing chilled food products include:

- Ingredients hygienic quality
- Product formulation/characteristics
- Processing parameters
- Intended use of the product
- Storage and distribution conditions
- Manufacturing hygiene
- Shelf life.

The Association represents many of the best-known UK chilled food manufacturers and campaigns actively on their behalf.

PACKAGING HYGIENE STANDARDS – Grocery stores and their product suppliers have a legal obligation to take all reasonable precautions and due diligence to avoid possible problems of contamination of a food product, whether this occurs during development, manufacture, distribution, advertising or sale of the product to the consumer.

Suppliers of packaging (and labeling) to food manufacturers or food retailers also share this obligation to ensure the consumer is provided with foods that are safe to eat. This means that they should have appropriate systems and controls in place. To this end, some retailers and food manufacturers have traditionally commissioned hygiene audits and/or inspected packaging and labeling premises to satisfy themselves that the necessary systems and procedures at their suppliers are in place. However, this often meant that a packaging or label supplier might have multiple inspections by different food manufacturers and retailers, perhaps with differing requirements.

To ensure that major retail grocers and food manufacturers work to a common standard, schemes, such as the BRC/IoP Global Standard – Food Packaging and Other Packaging Materials scheme in the UK, have been developed and introduced to ensure all packaging and label suppliers are aware of their obligations and how to meet them. A major feature of such schemes is that certification is provided by a third party audit body, which in turn is accredited to a national/international standard.

In the UK the Standard is jointly owned by the British Retail Consortium (BRC) and the Institute of Packaging (IOP) and covers:

- The adoption of a hazard assessment system
- A documented quality management system
- Control of the factory environment – standards, products, processes and personnel.

The Standard established hygiene management procedures to be used by manufacturers of food packaging and labels supplied to the food supply chain.

PACKAGING, NARROW-WEB – *See Narrow-web packaging*

PAD PRINTING – Although not a label as such, pad printing (or tampon printing) is an effective method of decorating both the curved surfaces of a very wide range of products and the surfaces of three-dimensional products. It is widely used to decorate and 'label' bottles, tubs and pots of almost every kind of material and product shape. Thanks to the flexible silicone rubber pad used, products with irregular shapes can also be printed.

There are two basic print methods used in pad printing: the stamping method, and the rotary printing method.

PANTONE – An international system of matching color for printing, designating unique colors by standard Pantone Matching System® (PMS) numbers. The Pantone name is known worldwide as the standard language for color communication from designer to manufacturer to retailer to customer.

Pantone is a world-renowned authority on color and provider of color systems and leading technology for the selection and accurate communication of color across a variety of industries. Developed by Lawrence Herbert in 1963, PMS is an innovative system for matching and communicating colors to solve the problems associated with producing accurate color matches in the graphic arts industry.

Pantone has also expanded its color matching

system to other color critical industries and continues to develop color communication tools, as well as adopting new digital technology to address the color needs of designers and production professionals. The Pantone color system lets designers access color libraries from inside mainstream systems such as PhotoShop.

Pantone was acquired by X-Rite, then by Danaher, owner of Esko, in 2012. *See also PMS.*

Pantone swatch

PAS 2050 – Considered to be the first global carbon standard set by the British Standards Institute in 2008, which further clarifies the life-cycle standards in BS EN ISO 14040 and BS EN ISO 14044. It specifies requirements for the assessment of the life-cycle GHG emissions of goods and services

PASS – A one-trip passage of a material through a press or other production equipment. Some constructions may require more than one pass before production is completed.

PASSIVE TAG – An RFID tag without a battery. When radio waves from a reader reach the chip's antenna it creates a magnetic field. The tag then draws power from that field and is able to send back information stored on the chip.

US paper name	Size	Metric to imperial (g/m2 to lbs)	Imperial to Metric (lbs to g/m2)
Bond	17" x 22"	0.266	3.760
Cover	20" x 26"	0.370	2.704
Tag	24" x 36"	0.614	1.627
Book	25" x 38"	0.675	1.480
1000 sq. feet (ft2)		0.205	4.831

Conversion of paper weight in the metric and Imperial systems

PAPER WEIGHT – The weight of paper is measured in grams/square meter in the metric system and in lbs/square foot or lbs/square inch in the imperial system, still used in the USA.

The imperial measurement is more complicated as different paper types are quoted in lbs weight for a given number of sheets (usually 500 sheets) in a popular size for that paper.

The table can be used to convert approximately the basis weights of paper from the imperial to the Metric System and vice versa.

PARTICLE GUMMED PAPER – Base paper, coated with particles of adhesive suspended in solvent to produce a remoistenable gummed paper with a matt finish. Also known as dry gummed.

PATCH FILM LABELS – Patch labels printed on OPP film in a single-web by flexo or gravure or on sheet-fed offset presses can provide a film alternative to paper labelling on most existing patch paper, cold-glue labeling equipment – without any machinery or adhesive modifications. Using standard water-based cold glues, patch film labels reduce glue usage by up to 50 per cent over paper labels, as well as providing excellent moisture resistance, eliminating curl, tear and wrinkling, and passing ice water/ice chest test procedures.

Clear patch films offer an alternative to self-adhesive labels for 'no-label' look applications, and can be reverse-side printed and applied at high speeds on magazine fed labelers.

Applications in which patch film labels are currently being used include beer, wine, spirits, food and household chemicals.

PATIENT COMPLIANCE PACKAGING – Ensuring that patients take their medication as prescribed has become an increasing challenge. Many patients forget to take their drugs, others loose track of what they have taken, and some stop taking them before they have been used up – all at significant cost to the economy and an estimated many thousands of deaths each year from non-compliance.

To address these issues, drug manufacturers are looking at introducing new and affordable means of patient compliance into their packaging and labeling, and a number of solutions have already been developed. At the forefront of these developments are solutions such as conductive inks and RFID technology – frequently incorporated into blister packs.

Conductive inks to print circuits or a sensor grid on to blister packs can subsequently enable a microchip to record the date and time that an individual blister is opened and a pill removed; active or passive RFID tags embedded into pharmaceutical packaging can also aid patient compliance, as well as be incorporated into a supply chain management system. RFID tags, used with specially designed readers, also have the ability to collect information about drug storage or the drug condition and provide relevant data to doctors.

Other developments include bottle closures incorporating an embedded RFID smart tag to remind the patient when medication needs to be taken and monitor that this has been done by recording the time the patient opens the bottle to remove the tablet – the doctor or pharmacist can then retrieve the recorded data. In this system, medication can be tracked without active input from the patient.

With RFID tags now being produced in extremely thin formats, and able to be incorporated into different pressure-sensitive label structures – some almost invisible to the naked eye – the use of patient compliance labels, closures and packaging is expected to grow rapidly over the coming years.

PATTERN COATED – Where an adhesive is laid down in strips parallel to the machine direction across the width of a pressure-sensitive label stock during manufacture. This is also called "strip coating". Also refers to an adhesive coating that is printed in a pattern not related to web direction.

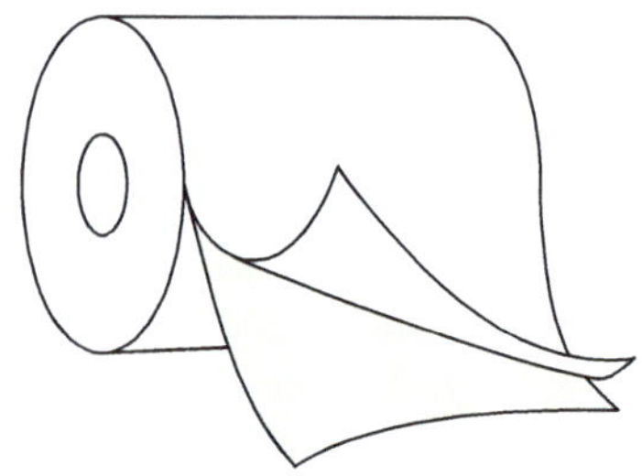

Pattern or strip coating parallel to the machine direction

PATTERN PRINTED ADHESIVE – An adhesive printed in a pattern, shape or strips on the printing press to provide adhesive-free areas or printed adhesive areas that can be used in the manufacture of special label constructions, leaflet labels or form/label combinations. The printable adhesive may be UV-curable on the press. *See also Stripe and intermittent coating.*

PATTERN VARNISH – Varnish applied to a printed surface or label in a specific pattern, area or shape. May be a gloss or matt varnish to enhance protection or image of the label.

PCB – Printed Circuit Board. A flat plastic or fiberglass board with circuits printed on the surface and chips mounted on it.

PCB LABELING – To automate the assembly, processing, testing and packaging of printed circuit boards (PCB) a barcode label is normally adhered to the board before soldering. This label must withstand harsh temperatures and chemical exposures and yet still meet all the necessary optical requirements for reliable barcode scanning.

The barcode label is normally fixed to the board when production commences on the bare board and, at each stage of production, the barcode is scanned to identify the board. The code may also include information about flux density and solder temperature, quantity and lot number, as well as a variety of test data. Reading the code may also help to reduce work in process inventory and reduce set-up times.

Successful PCB labels adhered before soldering must meet harsh environmental requirements, particularly thermal, chemical and optical requirements. Tamper-evident properties may also need to be incorporated to ensure that a serial number cannot be transferred to another board.

Solder temperatures up to 288 degrees Centigrade (550 degrees Fahrenheit) may be experienced in PCB manufacture; the labels are exposed to fluxing chemicals and defluxing solvents and must be firmly adhered to the board so that the liquids do not seep underneath the label; the adhesive must be resistant to the chemicals and solvents used; the barcode must give an accurate and reliable scan reading at every stage.

The selection of label materials for PCB labeling, whether before or after soldering, should be undertaken with care and must match all application and usage requirements. Materials used include polyimide and polyester facestocks, matched to thermal transfer ribbons and to any overlaminate used.

PCBS – PCBs (Polychlorinated biphenyls) belong to a family of organic compounds known as chlorinated hydrocarbons and were historically used in applications where stable, fire resistant, heat transfer properties were demanded. They were used in lubricating and cutting fluids, dye carriers in carbonless copy paper, plasticizers in paints and adhesives and in caulking compounds.

The manufacture of PCBs is now prohibited and their existing usage is being phased out. This is because they are believed to have harmful effects on human health and the environment.

Although PCBs are no longer made, many electric transformers and capacitors once filled with PCBs are still in service. Labeling regulations therefore require items such as PCB containers, PCB transformers, PCB hydraulic systems, PCB storage areas, PCB transport areas, etc, to be identified and labeled with their PCB classification. If the items were not originally labeled by their manufacturer, the owner must label all items that contain PCBs.

PCS – Print Contrast Signal. A measurement of the contrast between the bars and spaces in a barcode symbol, usually expressed as a percentage (%). To enable a barcode symbol to be scanned a minimum PCS value is needed. PCS values can be calculated and automatically displayed on suitable instruments.

PDF – Portable Document Format. A standard for digital file transfer, where the information about all elements on the page – format, text, font, attributes, graphics, etc, is delivered with the page.

PDF 417 – A high density, two-dimensional barcode with a high information capacity in a very small area. Normal codes may store up to 20 characters per symbol; PDF 417 symbols can carry up to 1,800 characters per symbol. Large amounts of text and data can be stored inexpensively and securely with the PDF 417 two-dimensional code and the printed symbol can withstand a certain amount of damage without causing any loss of data; a level of error correction which is far more advanced than normal one-dimensional barcodes.

PDF 417 2D barcodes can be printed from within any MS Windows application using a licensed package or site license from a variety of specialist supplier sources.

PEEL ADHESION – The force required to remove a pressure-sensitive adhesive label or tape from a standard test panel at a specified angle (usually 180º or 90º) and speed after it has been applied to the test panel under specified conditions for a specific time.

PEEL PLATE – A peeler plate or beak is a sharp-edged, flat piece of metal around which the release liner backing material is drawn in a label applicator or dispensing system so that the label is carried forward from the backing and can be applied to the product or pack. *See Applicators, self-adhesive.*

PEIAC – PEIAC is the Printing and Printing Equipment Industries Association of China.

PEN READER – A small hand-held barcode reading device. May also be referred to as a light pen or wand scanner.

PENETRATION – The migration of an adhesive into, or through, a face material.

PERFORATION – a line or row of cuts or tiny holes that enable a paper or web of labels to be folded, torn off or separated easily. Perforation may be horizontal or vertical, with standard or micro perforation (cut versus tie).

PERFORATING RULES – steel cutting lines to be inserted in perforation cylinders for making cross perforations.

PERMANENT ADHESIVE – An adhesive which, when applied to the specified surface, will display strong final adhesion and a permanent key within a short period of time. Generally difficult to remove without damaging the label or sticker. The actual strength of adhesion will depend on the nature of the substrate and the conditions under which the self-adhesive product is applied, consequently adhesion will build up over 24 hours and assessment of the effectiveness of a bond can show a marked difference with the passage of time. Tests can be made after one minute, one hour, one day and one week. *See also Adhesive and Adhesive, pressure-sensitive.*

PERMEABILITY – The property of a material that allows or resists a substance to pass or flow through it.

PERSONALIZATION – The use of variable data or text on demand, with possibly fixed information pre-printed or electronically printed at the same time and traditionally used to personalize address labels or label/form combinations.

With the advent of digital printing using electrophotographic and inkjet printing technologies it has now become feasible to personalize retail and branded products and increasing numbers of brands have initiated personalization of their products. Notable examples are Coca-Cola and Heineken.

PESTICIDE LABELING – Codes of conduct on the distribution and use of pesticides and guidelines on good labeling practice for pesticides, have been published by the Food and Agriculture Organisation (FAO) of the United Nations, as well as any requirements laid down by state, national or other international bodies.

Requirements set out by the FAO cover warning words, identification of contents, directions for use, safety precautions, antidote and first-aid instructions, environmental hazards, physical and chemical hazards, storage and disposal. References are based on the type of pesticide, its formulation, uses, toxicity and potential hazards.

Producers and users of pesticide labels are advised to consult the current national or international Regulations or Guidance for the country where the products are produced and used before undertaking labeling.

PET – Polyethylene Teraphthalate. A clear, tough, solvent resistant material used for some label face materials and for the manufacture of mineral water, soft drinks and fruit juice bottles, and for some laundry detergent and kitchen bottles. *See also Polyester face materials.*

PHARMACEUTICAL LABELING – The labeling of pharmaceutical/drug products provides more challenges to the label industry than almost any other label user market, both in terms of legislative requirements and in terms of the wide range of applications and solutions required. Overall, the pharmaceutical label market is made up of five main areas of application:

- Primary labels for ethical drugs
- Over-the-counter products
- Prescription labels
- Veterinary products
- Sundry healthcare products.

Regulations on pharmaceutical/drug labeling laid down by national, state or international bodies (such as the EEC or FDA) generally require labels to carry the name of the product, name and quantity of each active ingredient, strength and route of administration, name and address of the manufacturer, a registration or licence number, batch number, expiry date and any specific storage conditions.

The regulations provide for the correct particulars to be included on each label at each stage in the distribution process; that the label is consistent with the provisions of the Product Licence; that all labeling is carried out in a way that is indelible, clear, legible and can be easily read by the purchaser.

Regulations may also require the inclusion in the packaging of all medicinal products of a package leaflet for the information of users. This leaflet shall include: the identification of the product, name of the product, a statement of active ingredients, the pharmaceutical form, the name and address of the holder, therapeutic indications, a list of information which is necessary before taking the product, necessary and usual instructions for use, a description of undesirable effects, a reference to the expiry date, the date the leaflet was last revised, any symbols or pictograms designed to clarify certain information.

In terms of label application, the requirements may range from gloss-coated papers applied to glass bottles, to luminescent, detectable synthetic no-label look labels on small diameter syringes or ampoules. Where there are critical applications, such as infusion or blood bag labels or small diameter packages, the requirement can be even more challenging.

There is also a high requirement for quality and inspection, the risk of migration and the prevention of counterfeiting. 100% label count may be required by manufacturers, as well as specific manufacturing procedures.

In all cases, the current national, state or international regulations should be obtained and followed.

PHARMACEUTICAL LABEL PRINTING – Pharmaceutical labels are frequently printed in two or three colors and include house or brand colors. Generally there is not too much four-color halftone printing. Some applications may require the inks to be autoclaved or sterilized; others may need to be stored in chill or freezer conditions. The sector has a requirement for many small quantity short runs, often with short re-run times from standing plates. Mainly printed on paper but ampoules, vials and syringes may be filmic label stocks. Letterpress, flexo and some screen have traditionally been the dominant processes.

More recently, digital printing has become more widely used for the printing of short-run pharmaceutical, veterinary and healthcare labels.

End-user printing and ink requirements are: small quantities, some special colors and color matching, both paper and film usage, low or high temperature performance for certain product and, frequently, 100% accurate count of labels on the roll. Labels for different products must be kept separate throughout production and despatch.

Security of the label is essential in the pharmaceutical sector. The label must be correct in every particular and must be fully reconciled with the product. The extent to which this aim is pursued by drug companies and the pharmaceutical label producers supplying the drug companies, would probably not disgrace a printer of bank notes.

PHOSPHORESCENCE – Phosphorescence occurs when a compound, after excitation ceases, continues to glow for a finite amount of time. This period of continued light emission or 'afterglow' will range from a few milliseconds to some hours. This phenomena can be used to characterize phosphors or as a trigger in a machine readable system.

PHOTOINITIATOR – In light-induced curing or polymerization systems, such as the UV-curing or crosslinking of UV inks, varnishes, adhesives and silicones, photoinitiators are the chemicals which, when exposed to light, break certain chemical bonds to form active species. These then initiate the polymerization or crosslinking reactions. These chemicals are commonly referred to as catalysts.

Some UV-curable silicone systems are catalysed by a photoactive complex. UV radiation breaks down the complex and allows it to initiate the cure reaction.

PHOTOPOLYMER MATERIALS – Photosensitive materials that undergo a chemical change when exposed to UV light to form a tough, abrasion-resistant surface which becomes the inking medium when used in the manufacture of photopolymer printing plates.

PHOTOSHOP – Adobe's leading image manipulation program that is used to retouch, clean-up, manipulate and generally work on bitmap data files.

PH VALUE – A number on a logarithmic scale that extends from 0 to 14, which indicates the degree of active acidity or alkalinity of an aqueous liquid. Neutral solutions, which are neither acid nor alkaline, have a pH of 7. The lower the reading below 7, the higher the level of acidity. The higher the reading above 7, the stronger the alkalinity.

pH is a measurable property of a solution in much the same way as temperature. Just as we give the temperature of a solution on a scale of temperature, so we give pH of a solution a number on a pH scale. Each point on the pH scale represents a definite degree of acidity or alkalinity. The scale is arranged so that every time the acidity or alkalinity changes by a factor of 10 the pH changes by 1.0 pH units. For example, a solution of pH 6 has only one-tenth the acidity of a solution of pH 5. A solution of pH 2 is 100 times more acidic that a solution of pH 5. pH can be measured with indicators or by electrical methods.

The effects of varying pH are to be found in many processes within the paper, printing and label industry, e.g. the drying time of some letterpress or lithographic inks and the fountain solution in offset lithography.

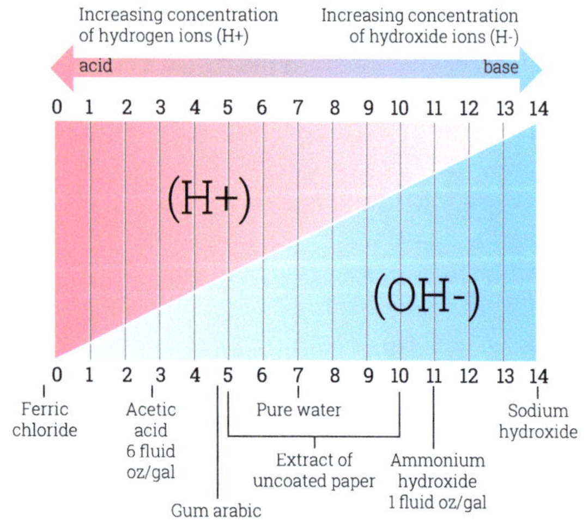

The pH scale of some common solutions.

PICK – The tendency of paper fibers or coating particles to be pulled away from the paper surface during printing when using tacky inks. These particles may remain on the surface of the printing plate or inking rollers and create faulty inked or printed images on subsequent prints.

PIGGYBACK LABELS – Pressure-sensitive label constructions that have two release coated liners, two layers of adhesive and a face material which enables the label to be applied – complete with the backing – for further or future application.

PIGMENT – The particles or substance that gives printing inks, dyes or paints their color. They exist as an insoluble suspension and absorb and reflect lighting, giving the appearance of colors to the human eye.

Pigments in printing inks both scatter and absorb light – so determining the final color and opacity of the printed image. To give the necessary color and performance characteristics, pigments should have a number of key properties. These include:

- Excellent light fastness properties
- Good color strength
- Adequate dispersion into ink vehicles
- Chemical and/or end-user product stability
- A particle size that meets a pre-determined range for the ink being manufactured.

PIN REGISTER – Accurate color registration of copy, film, plates and presses through the use of exactly positioned holes and special pins during pre-press operations and press set-up.

PIQET – Packaging Impact Quick Evaluation Tool. A web-based life-cycle analysis tool for product packaging developed by the Sustainable Packaging Alliance based in Sydney, Australia

PITCH – Pitch is a measure of the distance between the left-hand edge of the first label to the left-hand edge of the second label across the web.

PITCH DIAMETER – Refers to the measurement of a gear or cylinder and can be determined by dividing the circumference by Pi (3.14159).

A predefined diametrical position on the gear where the circular tooth thickness, pressure angle and helix angles are defined. The standard pitch diameter is a basic dimension and cannot be measured, but is a location where other measurements are made. Its value is based on the number of teeth, the normal module (or normal diametrical pitch), and the helix angle. It is calculated as:

$$d = \frac{Nmn}{\cos \Psi} \quad \text{in metric units}$$

PIVOT POINT – The central point of rotation, such as that found in a press dancer roll.

PIVOT POINT SENSOR – A sensor mounted at the pivot point of the dancer arm. The sensor determines in which direction the dancer is moving and where it is in its arc of travel.

PIXEL – The shortened version of Picture Cell. The smallest distinct element (bitmapped image) that a computer screen can display. Computer screens are made up of hundreds of thousands of pixels, each of which is controlled by information stored in the computer. The more pixels there are per inch or centimeter, the higher the monitor resolution.

One halftone, or process color dot, may spread over two pixels. Color monitors have pixels made up of three different dot combinations – red, green and blue. The color seen on the screen from each pixel is a combination of the light from these dots. *See Additive color.*

PLANOGRAPHIC – A printing process in which the printing surface carrying the image is on a flat plane. Both the inking (image) and non-inking (non-image) areas of the plate are carried on a common plane, such as in lithography. *See Offset litho and Label printing and production.*

PLASTICIZER – A substance which is added to certain plastics, primarily PVC, in order to obtain the desired degree of flexibility, moldability, softness, elasticity and dispensability. Plasticizers are also used in adhesives to obtain certain properties.

PLASTICIZER MIGRATION – Monomer plasticizers are an essential part of flexible PVC materials. Any plasticizer that migrates from a PVC label or substrate into the adhesive will cause excessive softening or degradation of the adhesive and weaken it. This phenomenon is accompanied by a simultaneous shrinkage of the PVC film.

PLASTICIZER RESISTANCE – Plasticizers used in PVC film stocks can migrate into the adhesive with potentially disastrous results. Rubber and hot-melt adhesives soften, become stringy and lose both adhesion and cohesion. The bond strength of acrylic dispersion adhesives is affected to a lesser degree. Specially formulated acrylic solution adhesives can provide adequate plasticizer migration resistance.

PLATE – The image carrier in letterpress, flexographic and offset printing presses. A flexible die in the manufacturing process.

PLATE CYLINDER – The cylinder in a printing unit to which the printing plate(s) or a plate sleeve is attached. Print cylinders may have the shaft as a permanent part of the cylinder body, or have a removable shaft.

PLATE ELONGATION – A physical lengthening of the flexographic image carrier (plate) material which tends to stretch as it is mounted around the plate cylinder, sometimes enough to distort images or text, e.g. a circle may be distorted to look more like an oval. The smaller the cylinder diameter the greater the potential elongation problem. Software to compensate for plate elongation can slightly distort images at the flexographic software application stage, or at the RIP stage before the film is imaged. *See Dispro.*

PLATE MOUNTING TAPES – Special adhesive mounting and cushion mounting tapes for affixing flexographic plates to printing cylinders have been developed by a number of commercial tape suppliers. These should enable the plate to be mounted easily without blistering, grip it tight to resist edge lifting and, after printing, remove so easily that plate back treatment is eliminated.

In recent years, foam mounting tapes have played a key role in helping the flexo industry achieve more consistent reproduction of high quality print design and in reducing dot gain. Foam material in the mounting tape, being softer than conventional filmic mounting tapes, compresses under the impression pressure. This enables the printing plate to remain stable and the print image to be reproduced without deformation.

The most important element of foam mounting tapes is the foam cell construction. An even cell structure will promote even compression and good recovery, consistently, throughout the print run – no matter how long the job is. An even cell structure is also critical over wider web printing widths,

The tight-gripping adhesives used on mounting tapes eliminate the work and time of sealing the plate edges and hold the plate securely throughout the press run. When finished, the plate needs to peel off easily without damage or kinking.

PLATE THICKNESS – The thickness of a printing plate measured from the back of the plate to the top of the image area on the surface. Most commonly used to describe the thickness of letterpress or flexographic plates.

PLUG-IN MODULE – Software developed by a third party vendor that enables a function to be used that is not available in a standard computer software application.

PLY – Each of the layers in a multi-layered structure.

PMS – The Pantone Matching System®, a book of standardized colors in fan format. Trade name of Pantone Inc's commonly-used method of color standards, color data, color reproduction and color identification. A system used for matching colors when printing on a press, which helps designers and print specifiers to describe and select the exact colors for the printer to match. *See also Pantone.*

POISE – The unit used to measure viscosity, expressed as one dyne per second per square centimeter.

POLISH LABEL ASSOCIATION – *See under SPPES*

POLITICA NACIONAL DE RESIDUOS SOLIDOS (PNRS) – PNRS is the solid waste policy enacted by the Brazilian government in August 2010. The legislation places responsibility for solid waste on the producer.

POLYCARBONATE FILMS – Polycarbonate films have a very high clarity and offer the versatility of acetate with the durability of polyester.

POLYESTER FACE MATERIALS – Polyethylene terephthalate, otherwise called polyester – or PET – is recommended for versatile, high-performing labels, tags and window decals. It has high dimensional stability, exceptional clarity, high tensile strength, strong tear and abrasion resistance, displays excellent resistance to solvents, oils and many chemicals, and has thermal stability at high and low temperatures. It is unaffected by water and practically impermeable to odors.

The durability, clarity and chemical resistance of PET makes it an ideal material for converting a wide range of niche products. These include nameplates and bar-coded labels for various security products, printed circuit boards, ID labels for automotive and aerospace parts, high and low temperature labels, autoclaved labels for sterilized medical applications and much more.

Thin PET films down to 36 micron (1.5 mil) or even 26 micron (1.1 mil) thick are adequate for most applications, including clear-on-clear labeling – i.e.: clear label, adhesive and backing. Overlaminating films used with optically-clear adhesives can be as low as 12 microns (0.5 mil) thick. Gloss clear grades are ideal for matching glass containers, syringes or ampoules, allowing visual inspection during packing. Although normally clear, PET films can also be pigmented white, metalized or produced in any color.

Polyester is generally easy to print and UV cure. The strength and dimensional stability of PET facestocks and liners virtually eliminate web breaks. Cutting dies should be of the sharp carbide-tungsten type. Top coating allows all label printing processes using solvent-based or UV-cured inks.

Polyester materials that show 'void' or similar messages when removed are a familiar tamper-evident device. Variations include destructible metalized polyester with patterned effects, sometimes incorporating security holograms. Typical advantage and benefit properties of PET face materials are shown in the table.

Advantages and benefits of polyester films

Printing and application performance	End-usage criteria
Resistance to heat	Range of products
Resistance to stretch	'No-label' look
Resistance to tearing	Resistance to abrasion
Dimensional stability	Resistance to solvents
Resistance to tearing	Dimensional stability
Resistance to solvents	Resistance to heat
Dispensability	Resistance to chemicals
	Resistance to moisture
	Higher cost*
	Conformability

Key: Good Poor or caution

*PET films are available in very much lower callipers than polyolefin films, which to some extent offset their higher cost.

POLYESTER RELEASE LINER – A silicone release coated polyester film with an excellent die-cutting surface. *See Release liner films.*

POLYETHYLENE FACE MATERIALS – Polyethylene (PE) is one of the polyolefin group of polymers and is usually produced using the traditional blown film co-extrusion process. Multi-layer films are made by using two or more extruders in the co-extrusion process.

Polyethylene is an environment-friendly PVC substitute and when disposed of by incineration merely releases carbon dioxide and water. Self-adhesive PE labels can be readily recycled with polyethylene containers.

The two main types of polyethylene face materials that apply to label production are:

Low or Medium Density Polyethylene (LDPE) which is transparent*, has good tear resistance and low stiffness. Corona-treated grades are used for squeezable and irregular-shaped containers. LDPE is also widely used for tape and packaging film, although medium density films are more likely to be used for labels

---

Advantages and benefits of polythylene films

Printing and application performance	End-usage criteria
■ Exterior tear resistance	■ Squeezability
■ Minimal shrinkage	■ Resistance to moisture
■ Printability	■ Resistance to chemicals
■ Low stiffness	■ Environment friendly
■ Prone to stretching*	■ UV resistance
	■ Transparency**

Key: ■ Good ■ Poor or caution

*during printing/converting.
**Standard PE films are not best for transparency, being slightly milky. However, companies such as ExxonMobil and Dow offer grades that are claimed to be sutiable for high clarity applications.

High Density Polyethylene (HDPE) has high stiffness, poor resistance but good barrier properties. Many types of plastic bottles and tubs are made with HDPE, although PET is challenging this material in certain beverage markets.

PE films are made in high-gloss white, super-white, matt white, transparent and metalized grades. They offer good resistance to water, chemicals and UV light, so are suitable for cosmetics, toiletry and household products, food and outdoor applications. Being free of plasticizers, PE's dimensional stability is good but it can easily be stretched. Minimal shrinkage avoids adhesive edge residues. It also has a high exterior tear resistance and is flexible enough to allow a printed label to resist creasing on squeezable products.

POLYMER – A long-chain molecular structure formed by the reaction and linking together of monomers.

POLYMERIZATION – The chemical reaction in which monomers and/or oligomers combine to form polymers. Polymerization may be initiated by light, heat or catalyst.

POLYPROPYLENE FACE MATERIALS – Polypropylene (PP) films derive from petroleum using cost-effective and high-yield polyolefin resins and have properties similar to those of polyethylene, but are stronger, more rigid and have a higher temperature resistance. The films are non-toxic, light, strong and odorless, with good moisture barrier properties. Efficient to produce, PP films require less energy and fewer raw materials compared to most other filmic substrates. As with PE, films are manufactured either by the blown bubble process or the cast extrusion process.

Depending on the process, PP films may be orientated (or stretched) in either one direction (OPP) as mono-axial film, or biaxially-orientated (BOPP) in both the machine and cross direction. Orientation improves the film's optical properties, while providing good stiffness and high tensile strength in the machine direction.

Co-extruded OPP labelstocks are available with several useful multi-layer finishes. For example, 'no-label' look grades are usually three-layer films of 57 microns (2.3 mil) in transparent gloss/gloss, and

---

Advantages and benefits of polypropylene films

Printing and application performance	End-usage criteria
■ Resistance to tearing	■ Resistance to abrasion
■ Dispensability of thin film	■ Resistance to chemicals
■ Die-cutting	■ Resistance to moisture
■ Printability	■ Dimensional stability*
■ Good film flatness	■ Clarity
	■ Low cost
	■ Conformability
	■ UV stability

Key: ■ Good ■ Poor or caution

*For best dimensional stability and clarity it is advisable to use oriented PP films (OPP or BOPP) rather than ordinary cast or especially blown PP.

---

transparent matt/gloss. White gloss and white matt give coated paper effects. Adding a white pigment to the original film gives an attractive appearance for many types of labels, including wrap-around labels on PET bottles and containers.

Overall, the market has seen a dramatic increase in the manufacturing and off-line coating of PP films for pressure-sensitives, reel-fed or cut sleeve wraps and in-mold labels, as well as filmic release liners. Key advantages and benefits of polypropylene label films are shown in the table on the previous page.

POOR TRAPPING – Found in printing when less ink is transferred to a previously printed ink layer than transfers to the unprinted paper.

POST CURE – A continuation of the cure reaction of UV inks or coatings resulting in increased degrees of cure after exposure to the thermal or UV-cure initiation energy. Some ink colors only surface cure during the initial radiation and take longer for the reaction to complete its cycle.

With some silicone systems, such as the tin catalyzed condensation reaction, post-cure may take as long as three weeks. Platinum catalyzed addition chemistry can range from a few days post-cure to virtually no post-cure.

The amount of post-cure depends on the degree of initial cure and on the chemical environment of the reactants.

POSTSCRIPT – A page description language created by Adobe Systems for defining the contents and layout of printed documents in precise detail, so enabling a computer to communicate with a printer. A coding language intended for output devices.

POT LIDS – Printed and heat-seal lacquer pre-coated lids used as the sealed top closure on yogurt, ice cream, water, desserts, snacks, and a range of other products.

During production the lids are coated with a heatseal primer/lacquer or protective coating, printed on a web-fed analogue or digital press. They may also be embossed with different surface textures and laminated.

Pot lids in shorter runs, versions and variation or customized, are seen as attractive options for narrow-web and digital converters to look to diversify.

POWER LEVEL – The amount of RF energy radiated from a reader or an active RFID tag. The greater the power output, the longer the read range. However, power levels are regulated by most governments to avoid interference with other devices.

POUCHES – Pouches, especially in the form of stand-up pouches, have become a replacement for many products previously supplied in a standard can or jar in markets such as shelf-stable foods, soft drinks, pet food, cereal, snacks, agriculture, pharmaceuticals, pesticide, medical, coffee packaging, dried fruit, lawn and garden products. They are often seen as a way of reducing the amount of packaging material.

Depending on the application and the products contained in the pouch, they may incorporate easy-tear notches, easy opening laser scoring, zipper locks for re-sealing, spouts with screen caps or tamper evidence. Many types of pouches are microwaveable and have an exceptional resistance to high temperatures, water, vapor and gas. *See also Stand-up pouches.*

PPF – Print Production Format (PPF) was the first attempt to standardize communication of digital data (often referred too as CIP3 data) between different parts of the print production process. Beginning with pre-press data and the automatic transfer of press-press information such as ink duct pre-setting, register control or color specification to the press, PPF developed into the next generation of job ticketing – JDF (Job Definition Format), the file that holds management and production details of a job.

PREFLIGHTING – Preflighting is a term used in the label and packaging industries to describe the process of confirming that the digital files required for printing are all present, valid, correctly formatted, of the desired type and are suitable for output.

The process of preflight checking a label or pack image file can take place manually within certain software applications or automatically within a workflow, which dramatically reduces the risk of output problems after a file has been rasterised (RIPped) – one of the major causes of print and plates are being output

Preflight checking of files is a fundamental part of the creative process and without it, invariably output problems can be missed. This then results in significant costs and delays. Preflight software generally achieves a significant Return on Investment as it saves employees' time as well as print materials and equipment.

PRESS DESIGN, NARROW-WEB – The technical machine design of a narrow web press today can draw on and incorporate a number of key areas into the building of a sophisticated press for high quality printing and converting, namely:

- drives, whether conventional, hybrid/ combined or single drive
- drives using servo drive technology
- machine controls, which may be conventional centralized machine control, centralized machine control or decentralized (modular) machine control.

production delays.

The preflight checking process looks for a range of potential problems that can include:

- Missing images and graphics
- Fonts that are inaccessible
- Corrupted fonts
- Font formats that are incompatible
- Incompatible image files
- Incompatible color formats (some RIPs have problems processing RGB images, for example)
- Resolution issues
- Presence of the correct color profiles
- Corrupted Images
- Incorrect page-layout, document size, margins, bleeds etc...
- Compatibility with the output device e.g. ink plates, colour separations
- Information all fits within the constraints of the output device and match the client specifications
- Confirm that the correct color separations or ink

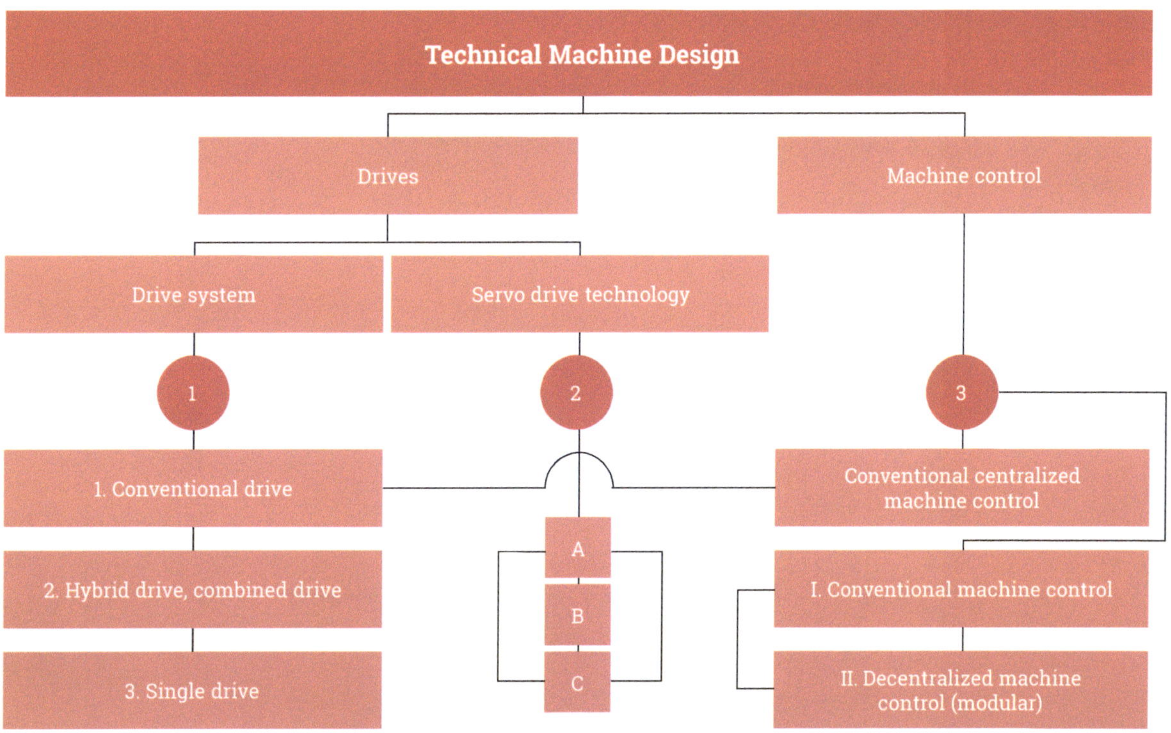

Technical Machine Design

Automation and operating concepts on label presses

AUTOMATION / OPERATING CONCEPT	CLASSIFICATION	DESCRIPTION
Automation by fully integrated closed-loop control functions	3/C/II	Fully integrated automation solutions for the centralized controls, e.g. fully integrated register control functions on the central HMI
Decentralized press operation	1/A,B,C/I	Control panels arranged to the individual press modules
Presetting of the entire label press	2,3/B,C/I,II	Presetting data can be recalled either out of the job-file or selected via pre-selection on the central operating panel. The presetting functions are: presetting of the web tension, register control (color sequence, position, register reference, register comparison, length and cross register), printing pressure adjustment, unwind and rewind tension

Different combinations of drive and control systems to meet different levels of press sophistication.

This general view of narrow-web machine design can be seen in the accompanying diagram, together with the flow diagrams showing the relationship between the different drive and control systems for label presses.

Drive and control systems. Three types of drive and control systems may be used with narrow-web presses:

1. Conventional driven presses, which consist of an electrical main drive motor which drives a main shaft. The force is transferred by gear boxes to subordinated modules, such as print units, rollers and cylinders. These presses use central control cabinets and decentralized control cabinets.

2. Hybrid/combined drive. Presses which are partially conventional driven (mechanical shaft) combined with single drive technology. These use central or decentralized control cabinets and decentralized control panels.

3. Single drive. Fully integrated direct drives using single module technology. Full multiple axes direct drive solution with assembly motors. There are no mechanical transmission elements (gearless). Decentralized control cabinets are preferred, with a centralized control panel.

Servo drive technology. Servo drive technology for hybrid and single driven label presses may be:

A. Axle driven with several gear ratios (backlash,

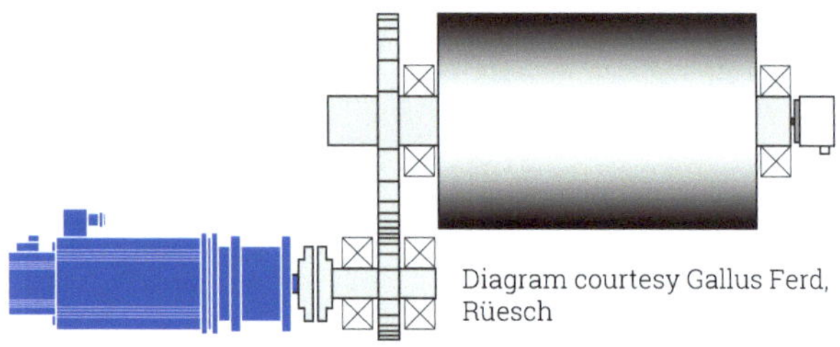

Diagram courtesy Gallus Ferd, Rüesch

Axle driven with several gear ratios.

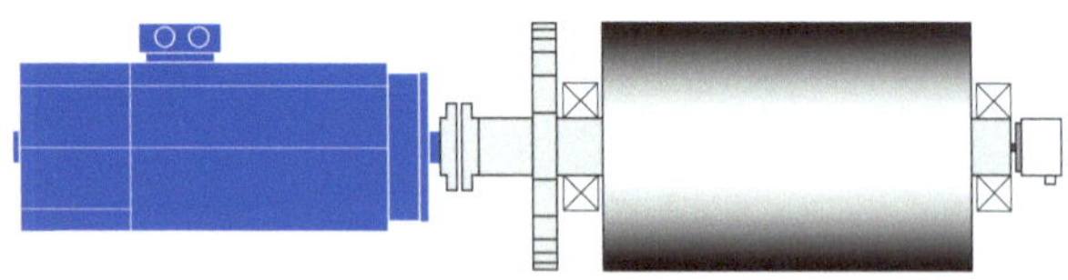

Diagram courtesy Gallus Ferd, Rüesch

Axle drive by mechanical transmission elements.

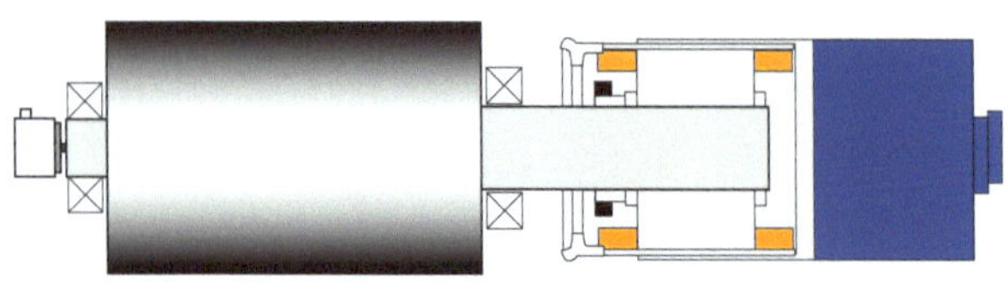

Diagram courtesy Gallus Ferd, Rüesch

Direct axle drive by assembled motor

noise, abrasion, lubrication, maintenance).

B. Axle drive by mechanical transmission elements, such as clutches (risk of deviations in angle-synchronism and axle misalignment).

C. Direct axle drive by assembled motor (no mechanical elements), highest accuracy in angle synchronization, maintenance-free, highest lifetime).

Machine control. Machine operating controls on narrow-web presses may be centralized, de-centralized or a combination of both methods:

i. With centralized press operation there is a central control panel for all the main press functions and digital entering of all the press settings (e.g. movable touch panel with digital hand-wheel).

ii. In de-centralized press operation the control panels are arranged to the individual press modules.

iii. With both centralized and de-centralized press operation there is a central control panel for the main press functions, and decentralized operating panels for the single press module functions.

By using different combinations of drive and control systems built into narrow-web presses it is possible to have a wide range of press automation concepts and systems to meet differing levels of press sophistication and requirements. Some examples of these combinations can be seen in the table on the previous page.

PRE-PRESS – Pre-press describes all the activities involved in setting up and preparing pages, documents and labels for printing. In conventional printing, several stages are involved – typesetting, scanning, separations, platemaking, etc. With digital technologies these stages are generally streamlined. *See also Digital workflow and Workflow.*

PRESSURE GAUGE – an indicator used to determine the amount of force applied to a cylinder.

PRESSURE-SENSITIVE ADHESIVES – Pressure-sensitive adhesives (PSAs) differ from other types of adhesives in that they are able to form a bond at any time, are permanently tacky, are capable of bonding to almost any surface and are adhesive above their glass transition (Tg) temperature. No activation by water, solvent or heat is required to exert a strong adhesive bond on materials as diverse as paper, plastic, glass, wood, cement and metals.

Four well-established pressure-sensitive adhesive technologies are currently used, with a fifth now beginning to find commercial applications:

Acrylic solutions. Solvent-based acrylic PSA solution formulations have been widely displaced by water-based and hot-melt systems for economic as well as ecological reasons. Solvent recovery and/or incineration are essential to meet clean air legislation requirements. Such equipment is expensive and can only be justified for large output operations. Acrylic solution adhesives are therefore not widely used today, except for speciality applications such as glass clear 'no-label' look PET labels or PVC labels where optimum water resistance is needed.

Rubber/resin solutions. These are also less

Properties of main pressure-sensitive adhesive types

	Rubber solutions	Acrylic solutions	Acrylic dispersions	Hot-melts
Tack	■ (green)	■ (yellow)	■ (yellow)	■ (green)
Peel strength	■ (green)	■ (yellow)	■ (yellow)	■ (green)
Shear strength, room temp.	■ (green)	■ (green)	■ (green)	■ (green)
Shear strength, high temp.	■ (yellow)	■ (green)	■ (green)	■ (red)
Die-cutting behaviour	■ (yellow)	■ (green)	■ (green)	■ (yellow)
Adhesive bleed/ migration	■ (yellow)	■ (green)	■ (green)	■ (yellow)
UV resistance	■ (red)	■ (green)	■ (green)	■ (red)
Heat resistance	■ (yellow)	■ (green)	■ (green)	■ (yellow)
Effect of plasticizers	■ (red)	■ (green)	■ (yellow)	■ (red)
Color/ transparency	■ (yellow)	■ (green)	■ (green)	■ (green)
Low temperature behaviour	■ (green)	■ (yellow)	■ (yellow)	■ (green)
Moisture resistance	■ (green)	■ (green)	■ (yellow)	■ (green)
Water resistance	■ (green)	■ (green)	■ (red)	■ (green)

Key: ■ (green) Positive feature ■ (yellow) Meets normal requirements ■ (red) Unsuitable for applications needing this property frequently used, because they are coated in solvent, except for high performance 'peelable' labels and specialities such as oil can labels.

Hotmelts. Hot-melt PSAs are a fast-growing sector, because their conversion performance is now very good and they are competitive. They can be used where an aggressive permanent adhesive is needed with high tack and some recipes have excellent performance in cold and wet conditions, particularly when labeling plastic surfaces. Hot-melts are easy to coat on compact equipment and are also the preferred choice for most in-house converters and printers.

Acrylic dispersions. Water-based acrylic PSA dispersions (emulsions) now represent the dominant technology for the label stock producer and can also offer a practical option for in-house coaters. The large label stock suppliers formulate their own adhesives, but ready-to-use formulations are commercially available for small and medium-sized coaters.

Radcure. Adhesives can be cross-linked (cured) by electron beam or ultra-violet radiation. This enhances some characteristics such as high temperature performance of hot-melt adhesives. Radcure adhesives are now used in some speciality tape applications and are also beginning to be used in limited applications by label or forms converters.

Pressure-sensitive label adhesives successfully meet an enormous range of demands. They can provide a permanent bond or can be removable. Some substrates are easy to stick to, e.g. paper or board, but PSAs can also be formulated to stick to 'difficult' surfaces such as plastic containers or moist glass bottles. Label service temperatures normally range from -20°C (-4°F), or lower for freezer packs; up to +200°C (392°F) or more for car engine parts. The cohesive strength of PSA adhesives can be reduced to allow labels to disintegrate to discourage tampering. The key properties of the main pressure-sensitive additives are shown in the table opposite. *See also Adhesive and Adhesive, pressure-sensitive.*

PRESSURE-SENSITIVE ENVELOPES – Traditionally sealed with remoistenable gum adhesives, envelope sealing has now adapted to the growing use of pressure-sensitive methods of sealing, both using a release liner and self seal options.

The pressure-sensitive adhesive stripe is conventionally protected by a paper or film release liner tape, which is peeled off and discarded. A more elegant alternative that is not only less costly but environmentally more acceptable is to eliminate the release tape by making the release liner an integral part of the product.

This is achieved by simply printing the silicone release patch on to part of the envelope so that it protects the pressure-sensitive adhesive stripe before use. The seal is effected by folding up the flap. Although deceptively simple, this method of sealing is quite complex and requires silicone formulations that can be coated on a wide variety of envelope papers –

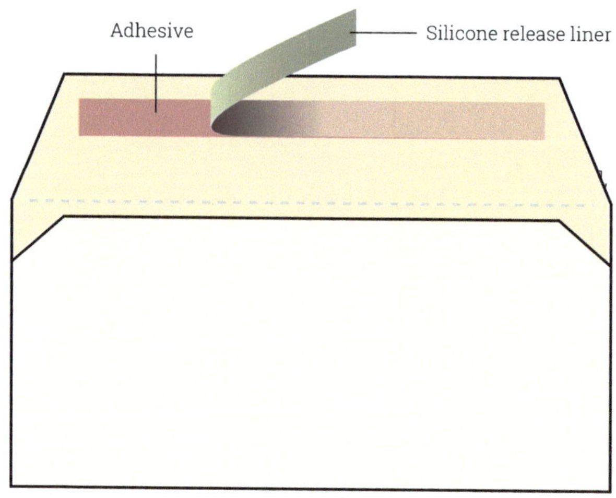

Pressure sensitive envelope seal with peel-off release liner

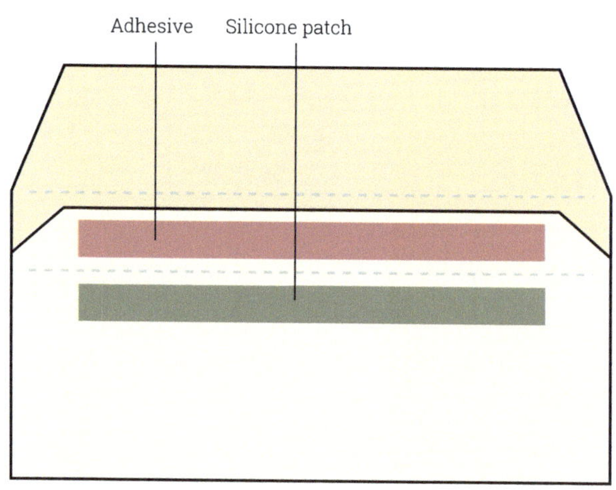

Pressure sensitive envelope seal with built-in release patch

including those made from recycled paper. The silicones must, therefore, be formulated to remain on any surface. Any silicone that is allowed to soak into the body of the paper is effectively wasted.

PRESSURE-SENSITIVE TAPE – *See Tape, pressure-sensitive.*

PRICE GUNS – Variously known as price guns, pricing guns, hand labeling guns or hand labelers, they all relate to hand-held machines that can print 1, 2 or 3 lines of fixed, variable or sequential information, with up to 14 digits per line on a range of small self-adhesive label sizes from 18 x 12 mm up to 37 x 28 mm. Used for printing price, code, date, marking or batch information, price and labeling guns are one of the key items needed by busy retailers, grocers and supermarkets.

Typical price/ hand gun

Price guns can be supplied with a variety of font faces, in various font colors, so as to print onto plain, pre-printed or colored self-adhesive die-cut labels which, in turn, may incorporate features such as security cuts so that labels are difficult to remove or substitute without breaking up. Some guns can print custom messages or logos or be used with pre-printed labels to enable companies to advertise their name, improve presentation, enhance clarity and, in some cases, satisfy legal requirements, codes of practise or industry standards – such as Best Before information.

Price guns and hand labelers incorporate alphabet, numeric or alpha numeric print wheels, an inking system and a dispensing head that will enable the printed label to be applied to the product being labeled. Labels supplied for use with hand guns may consist of different die-cut shapes to suit the label feed system used by different hand gun manufacturers.

Examples of price gun printed labels.

PRICE-WEIGHT THERMAL LABELS – Price-weight or catch-weight labels have grown to become one of the major self-adhesive roll-label sectors since the introduction of chemical-thermal/ direct thermal self-adhesive labelstocks around 1983/84. Once produced as premium-priced products – on premium-priced materials – thermal price-weight labels have now become a commodity product within the food/supermarket sector.

Like conventional food labels, price pressures, flexibility, ISO quality assurance, quick turnaround and product consistency are paramount to the price-weight label producer. Production of thermal price-weight labels is predominately by the flexo process, although both rotary letterpress and flat-bed semi-rotary letterpress are also used.

Special thermal-sensitive coated papers that change color under the action of heated styli are used to electronically print price-weight labels. *See Direct thermal printing and Thermal paper for more information on the special papers used.*

PRIMARY COLORS – The three primary subtractive colors of printing inks are cyan, magenta and yellow. When each of these colors is mixed with one or both other colors, a new color is produced.

The three primary additive colors of light are red, green and blue. Mixed in pairs they create the secondary light colors of cyan, magenta and yellow. White light is created when all three colored lights are mixed together in equal amounts.

PRIMARY LABELS – These are designed to be the main identification on a product or pack. They carry the product's brand name or image identification that will attract attention on the retail shelves and appeal to prospective buyers. The labels usually promote a specific logo or brand name, incorporate special brand or house colours and may carry a picture, drawing or representation (which generally may not hide, obscure or interrupt other matter appearing on the label). They are normally applied at the time of product manufacture.

For some applications the primary face label may be reverse side printed with secondary information. Used with clear liquids and clear bottles (glass or plastics) the information is read from the back of the container, through the container and contents, thus eliminating a separate secondary label. *See also Secondary labels.*

PRIMER – A coating applied to a label face material to improve its ink or adhesive key properties. A primer also serves to improve the scratch resistance of printing ink and reduces the tendency of adhesive components to migrate into and through the face material.

PRINTABILITY – Printability is the characteristic or ability of a label face material to permit high quality printed results. Although difficult to define precisely, good printability provides a uniformity of printed color, uniformity of ink transfer, good quality of 'black-on-white' image, and consistency of ink drying and setting.

Factors that influence the printability of paper label stocks include the surface smoothness and hardness, pick resistance, absorbency, strike through, loose or weakly-bound surface particles. Printability of film will also relate to smoothness, but absorbency or strike through, loose surface fibers or coatings will not be key factors.

A range of printability testing machines have been developed by various test organizations. Depending on manufacturer, they measure color, weight of the ink on the surface, density, picking, roughness, print penetration, mottle, uniformity of solids, darkness of printed areas, etc, and provide comparisons between various inks on different substrates for quality control and R & D purposes.

Where test strips are produced in the printability tester they can additionally be used for testing print quality, color matching, visual appraisal determination of ink coverage, scratch resistance, wear resistance, light fastness and resistance to chemicals.

Printability testers are most commonly used by ink, paper and board manufacturers, laminators, printers, plastics film manufacturers, lacquer and coating industries, and raw materials suppliers.

PRINTER FONT – The font stored in the computer system that enables a laser, ink jet or other printer to recognize the typeface when printing. If the designated font is not in the system then the printer will select another font.

PRINT QUEUE – Jobs transmitted from a computer and which are held in a queue awaiting subsequent processing by a printer(s).

PROCESS CHLORINE-FREE – A process being used for some recycled grades of papers

PROCESS COLORS – The four process color inks – cyan, magenta, yellow, plus black (CMYK) – which are used in the printing of four-color halftone labels.

PROCESS PRINTING – Printing from two or more process color halftone plates and inks to produce intermediate colors and shades.

PRODUCT & IMAGE SECURITY FOUNDATION – A non-profit making trade association originally formed under the name of Label & Tag Security International (LTSI) in the early 1990s. Today it operates with more than 100 members in 27 countries and is the only international forum for manufacturers, suppliers and users of labels, tags, packaging, identification systems, materials and systems for any form of product security, brand image protection or product traceability applications and technology.

The Foundation is managed by Governing Members/Trustees, each with extensive expertise in the worlds of security, documents, labels and printing.

PRODUCTION CONTROL – A management system used to ensure that label materials, inks, manpower, machinery, finishing, transport and other facilities are used efficiently and that will ensure a specific label order is produced to a specific quality and quantity, within a specified time schedule. *See also Statistical Process Control and TQM.*

PROTECTIVE COATING – A coating that is applied to the surface of a label to protect the print and and/or label stock from abrasion, chemicals, moisture or sunlight, or any combination of these.

PROTECTIVE HAND SHIELD® – Developed in response to the growing concerns regarding cross-infection in the food and healthcare sectors, Hand shield consists of a flexible, conformable substrate, die-cut into the shape of a hand with fingers and thumb spread out onto which a low tack, hypoallergenic adhesive has been coated.

When a hand is lightly pressed on a stack of Hand shields this will cause the topmost shield to adhere to the hand. A release coating on the reverse side enables the shields to be dispensed one at a time. After use, the shield is simply disposed of. The Hand shield therefore provides a low-cost, user-friendly pressure-sensitive alternative to protective gloves.

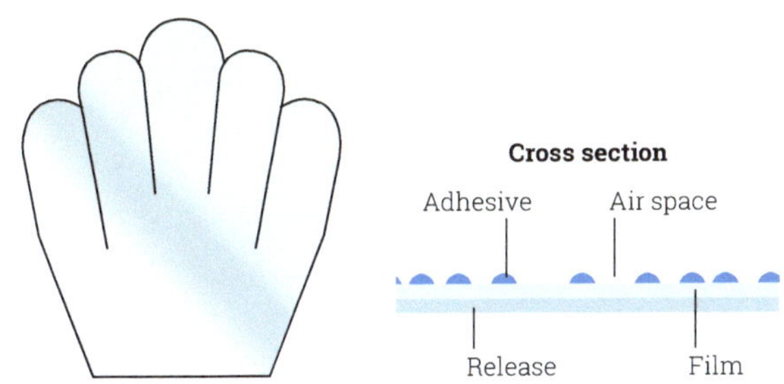

Structure of protective self-adhesive handshield

PROXIMITY DEACTIVATION – Also referred to as non-contact deactivation. The process by which an active tag can be rendered inactive without physical contact with the tag.

PSI – Pounds per square inch.

PSTC – The Pressure Sensitive Tape Council. Formed in 1953, the PSTC is a not-for-profit, 60-year old, North American Trade Association for tape manufacturers and affiliate suppliers, dedicated to helping the industry produce quality pressure-sensitive adhesive tape products in the global marketplace. It promotes the manufacture of high quality and performance pressure-sensitive tape products for industrial applications.

Its 25 tape manufacturer member companies produce over 90 percent of the total PSA tape manufactured volume in North America and represent $10 billion in annual sales.

PSTC publications provide vital material on current pressure-sensitive tape products, testing procedures, technological developments and background information on the tape industry. Details of products manufactured by members can be found in the Council's Tape Products Directory, including market applications, tensile strength, adhesion, thickness and elongation.

PUNCHED OUT LABELS – Refers to anvil cut or sheeted labels, rather than die-cut labels.

PVC FACE MATERIALS – Polyvinyl chloride (PVC) films are flexible and durable and display excellent weather and chemical resistance. They are available in gloss, matt, clear and pigmented formats. PVC films are extremely versatile and adaptable and used for a whole diversity of indoor and outdoor application. The material is noted for good printability, die-cutting and registration control.

Advantages and benefits of PVC films

Printing and application performance	End-usage criteria
■ Printability (Good)	■ Chemical resistance (Good)
■ Die-cutting (Good)	■ Weather resistance (Good)
■ Registration control (Good)	■ Flexibility/squeezability (Good)
■ Resistance to tearing (Good)	■ UV stability (Good)
■ Dimensional stability (Good)	■ Scuff resistance (Good)
■ Waste disposal (Poor or caution)	■ Environment friendly (Poor or caution)
	■ Waste disposal (Poor or caution)

Key: ■ Good ■ Poor or caution

Nevertheless, its use for prime-product labels has declined in recent years due to the aversion to using chlorine-based products in an increasingly environmentally-conscious age. Packaging grade PVC was seen as a toxic material because of the vinyl-chloride monomers associated with its production and disposal, especially when burnt.

However, PVC's decline as a label stock material is countered by its established place in industrial and exterior applications. These include point-of-sale material, outdoor promotional and warning signs.

Vehicle liveries and fleet markings, usually made with screen printed vinyl sheets, are finished on computer-aided cutters. Wide-format ink jet printers, producing high-quality results, have made major inroads into the screen and display markets.

Laminators have also formulated many versatile PVC products to meet changing market needs. Factors include good long-term UV stability, compatibility with specified adhesives, and flexibility.

Flexible PVC is soft and malleable and is highly resistant to scuffing and most chemicals. Semi-rigid vinyl's have similar properties, but are more dimensionally stable. The destruct variety is brittle and cannot be removed intact after application. Applications include tamper-evident container seals, also computer-imprint able face stocks where security demands non-removable labels. *See also Vinyl.*

Q

QR CODE – *See Quick Response code.*

QUALITY CONTROL – Also called Quality Management or Total Quality Management (TQM). A management system in which tests and/or inspections are applied at various stages during label production and printing to ensure that the end product meets pre-determined specifications and standards of quality. *See also Statistical process control, TQM, Production control and Quality Management.*

Management and Six Sigma. Each of the quality management/control systems is briefly described below:

BS EN ISO 9000. Implementation of the BS EN ISO 9000 quality standard involves making a large number of small quality improvements across a wide range of business processes. Activities can be managed as a process, with each activity adding value to the last, all ultimately leading to customer satisfaction. Simple, user friendly, documentation can be created around a company's existing business practices.

Statistical Process Control (SPC). A system by which printed label results can be controlled, Statistical Process Control (more commonly known as

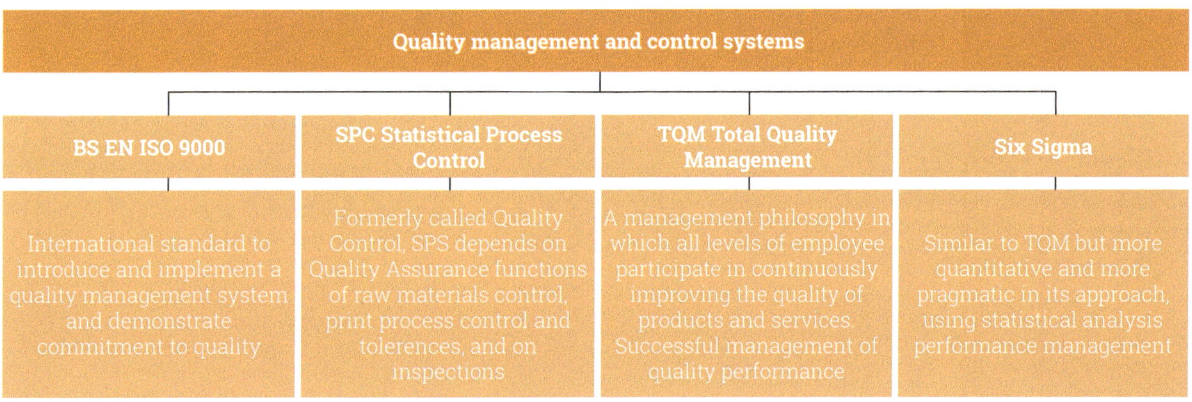

QUALITY MANAGEMENT – Quality can be defined as the ability to satisfy, or exceed, the needs and expectations of customers.

Label printing, whether sheet-fed or web-fed, is a manufacturing process that uses variable components, such as paper, inks, plates, films, etc, and which is carried out on machinery that will eventually show signs of wear, or can be set up differently by the same or different operators. As such, all these possible variable elements need to be quality managed and quality controlled wherever possible.

In the label industry, BS EN ISO 9000 has become the standard by which key label users, brand owners and buyers assess whether a system and procedures are in place to manage and control quality. More sophisticated quality management and quality control systems now being adopted by the label industry include Statistical Process Control, Total Quality

Quality Control in the past), depends on the Quality Assurance functions of raw materials specifications and controls, printing process control, on acceptable standards and tolerances and on inspection.

Total Quality Management (TQM). The basis of TQM is that any product or service can be improved upon and that improvements reduce costs, provide better performance and give higher reliability. The approach of TQM is that any company can continuously improve the quality of its goods and services – manage its quality performance – through the participation of all levels of employees.

Six Sigma. Six Sigma and TQM have many characteristics in common. They both focus on quality control, both address quality problems at their roots, both focus on solutions geared to the overall goal of the business, both rely on data as evidence, both require senior management to buy into the quality program and both aim for a lasting cultural change throughout the whole organization. Six Sigma however, is more quantitative and more pragmatic in its approach, using statistical analysis performance management. Six Sigma can be incorporated within an existing TQM environment.

Starting with ISO 9000 as an internationally recognized quality management standard, it is possible to bolt on any or all of statistical process control, total quality management and Six Sigma systems to eventually rise to the highest levels of quality management and customer quality satisfaction. Further information on the above quality management systems can be found by visiting ISO 9000, Statistical Process Control, TQM and Six Sigma.

QUARKXPRESS – A printing/publishing software package that enables pages to be designed and made-up to incorporate images and text.

QUICK RESPONSE (QR) CODES – Printed codes which interact with smart phones and dedicated QR reading devices to access related on-line content.

QR codes can be scanned vertically or horizontally and where a standard UPC or EAN barcode can store up to 30 numbers, a QR code can store 7,089 numbers.

Applications include adding traceability or interaction to consumer goods, and in industrial supply chain track, trace and product ID applications.

Quick Response (QR) code

QUIET ZONE – The light margin or clear area that surrounds a barcode symbol if it is to be read successfully. *See also Barcode symbol and Barcode.*

R

RADCURE ADHESIVES – Adhesives can be cross-linked (cured) by electron beam or ultra-violet radiation. This enhances some characteristics such as high temperature performance of hot-melt adhesives. Radcure adhesives are now used in some speciality tape or in-house label applications, but do not yet offer a practical option for most label or form converter requirements.

RADIATION CURABLE SILICONES – Radiation-cured silicone release agents provide opportunities for new applications that are not easily met using conventional thermal curable silicones. Radiation cures at, or only slightly above, room temperature. It allows release liners to be based on thin, thermoplastic film substrates, such as low density polyethylene, polypropylene or PVC which cannot easily be processed through long web path, high temperature curing ovens.

Advantages of Radiation Curing of Silicones

ROOM TEMPERATURE CURE	PROCESSING
No substrate restrictions	Warm application possible
No paper re-moisturizing needed	Compact - little space
No paper shrinkage productivity	High line speed - higher
	Low energy consumption
Absolutely lay-flat	100% product - no solvents
No post-cure line speed	No misting - even at high
Thermal stability	
Long pot life	

Moreover, paper substrates are not dried out, so that complex and expensive re-moisturizing units are no longer required and good lay-flat properties are ensured – an essential prerequisite for labelstocks.

Radiation-curable silicone release agents are 100% solids systems. They require no energy to drive off water or solvent and emit no organic or other volatiles into the atmosphere, thus offering a number of advantages over conventional cure silicones.

While the use of solvent-based silicone systems is declining, radcure technology is unlikely to displace thermally-cured silicones or emulsion silicones from existing processes. Rather, they are opening up new opportunities in areas where thermally-cured silicones are difficult, such as thermoplastic film liners, complex 'piggyback' constructions and especially in-house coating for labels and forms converters.

RADIO FREQUENCY IDENTIFICATION (RFID) – A method of electronically identifying (and sometimes changing data in) unique items at a distance without a line of sight using radio waves. In RFID systems a reader will typically communicate with a tag, which holds digital information in a microchip.

An RFID system basically consists of three separate components – an antenna or coil, a transceiver (with a decoder) and a transponder (the RF tag) that is electronically programmed with unique information. The antenna sends out radio signals to activate the tag and read it or write data to it.

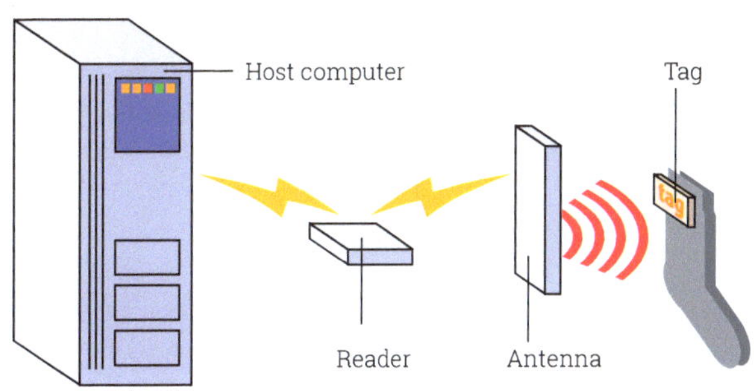

Reading and analysing data from an RFID tag

Antennas used in RFID systems come in a variety of shapes and sizes and may be packaged with the transceiver and decoder to become a reader/interrogator that can be configured either as a hand-held or a fixed mount device. The radio waves emitted from the reader are in the range of one inch to one hundred feet or more, depending on the radio frequency used and the power output. As the tag passes through the electromagnetic zone it detects the reader's activation signal. The reader then decodes the data encoded in the tag's integrated circuit (chip) and the data is passed to the host computer for processing.

Like antennas, RFID tags also come in a wide variety of shapes and sizes and can be categorized as either active or passive tags. Active RFID tags are powered by an internal battery and are typically read/write, allowing data to be re-written or modified. Memory capacity of active tags will vary according to the particular application it is being used for.

Passive RFID tags operate without any separate external power source, obtaining their power from the tag reader. They are therefore much lighter than active tags – and less expensive – and have a virtually unlimited operational life. However, they have a shorter read range than active tags and require a higher-powered reader. There are also chipless RFID tags that use material to reflect back a portion of the radio waves beamed to them. *See Chip tags; Chipless tags; RFID tags.*

RAID – Redundant Array of Independent (or inexpensive) Disks. RAID provides low-cost, reliable storage by saving data on more than one disk simultaneously.

RASTER – Systematic scanning of a screen by a beam of light or electrons. The technique is also used in the graphics industry to detect and scan images, turning a continuous tone image into a multitude of dots of varying size, but of an identical distance apart. The size of the dots controls the amount of printing ink printed on to the substrate.

A coarse raster produces dots that may be visible to the naked eye; a fine raster decreases the number of discernible tonal values, especially in the shadow areas. *See also Scan.*

RASTER IMAGE PROCESSOR – *See RIP.*

RASTERIZATION – The converting of image data into a pattern of dots, usually performed by an image setter for the production of a negative or positive image on paper or film. The process by which Printer Description Language (PDL) is turned into bitmaps that can be outputted by a printer.

RATE OF SET – Refers to the time required for an adhesive (under specified conditions) to reach a fiber tearing bond.

REACTIVATION – The process of rendering inactive tags to an active state.

REACTIVE-CODE LABELS – These work on the key-lock principle of genetic engineering, with invisibly printed features only made visible by means of precisely defined antibodies. This means that authenticity can be verified by using handy coding pens or hand-held reading devices.

Special laser equipment is used to apply markings, images, text and other features on the labels, making them visible only when exposed to UV-light. The markings cannot be removed without destroying the material. Secret personal codes or identification numbers can also be applied in a tamper-proof manner underneath a scratch coating. Such numbers may be used for enabling technical devices or verifying authenticity.

READ AREA – Used to describe the scanning path or scanning area of a barcode symbol.

READER – May also be called an interrogator. Used in RFID systems, the reader communicates with the RFID tag via radio waves and passes the information in digital form to a computer system.

Also refers to the device used for 'reading' data contained in barcode, MICR or OCR symbology.

READ-ONLY TAGS – RFID tags that contain data that cannot be changed without the microchip being reprogrammed electronically.

READ RANGE – In RFID systems, the distance from which a reader can communicate with an RFID tag. Active tags have a longer read range than passive tags because they use a battery to transmit signals to the reader. The read range of passive tags is influenced by frequency, antennae design, reader output power, and the method of powering up the tag. Low frequency tags use inductive coupling in which the RFID tag is required to be within a few feet of the reader.

READ-WRITE TAGS – These are RFID tags that can store new information on their microchip. Such tags are more expensive than read-only tags and consequently have a limited use in supply chain tracking.

REAL TIME LOCATING SYSTEMS (RTLS) – Fully automated systems that continually monitor the locations of assets and personnel. An RTLS solution typically utilizes battery-operated radio tags and a cellular locating system to detect the presence and location of tags. The locating system is usually deployed as a matrix of locating devices that are installed at a spacing of anywhere from 50 to 1000 feet. These locating devices determine the locations of the radio tags.

The systems continually update the database with current tag locations as frequently as every several seconds or as infrequently as every few hours for items that seldom move.

The frequency of tag location updates may have implications for the number of tags that can be deployed and the battery life of the tag. In typical applications, systems can track thousands of tags simultaneously and the average tag battery life can be five or more years.

REAL-TIME PROCESSING READERS – The ability of a computer to process data so fast that the user perceives no delay between a command and the result.

RECYCLABLE – Materials that are able to be reprocessed and used again.

REDUCERS – Refers to materials such as varnishes, solvents and greasy or oily compounds that are added to printing inks to reduce the body and/or viscosity of them for printing.

REEL – A finished roll of film, paper, laminate or labelstocks.

REFLECTANCE – Reflectance is the amount of light which is returned from an illuminated surface.

REFLECTION DENSITOMETER – Equipment that measures the optical density, or light absorbing qualities, of an ink or surface. In application, a reflection densitometer exposes a print to a source of light and then measures the reflection directed back from the print as a percentage. This is done with a photoelectric cell that converts the reflected light into an appropriate electric signal; the densitometer converts the electrical signal into density values. The density scale gives a linear relationship, equating to the method by which the human eye determines light, but much more accurately.

By placing complimentary color filters between the substrate and the photoelectric cell it becomes possible to measure the density of all the process colors, which gives a simple and effective method for maintaining process color control on press.

For transparent materials a transmission densitometer is used. Instead of measuring the light reflected back from the substrate, it measures light projected through the substrate.

REGISTER – The correct placement of one color in relation to the next during printing, as well as to the placement (registration) of printed images with subsequent operations such as die-cutting, perforating, embossing, sheeting, etc. When colors or subsequent processing operations are not placed in the correct relationship to each other this is referred to as miss-register.

REGISTER CONTROL SYSTEM – A system on a press that is responsible for maintaining the accurate positioning of each color being printed. The system has to be capable of dot-to-dot accuracy between each print unit, print-to-embellishing and print-to-die-cutting/converting, throughout the press run.

Register may be manually or electronically controlled and the system has to be able to allow four adjustments of the print cylinders/unit whilst the press is running i.e. side movement left – right and circumferential advance and retard.

Manual systems are totally dependent on the operator to make all the necessary adjustments, but with automatic register control, the system will read the position on each printed colour and automatically make any adjustments that may be required.

RELATIVE HUMIDITY – Relative Humidity (RH) is the amount of moisture in the air compared with the amount of moisture needed to saturate the air at the same temperature, with the result expressed as a percentage.

Changes in RH in a printing plant may cause printing paper to take up or lose moisture, so causing the paper to curl, wrinkle or change size, leading to printing or register problems. Relative humidity can also affect sheets of die-cut or butt-cut self-adhesive labels converted from rolls of labelstocks for subsequent use in laser and ink-jet printers. Excellent protection from changing humidity is afforded by the use of LDPE packaging in the form of sealed wrap, bags or stack covers.

RELEASE – The act of freeing or separating a pressure-sensitive face material from its backing; the degree of ease with which the print carrier can be separated from the silicone backing paper. Determined primarily by release additives in the silicone coating, adhesive and silicone coat weights plus, the strength of the adhesive.

Most paper and film labels feature a relatively light release, especially those intended for application in automatic labeling machines. Terms used to describe different release levels include:

- Low forces = low, easy, free.
- High forces = tight, hard, high.
- Force oscillation = noisy, zippy, raspy.

May also refer to the force required to separate a pressure-sensitive label or tape from its release liner under specified conditions of peel angle and speed.

RELEASE AGENT – A material used to facilitate the removal of molded parts from a mold, and which may cause labeling problems when applying labels to molded components, e.g. to glass bottles or plastic containers.

RELEASE COATING – The release liner treatment material – usually silicone – that is coated onto the backing paper or film and which enables pressure-sensitive face materials or labels to be readily removed from the protective release liner prior to application. *See Silicone release coatings.*

RELEASE FORCE – The force required to separate a pressure-sensitive adhesive coated material from its backing or protective sheet at an angle of 180 degrees. High speed release, as described in, say, FINAT Test Method FTM 4, simulates die-cutting and waste stripping label materials at jaw separation rates between 32 and 984 feet per minute (10 and 300 meters per minute).

Slow speed release testing simulates label dispensing and application *(see FINAT Test Method FTM 3).* Laminate is peeled apart at 300 millimeters (12 inches) per minute and the separation force measured.

RELEASE LINER – The release liner or backing is that component of the pressure-sensitive laminate construction which functions as a protective carrier for the pressure-sensitive adhesive. It is separated from the adhesive and the face stock prior to the label being applied to a container, pack or other surface.

RELEASE LINER FILMS – Although the majority of release liners are still paper-based, film liners have been showing good growth in recent years. One reason for this late development is the perceived high-cost of film compared to paper, although this cost differential depends on which film is being considered.

Another reason is the difficulty of processing heat-sensitive films through the high temperature cure ovens of a silicone coating line.

The development of low temperature (70ºC or 158°F) thermal cure silicones, and especially the room temperature cure UV and EB cure systems, now enables thermoplastic films to be coated with silicones without difficulty. Solvent less reducer silicones can be coated and cured on very compact, short web path equipment at high speed. Filmic release liners offer the converter and end-user some unique properties, which conventional paper backings cannot match, namely:

- A smooth, maximum hold-out surface
- Uniform caliper
- Excellent dimensional stability

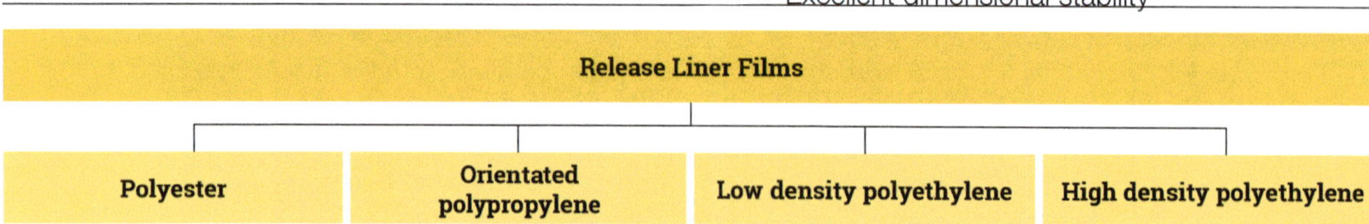

Main types of release liner films

- Transparency
- Gloss
- High tensile and tear strength
- Impervious to damp or highly humid application environments that affect paper backings
- Resistance to die-cutting penetration of the backing that causes web breaks
- Potential for easier recycling of plastic than paper backings
- High internal delaminating resistance
- Tailored flexibility or stiffness
- Low compressibility.

When slitting self-adhesive materials with a synthetic backing it is very important to ensure a clean cut edge, for if reel edges are even slightly ragged, die-cut label reels will tear causing web breaks on the label applicator. For this reason scissor cutting is preferred to crush cutting, when slitting these reels. A number of different films are now used as release liner substrates including:

Polyester (PET) films in the 12 – 125 micron (0.5 mil – 5.00 mil) thickness range, which offer dimensional stability, low heat shrinkage, maximum tensile strength at minimum gauge, and outstanding clarity and smoothness. They are ideal as release liners for 'no-label-look' self-adhesive labels.

Orientated polypropylene (OPP) films are widely used in the release layer of repositionable fastening tape on diapers, and are becoming a strong contender in the label release sector. OPP films in the 30 – 75 (1.20 mil – 3.00 mil) micron range provide good conformability, excellent dimensional stability, good tensile strength and clarity, at a lower cost.

Low density polyethylene (LDPE) film liners in the 25 – 250 micron (1.00 mil – 9.90 mil) thickness range offer low cost, excellent flexibility and conformability, elongation and tear resistance, have poor clarity, low temperature resistance and tensile strength, and relatively poor calliper control.

High density polyethylene (HDPE) film is slightly stiffer, provides good conformability and dimensional stability, with good tensile strength and delaminating resistance. HDPE liners are used in self-wound tapes, medical applications and for self-seal bags and pouches. *See also Release liner papers.*

RELEASE LINER PAPERS – Paper liner substrates have dominated the release liner market from the early days of self-adhesive laminate development, and will continue to do so for the foreseeable future. Within the market, glassine calendared krafts account for around two-thirds of demand, clay-coated around one quarter, and polymer coated papers around 10 per cent. Filmic liners are still a relatively small, albeit faster growing sector.

Calendared krafts. Release papers based on glassine calendared krafts are used in practically all roll label self-adhesive applications. Suitable papers have to be smooth and dense and are made by heavy calendaring and by using suitable sizes. Refining of the pulp raw materials is closely controlled to provide adequate strength for all label conversion activities, plus optimum transparency and infra-red transmission allow the gap between labels to be scanned when labels are being automatically applied or dispensed.

The reasons why glassine calendared krafts maintain their position is their attractive property spectrum, especially for rolls: smoothness, density, even calliper, translucency, silicone hold-out, mechanical strength and adequate dimensional stability for roll applications.

Clay-coated papers. These have become well-established over the past twenty-five years, particularly for sheet labels and heavyweight vinyls and decals. They have good lay flat characteristics, but their lack of transparency excludes them from many roll label applications. By clay coating the surface that will receive a silicone release coating, laminators stop unwanted penetration of expensive

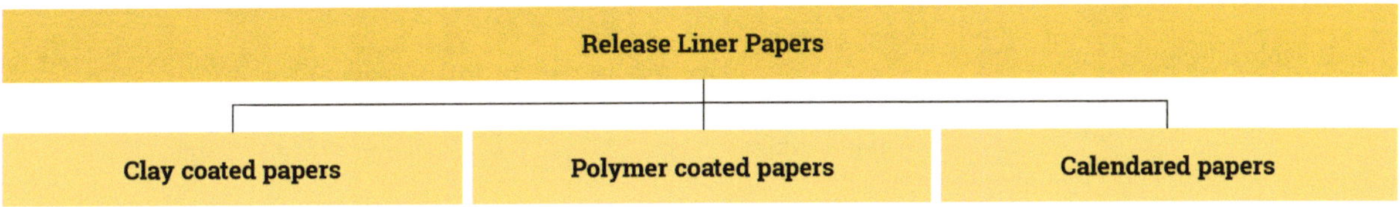

Main types of release liner papers

silicone into the backing paper, ensure a uniform coating pattern of around 1g/m2 and a marginal cost saving over glassine, at comparable paper thickness.

Clay coated papers have much greater tear resistance, but inferior tensile strength, as compared to comparable glassine backing papers.

Polymer-coated papers are a further important base material for release liners. They combine the strength of long fibered papers such as MG krafts with the smooth, non-absorbent surface of plastic film. However, such papers are significantly more expensive than glassines, calendared krafts or clay coated papers. They are therefore used primarily in speciality applications, such as self-adhesive sheet labels, document pouches, building and insulation products, covering films, tapes, etc.

RELEASE LINERS, NON-LABEL APPLICATIONS – Label-related applications account for around three-quarters of the total market for release liners, yet the non-label related end-uses represent a major and important market segment that cover a very wide spectrum of applications and can be broadly divided into nine distinct groups:

Graphic arts markets. These include decals and stickers using clear films, vinyl's and metalized films; truck and fleet marking for advertising or identification mainly using cast or calendared vinyl; name plate applications including appliance marking, but also the growing back-lit instrument panelling and membrane switch markets; self-adhesive reflective sheeting for road signs and fleet graphics.

Automotive and boats. Liners are used with laminates comprising adhesive-backed carpet panels and sound-deadening matting and foam used for mounting trim moldings.

Tapes. One-side and especially two-side coated tapes comprise an important market for release liners. Also includes transfer tapes used for bonding and joining.

Sealants and caulks. Sealants supplied on release liners are used for sealing windows to improve thermal insulation, and for sound and weatherproofing in the automotive and other industries.

Medical, surgical and sanpro markets. Release liner applications include bandages and wound dressing materials, electrodes to monitor heart and brain functions, colostomy bags, transferral drug delivery patches and surgical incise drapes that protect surgical entry wounds from infection. The sanitary protection sector includes feminine hygiene products, baby diapers and adult incontinence pads.

Building and construction. Applications for release liners in this sector include solar screens, bit mastic waterproof membranes, some floor tiles, wall coverings and shelf-liners, as well as pressure-sensitive adhesive-coated abrasive sanding discs and foam draught excluder strips for doors and windows.

Casting and interleaving. Covers release liners used in the processing of cured and uncured rubber and the high pressure laminating of Formica-type kitchen tops.

Fabric and apparel applications. Although a relatively small market for release liners, it comprises foam inserts, supports and items used in shoe manufacturing, as well as iron-on decals and screen printed apparel transfers.

Composite and pre-peg. The processing of carbon fiber by the aerospace, military and commercial aircraft and sporting goods industries offers a small, but very high added-value market for release liners. Examples include helicopter rotor blades, yacht masts and tennis racquets.

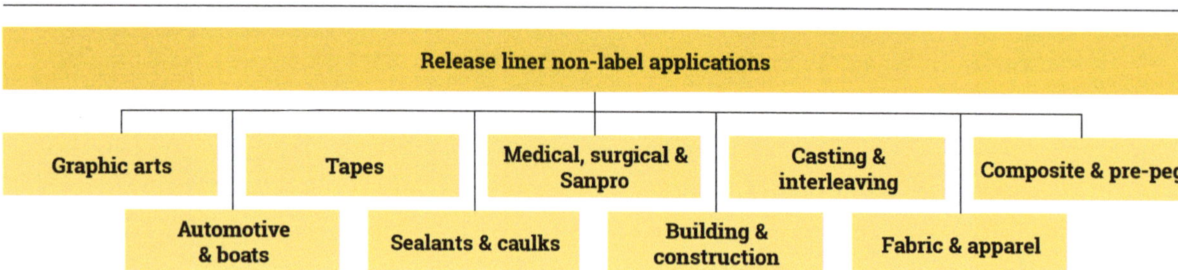

Use of release liners in non-label applications.

RELIEF PRINTING – Printing in which the image, ink-carrying area on the printing plate is raised above that of the non-image area. Letterpress and flexography are both relief-printing processes. *See also Label printing and production, Letterpress printing and Flexography.*

REMOTE DIGITAL PROOFING – May also be called remote proofing, digital proofing or just soft proofing, and can refer to either viewing a proof on a screen (soft proof) or to a hard copy proof viewed remotely in a customer's office.

There are two categories of soft proofing:

Color-critical in which a calibrated monitor that accurately mimics CMYK output on-screen is required for accurate color reproduction.

Review of content only in which a simple PDF on any screen is suitable.

In hard copy remote digital proofing the proofing equipment is placed in the customer's office. Mainly used when the printer and client require a lot of proofs that require color accuracy. It takes time and money to set up but saves the time and costs involved in sending hard copy proofs by courier.

Remote proofs can be output at the client's site and give consistent, accurate color. No prior color management knowledge or expertise is required by the client as the system is all controlled by the printer.

The major benefit of remote digital proofing when compared to more conventional proofing methods is the time saving achieved at the proofing stage, in the proof correction and approval process. *See also Remote proofing.*

REMOTE PROOFING – The act of viewing a proof on a monitor or a digital proofing device which is calibrated by software to a 'master' image, file or proofing device located at a different site. May also refer to a secure internet site where the same master image can be viewed by stake holders at different geographical locations. *See also Remote digital proofing.*

REMOVABLE ADHESIVE – A pressure-sensitive adhesive which, when it has been applied to a specified surface, may be easily removed at a later date. A pressure-sensitive adhesive displaying relatively low final adhesion.

Although it adheres firmly to the substrate, it can be removed again for quite a long period of time. Ease of removal depends on the nature of the substrate, the ambient conditions at the time of application and the environmental conditions during the service life of the label.

Ease of removal varies with the quality of the label material. Clearly, a strong label material will not tear readily and will aid label removal. Some removable adhesive labels have an adhesive key coating to which the adhesive coating will have a greater affinity than the surfaces it is designed to remove from.

Lowering the coat weight of an adhesive can improve the performance of a removable label, as can directly coating the label material with adhesive, rather than the preferred transfer coating method.

Most removable adhesives become more permanent with the passage of time and storage. Environmental conditions that the labeled product has to contend with during its lifetime will affect this deterioration. *See also Adhesive, Adhesives, pressure-sensitive and Pressure-sensitive adhesives.*

REMOVABLE BLADED TOOL (SHEETER) – A rotary cutting tool, based on a steel body, which has an even number of channels machined around its circumference. Perforating or cutting rule can be fixed into a balanced number of these channels, so giving the tool a degree of variable or flexible options.

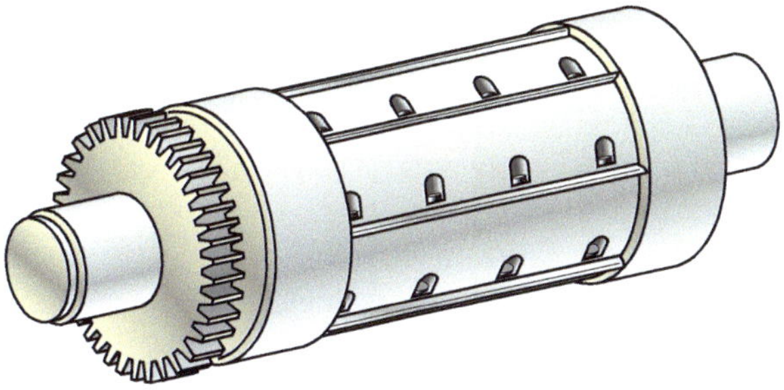

Removable bladed cutting tool for butt cutting, sheeting or perforating.

RENEWABLE – A process is renewable if you can do it more than once

RENEWABLE ENERGY – Energy that is naturally sourced and used to displace fossil fuels, such as the sun's light, wind, rain, tides and geothermal heat. It can also be energy created from waste.

REPEAT LENGTH – The circumference/printing length of a plate cylinder onto which the printing plates are mounted. Repeat length is determined by one revolution of the plate cylinder gear.

REPOSITIONABLE ADHESIVE – Repositionable pressure-sensitive adhesives are constructed with an adhesive coating that enables a label to be removed and then repositioned within a short, finite, period of time after application. Beyond that time, the adhesive bond starts to become permanent, except in those cases where a removable adhesive has been used. Because of the variety of surfaces that are labeled, there are several means to achieve label repositionability, depending on the label volume and whether the eventual bond needs to be fiber-tearing permanent, or an adequate adhesion that could be removed.

For small volumes, a removable adhesive can be ideal, providing that the end user is happy that the label is not designed to be permanent. Some removable adhesives become more permanent on some surfaces quite quickly anyway.

Lower coat weight permanent adhesives can be used, as can standard coat weight, moisture tolerant, freezer grades that are semi-permanent at comfortable room temperatures.

Harder acrylic permanent grades usually have a working temperature range down to +5ºC (41ºF), but at this extreme would not be as tacky as at +20ºC (68ºF), so can often be repositioned on surfaces with a temperature of +5º (41ºF) to 10ºC (50ºF). Permanence is achieved with time, or if the temperature of the labeled product subsequently increases.

With all repositionable label requirements, the surface type and shape, the labeling surface and environmental conditions, plus the length of time that repositionability is needed, must be considered and trials in the actual labeling area, as with all self-adhesive label applications, are vital.

RESHARPENING/REFURBISHING/REPAIR – Under normal circumstances, rotary dies, anvil rollers and magnetic cylinders can be repaired if they are damaged. Rotary dies have to be cylindrically ground to the right specifications, sharpened (machine-sharpened / manual sharpened) and tested in the right material.

Anvil rollers also will have to be cylindrically groundwh

Magnetic cylinders have separate bearers that have to be replaced. Sometimes, even the body of the magnetic cylinder has to be repaired . In all cases, the total magnetic cylinder (bearer & body) will have to be cylindrically ground to the right specification.

RESIDUE – Usually refers to an adhesive label that has been removed, but leaves a portion of the adhesive behind on the previously labeled surface. The adhesive label does not remove cleanly.

RESISTIVETY – The ability of certain materials to resist the passage of an electrical current, either through its surface or the bulk or body of the material. The unit of volume used to measure resistivety is the ohm-cm.

RESOLUTION – The crispness or sharpness of an image or typeface, either on a monitor or an output device. A measure of the 'fineness' of an image that can be reproduced by a given system. The scale of the resolution depends on the device being measured. Scans are measured in samples per centimeter or per inch, or pixels per inch/centimeter. Halftone screens can be measured in lines per inch (lpi) or lines per centimeter (lpc). In all cases, the higher the resolution, the higher the quality and more detailed the image.

RESOLUTION, INKJET – *See under Inkjet resolution.*

RETARDERS – Refers to materials , usually combinations of liquids, solvents, extenders, etc, that are added to an adhesive, ink or coating to slow down the drying rate of the material.

RE-USABLE EAS TAGS – An electronic article surveillance (EAS) tag intended to be removed from the merchandise. Such tags are most commonly used for soft products, e.g. clothing, soft furnishings.

RE-USABLE ELECTRONIC TAGS – RFID transponders which can be re-used to tag completely different objects simply by erasing the corresponding message and storing a new one that relates to the newly tagged object.

The memory chip of the electronic tag or label only stores a unique number, which is all that is needed to identify the tagged object, and communicates this number to the reader.

REVERSE AUCTION – A reverse auction is a B2B Internet-led auction in which prices are bid down, rather than up as with a traditional auction. Also, unlike normal bidding auctions, it is the buyer that initiates a reverse auction, not the seller. Quite simply, the buyer details his product requirement and the maximum price he is prepared to pay for it; suppliers of that product then post their bids in a reverse manner or descending price order.

At the close of the auction the buyer accepts one or more of the listed bids (not necessarily the lowest bid) and, at that point, the auction is stopped and the transaction is considered complete. The reverse auction process can often lead to buyers saving some 10 to 30 per cent on their purchasing.

Perhaps unsurprisingly, reverse auctions tend to work better for buyers than for sellers. Buyers love reverse auctions; suppliers usually hate them. Certainly many label converters claim to have had bad experiences of the reverse auction process. The process frequently alienates suppliers, can undermine the trust needed to manage complex products and services, and may end up reducing the number of surviving suppliers. Indeed, the view of some label trade associations is that reverse auctions are counter-productive to the procurement of labels.

Unfortunately, many label converters who have taken part in reverse auctions feel they have done so under duress, particularly in those cases where an existing customer holds a reverse auction on a job the converter is already producing or has produced in the recent past. Converters with excess capacity may also participate in a reverse auction to cover costs or make very little profit rather than having machines standing idle.

Reverse auctions can of course also be the wrong tool for a buyer to use, and do not always work successfully – particularly if the volume of purchases is too small, or it costs more to operate the auction than the potential savings that can be achieved. Also, buyers will not get useful bids unless they prepare their Request For Quotations (RFQ) well, detailing carefully the required products or services, quality, quantity, service levels, method of delivery and payment terms. This should be sent to potential bidders well in advance of the auction.

Reverse auctions probably work best for buyers in markets where there is over-capacity, and least well where there is insufficient competition amongst suppliers. They are also less effective in situations where the cost of switching suppliers is too high. Low-cost suppliers may be more inclined to welcome reverse auctions as they play to their competitive strengths and they do not need to include a margin for marketing or sales personnel.

To achieve the best results with reverse auctions, buyers should preferably pre-qualify the participating suppliers; train them in the concept of reverse auctions and how they should participate; only award business after a quality/performance assessment of the bid winner; define the rules and ethics of how the auction will operate – and make sure that everybody understands them.

REVERSE ANGLE DOCTOR BLADE – *See Doctor blade.*

REVERSE OUT – An image that has been reversed by photographic or computer processes. A solid image for example, may be reversed from black on a white background to a white image on a black background.

REVERSE PRINTING – Printing on to the underside of a transparent film, with the printed image then read from the front side of the film. The printing plate(s) need to have the image areas reversed so that the image can be viewed the correct way around from the front side.

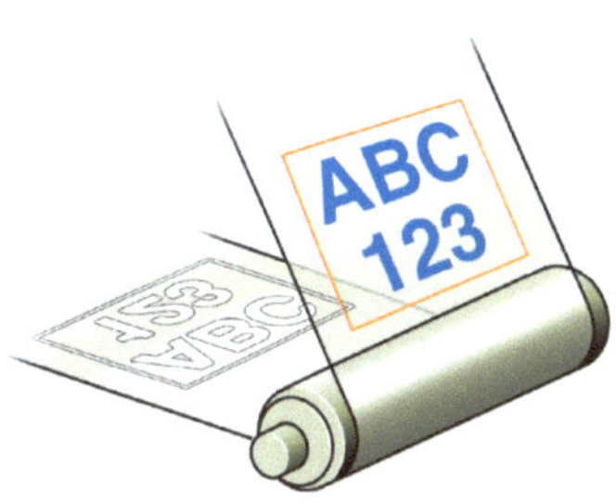

Printing on the underside of a transparent film, with the printed image then read from the front

REVERSE ROLLER COATING – The coat weight of adhesive or silicone can be metered between two rollers, moving in opposite directions. The metered coating on the second roller is then transferred to the paper or film at a nip with a third roller which moves the coated material forward, for drying or curing. The amount of coating transferred is controlled by the gap between the metering and application rollers, as well as by the relative speed of rotation between the same two rollers.

Reverse roller coating is also used for some paper coatings.

REVERSIBLE INKS – Most often used in beverage packaging, reversible inks have a temperature range between -20°C (-4°F) and 50°C (122°F). With cool drinks, thermochromic reversible inks have been developed which, for example, change from white to blue and reveal the words 'ready to serve' when the drink is at the appropriate temperature. When the product warms up, the message will disappear.

REWINDING AND INSPECTION – The operation of winding printed die-cut labels on a press roll through a machine that examines and inspects them for defects, web breakages or other faults, slits them to the required widths for subsequent application, counts the number of labels on each roll and rewinds the various inspected, slit, counted labels to the required roll tension and diameter in rolls ready for end-user application. Any out of specification or damaged labels can be removed during this operation. Where there are web breakages or sections of damaged or faulty labels removed, the web can be spliced before rewinding. *See also Label inspection and Splice.*

REWIND SHAFT – The take-up spindle, mandrel or shaft on a roll-label label printing and converting line.

RFID INVENTORY TRACKING – enables brand owners and retailers to interact electronically with suppliers, generate forecasts, control inventory costs, share information and transmit information with and between stores.

RFID ROLL LABELS – The incorporation of RFID transponders – consisting of a chip and antenna on a thin polyester film – into a traditional roll label construction. Transponders embedded inside roll label stock help solve problems in product identification, control tracking and security and can be used in a wide variety of applications. They can be delivered in roll format or die-cut to size.

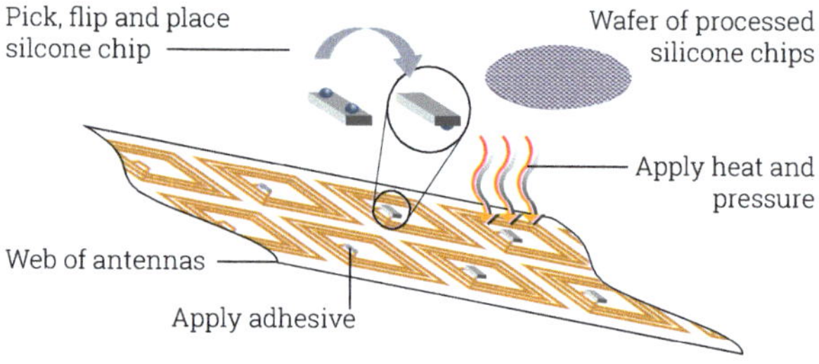

Construction of RFID label, based on illustration by Avery Dennison.

RFID roll labels are designed for mass production using a roll-to-roll process.

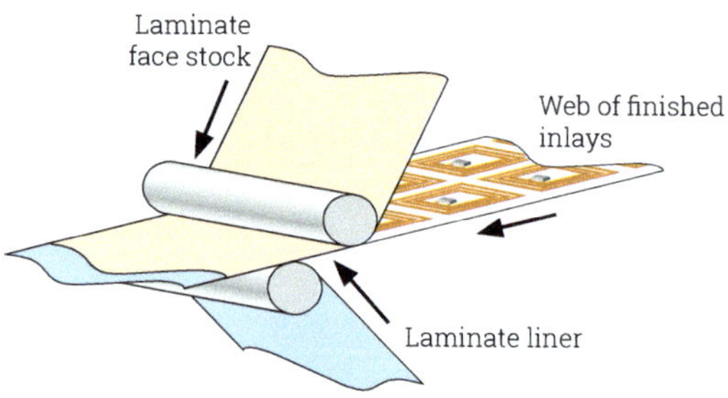

Converting a web of finished inlays into a label laminate structure based on illustration by Avery Dennison.

RFID TAG – A tag containing a microchip attached to an antenna that is able to pick up signals from and send signals to a reader. The tag contains a unique serial number and may also contain other information, for example, a customer's account number. Tag readers use the message they receive from the tag in accordance with their programmer's instructions.

Generally, these readers must be no more than 2 metres (6 feet) or so away from the tags to be read. Unlike barcodes, RFID tags do not require line-of-sight reading, so a pallet load of RFID tagged goods can be read in a second or so.

Rather than a generic code, which relates to a specific product line – as with barcodes – each RFID tag has its own specific code. This offers retailers the opportunity of tracing and recalling certain batches of goods very quickly in the case of safety or quality concerns. However, at the current time, RFID tags are too expensive to use with low-cost items, although this is expected to change with the forecast volume usage of smart labels and tags over the next 5-10 years.

RFID tags have the potential to offer significant benefits for brand owners, retailers, logistics and manufacturing businesses. In particular:

- Tracking batches of products throughout the whole logistics/distribution supply chain
- Minimizing inventory and stock levels
- Reducing scanning time at checkouts as full trolleys are read instantly
- Preventing shoplifting and other forms of store shrinkage
- The development of 'smart shelf' technology to identify movement of goods from shelves, so triggering shelf replenishment
- Airport luggage handling & security improvements
- Location of assets
- Event management and access control
- Remote servicing indication and failure prediction for equipment

RFID tags come in many different formats, including smart labels, swing tickets, smart cards, key-chain wands, toll boxes, etc. RFID tags are available in active, passive and semi-passive forms. *See also Active tags; Passive tags.*

RGB – The acronym or abbreviation used for the primary colors of light (red, green, blue) – the additive primary colors) that are used to form all other colors in computer monitors and photographic reproduction. All color monitors use red, green and blue input signals to create the colors that the operator sees on the screen. If the images are to be printed the values must pass through an RGB to CMYK color transformation. *See Additive color.*

RHEOLOGY – A branch of physics in which the flow and deformation of liquids is studied. This includes the study of printing inks, which are usually described in terms of consistency, tack, flow, length, stiffness or viscosity.

Differences in flow speeds of inks or varnishes can be attributed to friction in the particular liquid. The level of friction is in turn determined by the viscosity.

RIP – Raster Image Processor. A RIP is the link between the computer and the printer. Software in the RIP converts the image (rasterizes it) into an electronic bitmap, which indicates every spot position that the printer can use to print in various resolutions, which may be from 400 up to 2,540 dots per inch (dpi) or from 155 up to 1,000 dots per centimeter (dpc). The laser output device records the visual information received from the RIP onto film or plate material. The larger and more complex the files, the longer they will take to process through the RIP.

ROCKWELL SCALE – The Rockwell scale is a hardness based on the indentation hardness of a material. The Rockwell test determines the hardness by measuring the depth of penetration of an indenter under a large load compared to the penetration made by a preload. There are different scales, denoted by a single letter, that use different loads or indenters. The result is a dimensionless number noted as HRA, where A is the scale letter

ROCKWELL HARDNESS – A method of determining the surface (or 'Shore') hardness in degrees of a substance, such as an inking roller or rubber printing plate.

ROLL COATING – Any method of coating a web of paper in which the coating is applied directly to the paper surface by transfer from an applicator roll, which carries the coating material on its surface. The applicator roll may rotate in the same direction as the paper web or in the reverse direction (reverse roll coating).

ROLL-LABEL – Pressure-sensitive labels that are printed and converted in a continuous roll form for subsequent automatic application. *See also Rewinding and inspection.*

ROLL-ON, SHRINK-ON LABELS – Roll-On, Shrink-On (ROSO) labels were developed for use on traditional reel-fed wrap-around label applicators, but make use of a shrinkable BOPP film rather than paper, so providing a shrink capability which is ideal for the decoration of contored bottles or cans. Rather than being formed into a tube prior to application – as with shrink sleeve labeling – ROSO film labels are wrapped around the container and sealed with hot-melt adhesive. The container then passes through a shrink tunnel which shrinks the film to the precise contor of the container. Heat in the shrink tunnel is only applied to the areas where shrinkage is required.

The shrink capability of ROSO films is not as high as shrink sleeve films, being typically in the range of 18 – 24 percentage shrinkage. With clear films the labels can be reverse-side printed. White films provide an alternative to paper labels, offering improved gloss, excellent dispensing and better scuff and humidity resistance. If a design or copy falls within the shrink areas then allowance needs to be made in the design and printing for distortion.

ROSO labels are used for the decoration of such products as soft drinks, aerosols, drinkable yoghurts and flavored milks.

ROLL-TO-ROLL LABELS – Labels that are produced by unwinding a roll of labelstocks, printing and converting the unwound web, and then re-winding the printed and die-cut web into a roll at the end of the press. May also be referred to as reel-to-reel.

ROLL-TO-SHEET LABELS – Labels that are produced by unwinding a roll of labelstocks, printing and converting the unwound web and then cutting the printed and die-cut web into sheets as it exits the press. Laser, copier and ink jet labels used with desk-top printer systems are sheeted labels.

ROM – Read Only Memory. Memory permanently loaded during the manufacturing process and programmed into the chip. A computer can use the data or instruction in the ROM, but cannot alter the data by writing or switching off.

ROOM – Rip Once Output Many. A pre-press system in which the master file is Ripped once, then the same data is used to drive multiple output devices, such as digital proofers, plotters and CTP devices. Prevents the possible introduction of data anomalies from multiple, device-dependent RIPS.

ROSETTE – A rosette pattern that is created when all four-color halftone screens are placed at the traditional prescribe angles to each other. *See also Screen angles.*

How the rosette pattern is built up in four-color process printing.

ROSO – *See Roll-On, Shrink-On labels.*

ROTARY PRESS – A press that normally prints and converts label material from a roll and then rewinds it on a roll. Printing may be by any printing process or combination of processes. The web passage and printing units in a rotary press may be arranged in a variety of ways – in-line, common impression or stack. *See also roll-to-roll labels.*

ROTARY PRINTING – Printing operation that uses a cylindrical impression cylinder and a cylindrical printing plate (or sleeve). The web or sheet of labelstocks passes between the two cylinders to be printed. *See also Rotary press.*

ROTARY SCREEN PRINTING – *See Screen process printing.*

ROTARY TOOLING – The tooling used on rotary label presses to cross-cut, cross-perforate, punch holes or die-cut a web of pressure-sensitive material so as to produce finished cut-to-size, cut-to-shape, sprocket-punched or cross-perforated labels.

Rotary punching tools incorporate pierced hollow punches around the periphery of the tool to punch out holes. An air supply connected to the cutter blows air through the cavity and ejects the punched-out waste in a particular direction. A vacuum completes the action, ensuring that the waste is carried to a collection box close to the press. With all rotary punching it is important to ensure that the waste is under control at all times.

Rotary die-cutting tools are available in different formats: those in which the cutting edge is engraved from a bar of solid steel; those which are machined

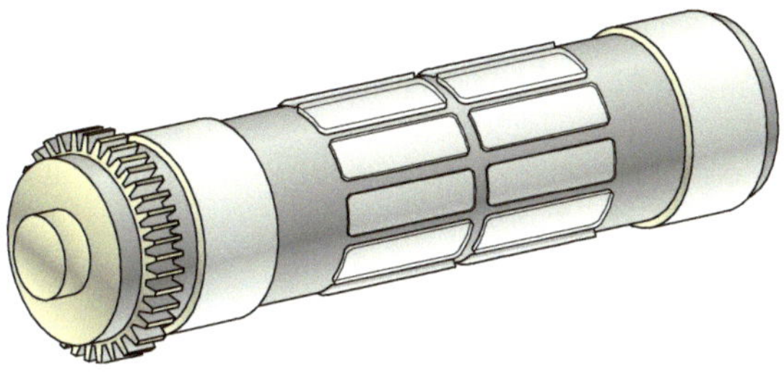

Standard cutter for producing pressure-sensitive labels.

from a steel bar by electronic discharge; or those produced in the form of thin steel plates that are mounted for use by wrapping the thin steel around a magnetic cylinder.

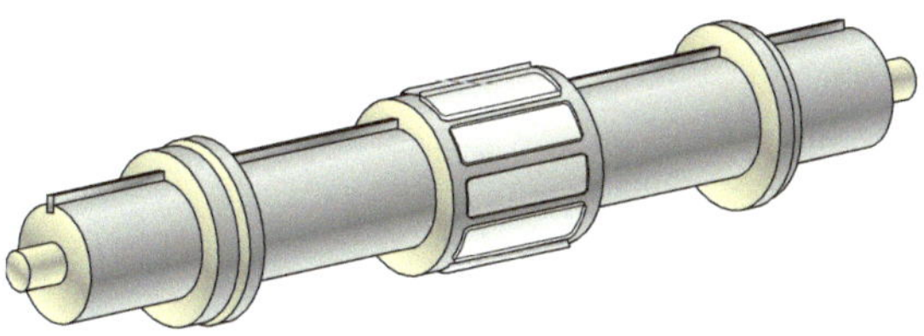

Slide-on cutter assembled on a parallel shaft.

Cross-cutting. With cross-cutting or cross-perforating rotary tools, the cutting or perforating blades are mounted individually and held in place around a special cylinder.

See also Die-cutting and Removable bladed tool.

ROTOGRAVURE PRINTING – A printing process that uses print cylinders in which the image areas have been etched or engraved into the metal surface. The recessed image areas of the cylinder hold the thin, fluid, ink used in the process. As the inked printing cylinder comes into contact with the label substrate, a rubber roller provides pressure to transfer the ink to that substrate. Electrostatic means may be used to aid the transfer of the ink to the substrate. Rotogravure is an intaglio process. Also known as gravure printing. *See also Gravure printing and Label printing and production.*

RTF – Rich Text Format. A universal means of transferring text between different software programs developed by Microsoft.

RUBBER-BASED ADHESIVE – A pressure-sensitive adhesive that is based on natural or synthetic rubber. The adhesive may be coated as a hot-melt, solvent or emulsion. *See also Adhesive, pressure-sensitive.*

RUBBER-RESIN ADHESIVES – These are blends of elastomers and tackifying resins, which provide a wide range of pressure-sensitive adhesives for different applications.

Rubbers range from natural, through styrene butadiene to styrene-butadiene-styrene (SBS) and styrene-isoprene-styrene (SIS) block copolymers. Rubber-based pressure-sensitive adhesives can be formulated as solvent-based, emulsion or hot-melt systems.

RUB RESISTANCE – Testing of label and packaging inks for rub resistance can be carried out using a number of different rub resistance or rub tests. The principle of them all is that a printed sample is placed face down onto an unprinted sample of the same substrate and then the two surfaces rubbed together under controlled pressure for an agreed number of cycles. The degree of transfer of color from the printed to the unprinted sample is inversely related to the rub resistance.

Rub resistance testing is mostly used for printed paper samples rather than for flexible films.

RULE CUTTERS – Rule cutters are used for flat die-cutting in presses where the web is momentarily halted during the actual cutting process. Consequently, the output speed is somewhat slower than when rotary tooling is used.

In its most simple form, the flat ruled cutter may be produced by working to dimensions, obviating the need for a drawing when the shape is a simple

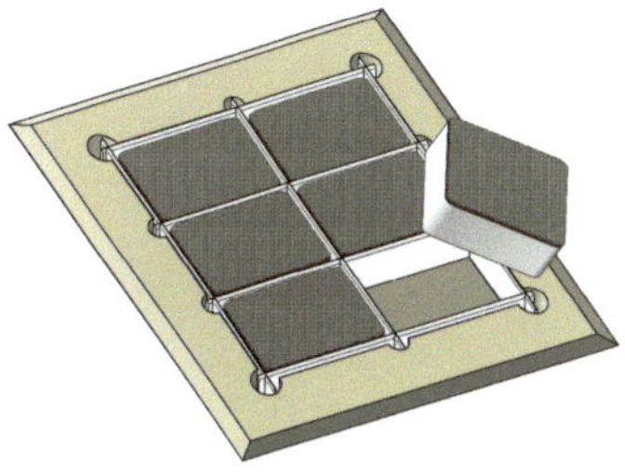

Construction of a flat ruled cutting die showing the cutting rule bent to shape, the ejection rubber and the assembled die.

square, rectangle or circle. For more complex cutting shapes a key line can be drawn by hand or created by some form of electronic origination at the same time as the label design is created.

To make a flat ruled cutting die, the desired label shape is cut in a base by means of a jigsaw, a vertical miller or a laser beam. Cutting rule is then cut and bent to fit the cut-out label shape and inserted into the base. Wherever practical, for facilitating removal of the waste, rule joints should be positioned so that they run with the web direction and not across the web.

Ejection rubber is fitted to ensure the die-cut web separates quickly and cleanly from the cutter. The amount of rubber required will vary according to the complexity of the label shape.

RUPTURE – The result of physical stress when an adhesive film is broken.

S

SACHET ENVELOPES – Used to contain scented herbs, flowers or other fragrances, potpourri sachet envelopes are also called wardrobe sachets.

SACHETS – Sachets are a popular method of packaging single-use quantities of a wide range of food, medical and pharmaceutical, industrial, consumer or toiletries products. They are commonly opened by making a small tear in part of the package and then squeezing out the contents. Depending on the product contained in the sachet, other methods of opening and removing the contents may be used.

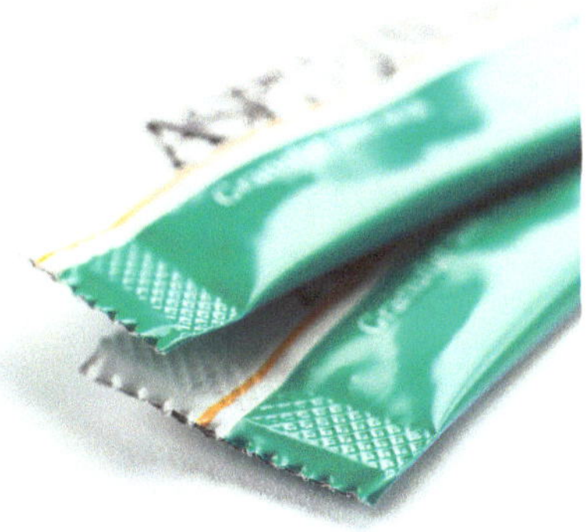

Sachets. *Source: Constantia Flexibles.*

Key end-use applications for sachets today include:

Foods – ketchup, mayonnaise, salad cream, sauce, vinegar, mustard, relish, tartar sauce, sugar.

Medical and pharmaceutical – hand wipes, impregnated wipes, creams, lotions, powders, hygiene products, swab sticks, patches.

Personal care – creams and gels for hair and hand care, mud masks, wet wipes, talcum powder.

Industrial – grease, industrial wipes, high tech surface wipes.

Veterinary – Powders for veterinary use.

Narrow-web printing of sachets, whether flexo or digital, enables product manufacturers to personalize them for branding and promotional purposes in hotels, restaurants, bars and clubs, events, etc., where each sachet provides the consumer with a new, clean and fresh product each time – presented in a convenient, distinctive and affordable portion.

There are a range of substrates available for sachet production, including unsupported films, barrier films and food grade films – depending on requirements. Printed substrates are converted into sachets through the use of form, fill and seal machines. *See also Narrow-web packaging.*

SALMA – The Self Adhesive Manufacturers Association, New Zealand, has a membership of 42 companies which includes approximately 90 per cent of the country's label industry.

SANDWICH CONSTRUCTION – Whilst normal self-adhesive label laminates are of two plies, so called sandwich, or multi-ply, constructions have developed, most commonly of three plies with two adhesive coatings.

Different face and middle-ply materials are utilised, which are die-cut down to the third ply. This rear part is a backing from which the first and second plies can be automatically applied onto a moving web of business forms, or an industrial product, as a 'piggy-back label'.

When applied, the second ply, which can be self-contained carbonless paper, translucent glassine or clear film, has an adhesive coating which adheres the piggy-back label to the labeling surface, leaving the top ply as a peel-off label for subsequent use by the end user.

Innovative sandwich constructions incorporating adhesive and silicone strip coating, plus different performing adhesives and dry-peel plies, have also been produced. A non-label hairdresser's application was reported as utilising seven plies, with six adhesive layers.

SATURATION (COLOR) – Color is measured in terms of hue, saturation and luminance. Hue is the predominant wavelength of a color which distinguishes it from other colors, such as red, blue or green; saturation is the color intensity (from white to pure color); and luminance is the brightness of the color (from light to dark).

Saturation can be defined as the amount of primary color applied on a specific area as a percentage of the maximum that could be applied to that area.

SATURATION (IMPREGNATION) – The addition of

various materials, such as latex or rubber, to a face material so as to improve its physical properties.

SCALABLE FONT – A font that can be resized on the screen, or when printing, to any point size without it altering its appearance or becoming distorted. PostScript and TrueType fonts are scalable.

SCALING – Changing the size of an image by its dimensions in both the X and Y directions. In the label industry this may be known as dispro. *See Dispro.*

SCAN – To search for a symbol that is to be electronically recognized; a search for marks that can be recognized by the recognition unit in an optical scanning device.

SCANNER – An electronic device for scanning a picture of the image, breaking it up into dots (digitizing) and then sending the digital information to the computer. Also relevant to four-color process reproduction.

SCANNER RESOLUTION – This is an important property of scanners, being the number of pixels the scanner can resolve. Resolution is inversely proportional to the speed of the scanning.

SCENTED PRINT – The preservation of fragrances on printed matter using a technique of micro-encapsulation. The scent is released by rubbing the printed surface. In use, fragrance capsules are mixed with an overprint varnish and applied with an additional print unit on the press, so offering an economical solution for printing scents in combination with high quality printing. Being encapsulated, the scent will not evaporate until the capsules are rubbed.

Scented print is used with some direct mail, in some printed promotional material and on some printed perfume samples or labels.

SCORE – A score is an indentation or partial cut in a material, such as carton board, to enable the material to bend, crease or fold in a defined place. Scoring rules can be incorporated into die-cutting forms for carton production. Machines using male and female rollers are also available for scoring materials.

SCOTCH TAPE TEST – A potentially imperfect test for ascertaining ink key on a printed surface. In use, a length of standard Scotch tape is placed on the print and smoothed down with the finger. It is then slowly pulled back for half of its length before being snatched back rapidly for the remainder. Both the print and tape are examined for any evidence of ink transfer. The test is used for flexible films and foils.

Although quick and easy, care must be taken not to leave the tape on the print prior to removal as the tape components can soften the ink film. Whilst the original test, using only Scotch brand tape, ensured that there was some uniformity of tape tack, nowadays there are many self-adhesive tape brands, using different types of adhesive. To achieve a meaningful result consistently, it is vital to use the same brand of tape and one of uniform quality. The test is supplemented by examining the flexibility. This is done by holding the film between the thumb and forefinger of each hand and flexing the film 10 times before examining it for ink removal.

SCRATCH OFF – An opaque coating or ink applied to a specific area on printed matter to hide the details underneath. The hidden details are revealed only when the coating/ink area is removed by scratching or rubbing with a fingernail or coin. Used on direct mail and mailer type documents, labels, tickets, etc, in the form of competitions.

SCREEN ANGLES – The angles at which halftone screens are positioned in relation to one another to avoid moiré patterns. To achieve this, screen angles for each color must vary by 30 degrees. Black is always at 45°. Traditionally, the most commonly used screen angles are in steps of 15 degrees, that is: black – 45°, cyan – 105°, magenta – 75°, and yellow – 90°.

Alternatively, with black always at 45°, the other screens would have to be printed at 15, 75 and 90 degrees. However, the fourth color would then have the same angle as the first, which doesn't work. The solution is then for one color screen to be out of 30 degree variance with the rest – so giving yellow a screen angle of 0 degrees. Yellow being much lighter than the other colors, any moiré it creates is less noticeable.

A rosette pattern is created when all four-color halftone screens are placed at the correct angles to each other.

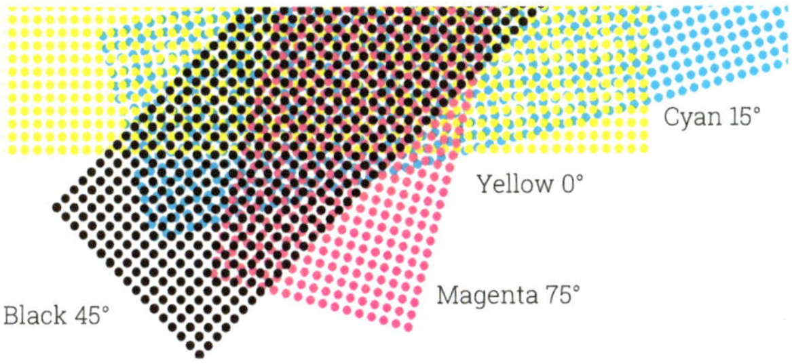

How the screen angles combine to create the rosette pattern in four-color process printing.

SCREEN PRINTING RETURNABLE BOTTLES – Screen printing has been used for point-of-sale, label and industrial printing for many years. In the industrial field, screen printing is used for clock faces, instrument panels, keyboards, signs, etc, as well as for the printing of plastic 'squeeze'; bottles for household products, toiletries, chemicals, agro-chemicals and similar applications.

Screen printing is also used for decorating returnable (or premium brand one-way) glass bottles and, more recently, PET bottles – including Coca Cola and Pepsi bottles. For returnable bottles the bottle is screen printed at the point of manufacture with a durable, scuff resistant, image that will endure numerous round trips, including washing operations prior to re-filling.

Screen printing of glass or plastic bottles is relatively slow at around 6,000 bottles per hour and could not be viably carried out on a bottling line. However, if used for decorating returnable bottles, where there may be up to 20 trips per bottle, the process becomes a viable operation and can compete with labeling. For non-returnable bottles the process would seem to be uneconomic.

By using different thickness gage meshes, mesh spaces and inks it is possible to print ink films thicker than any other process, so creating a tactile feel and image to the decoration. With some stencils it is possible to achieve an embossed feel to the decoration.

Considerable interest and development is now being undertaken with a view to directly printing bottles with inkjet.

SCREEN PROCESS PRINTING – Originally called silk-screen printing, screen process is one of the oldest methods of printing. Today, it is mainly used for the production of self-adhesive labels for cosmetics and toiletries, pharmaceutical and industrial and outdoor applications and is most frequently found in combination in a roll-label press line with letterpress, UV flexo or offset.

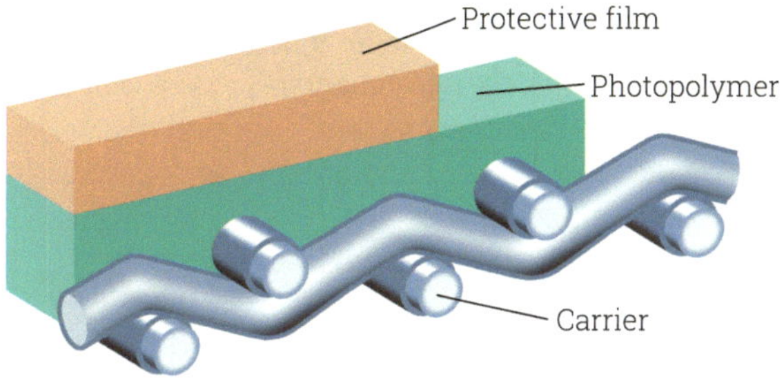

Structure of a screen carrier mesh and photopolymer coating (Illustration courtesy of Gallus Ferd. Rüesch)

The screen process has the ability to print a smooth, controllable lay-down of ink and to provide durable, high quality labels where good ink coverage is required for resistance to weather, moisture, chemicals and abrasion. It can also print a good opaque white image, something which other label printing processes find difficult to achieve.

The image carrier for screen printing is no longer a silk mesh, but today is most often made from a nylon, polyester or fine wire mesh onto which a photo-sensitive coating has been placed. Exposure to light through a film positive of the label image causes the coating to harden in the non-image areas enabling the unexposed (unhardened) areas to be washed away. This leaves a negative image of the label remaining on the screen mesh through which ink can flow.

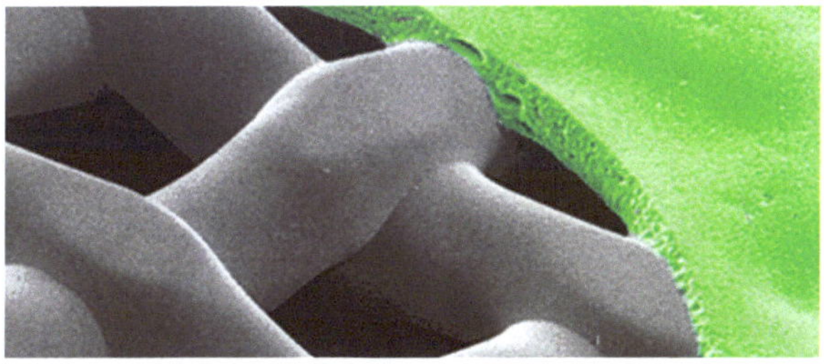

Magnification of exposed and washed-out screen image on a metal mesh (Illustration courtesy of Gallus Ferd. Rüesch)

In the printing operation, the screen is mounted on a frame and placed in position on the screen printing and die-cutting press. Ink is pressed through the open areas of the mesh by a blade called a 'squeegee' to produce a print on the label substrate – which in screen process can be virtually any material. Ink drying is mainly undertaken by UV-curing.

The screen printing process is capable of laying down a very thick film of ink and is the only process that allows a light color to be printed over a dark material with satisfactory results. Very high quality work is possible and fine line or tonal work can be produced with equal success.

Roll-label screen printing presses used for self-adhesive label production either use a flat-screen in a stop and go motion, or use rotary screens in which the squeegee is mounted inside the cylinder. Flat screen printing is a relatively slow process as it requires the web being printed to come to a stop for each print cycle. Although the rather slow, flat style of printing unit gained partial acceptance with label printers, the move towards building presses in which a combination of printing processes, including screen, demanded something more – rotary screen, which is faster, has a continuous motion and works well together with other printing processes.

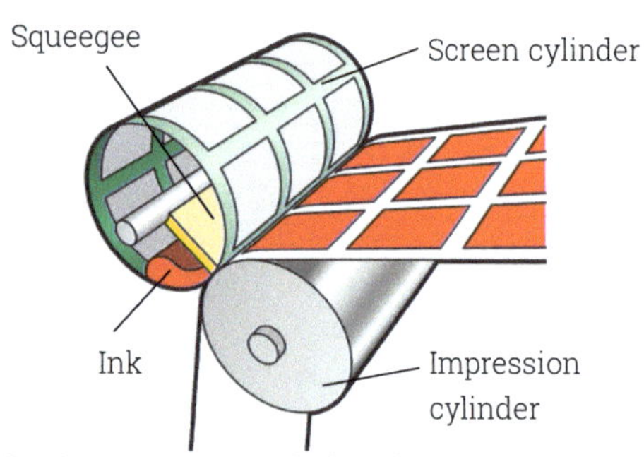

Principle of rotary screen printing showing the ink and squeegee inside the screen cylinder (Illustration courtesy of Gallus Ferd. Rüesch)

The rotary method of screen printing has been used for many years for the printing of fabrics and wallpaper and this technology was developed to create narrow-web rotary screen units. These could be interfaced with letterpress, flexographic, offset, or today even gravure units, for high quality and special effect multi-process printed self-adhesive labels. In the rotary format, both the mesh carrying the photographic image and the label substrate come together as cylinders, which reduces the possible area of contact and sharpens up the image being printed. The supply of ink is retained inside the rotary screen and the squeegee is also located inside the cylinder.

Rotary screen or screen combination narrow-web presses make up around 10 – 15 per cent of all narrow-web roll-label presses installed each year.

SCREEN RING – a support system for screen mesh used in rotary screen printing.

SCREEN RULING – The number of halftone dots per linear inch or centimeter. Also known as screen frequency.

SCROLLING – A technique used to recall information from a VDU display memory. As each line is recalled by scrolling, existing lines on the screen move up or down by one line to make room for the next.

SCUFF – Printed image or substrate damage that occurs through rubbing against a surface under pressure. *See also Rub resistance.*

SEALER – A surface coating designed to prevent the passage of a substance through a label or packaging material. A barrier coating. *See also Barrier coat.*

SECONDARY LABELS – Secondary labels, also described as 'Information' or 'Variable Information' labels. Where primary and secondary labels are adhered to a product, the secondary label is usually smaller and positioned so as not to detract from the primary brand label. They are used to carry information such as a list of ingredients, health or safety requirements, nutritional details, instructions for use, EAN code, warnings, manufacturer or supply name and address or registered office of the manufacturer or packer, prices, volume, weight or quantity and possibly promotional or special offer deals. If appropriate, the label may also contain details of any special storage conditions or conditions of use, and particulars of the place or origin.

Information contained on primary and secondary labels is normally required to be clear, legible and indelible, readily discernible and easy to understand and is likely to have to conform to one or more of the regulations, standards or legislations that relate to labels and to labeling.

With one-piece wrap-around can and bottle labels the primary and secondary elements are all incorporated into the one label, rather than being two separate labels. *See also Primary labels.*

SECU-LABEL™ – An innovative anti-counterfeiting product in which a special self-adhesive label is impregnated with a covert completely invisible mark, thereby creating a unique identification signature. The special security marking inks used (under licence) employ a method of scattering in which an electromagnetic wave is scattered to create the unique 'signature' in the label.

SECURITY DESIGN – An ever-more sophisticated and ingenious range of security design features have been developed and implemented in the world of security printing in recent years, all with the aim of providing deterrents against counterfeiting and forgery. Many of these features have begun to find applications in brand protection, anti-counterfeiting and security labeling.

Originally rooted in the traditional security printing sector where skilled hand and/or machine engravers and origination artists created sophisticated background designs, security designs for documents, tickets, tags, packaging and labels. Techniques advanced significantly with the introduction of sophisticated security design software and computerized origination systems.

When these complex computer-created security design features are used in conjunction with specialized security papers with watermarks or metalic threads, security inks, holograms, etc, the task of the counterfeiter is made significantly more difficult. Such computer-generated security design elements can generally be categorized as either overt or covert features and as security elements designed for the validation of authorized products or the detection of unauthorized products.

Security design features that have been developed include:

- Guilloches
- Linework fractals
- Registered images
- Relief
- Special rasters
- Variable line thickness
- Micro texts
- Screens

These security design features are generic, rather than proprietary and can be offered by those printers that have acquired the appropriate software and origination equipment.

SECURITY HOLOGRAMS – The use of holograms to provide security features began with credit cards in the early 1980s, with them experiencing relatively slow growth for much of that decade. More recently – from the mid 1990s onwards – growth has escalated quite rapidly and holograms are now quite widely used for both document protection and brand protection.

More precisely defined as 'security OVDs' (Optical Variable Devices) when their primary role is protection against counterfeiting/forgery or authentication – rather than when they are used for decoration or embellishment – security holographic OVDs provide the most effective yet economical solution to the problems of counterfeiting and forgery.

In brand protection applications, security holographic OVDs can now be found on labels and packaging for markets as diverse as pharmaceuticals, automotive, software, audio-video, fashion goods, wines and spirits, electronic products, chemicals and lubricants. Outside of brand protection, applications cover banknotes, passports, licences, checks, tickets, vouchers and payment cards.

There are many different techniques available to create security holographic OVDs, with complex, proprietary origination processes offering the highest level of protection. Images created by these processes cannot be absolutely or easily re-originated or counterfeited. Security holograms offer:

- Dynamic visual effects which allow immediate recognition and identification
- Unique covert information which can be incorporated into the OVD and interrogated for additional/absolute verification
- Unique optical elements that cannot be reproduced by color copiers and scanners, or by conventional photographic and printing processes
- Designs which complement the printed label or

package to which they are applied.

Optical images on security holographic OVDs should meet the key performance criteria of bright and clear images, tamper evidence, method and speed of application and be able to meet any necessary usage criteria such as, durability, chemical resistance, etc.

Security holographic OVDs are available as threads and tapes, laminating films, hot stamping foils, transfer films and tamper-evident labels or seals. *See also Holograms and OVDs.*

SECURITY INKS – Inks are a perfect medium to provide authentication or brand protection features on labels and packaging because almost all packs and labels will be subjected to a printing process of some sort. Thus, inks that provide some kind of special security or authentication features can offer an effective solution to security problems. Close co-operation at an early stage between package and label designers, printers and a specialist security ink manufacturer can be a cost-effective way of protecting packaged and labeled goods.

Security inks, of which there are many types, can be printed by almost all printing processes, including offset, letterpress, screen, flexographic, gravure or ink jet. It is important to use the correct method of printing as the printing process used can help in producing a good security effect. This is because many types of security inks rely on the thickness of the ink film – the thicker the ink film the easier it is to see the effect.

There are a great variety of different types of security inks available, which include:

Reactive inks. These are of several types which can be either reversible or irreversible. One type has a chemical reagent within the ink, which will change color when subjected to another chemical reagent – which could be with a felt tip pen or on a stamp pad. Another type is a black ink which will bleed yellow when subjected to water.

Water fugitive inks. Used mainly for security backgrounds on documents, these inks react with water and will show if a document has been attacked with a water-based reagent.

Solvent sensitive inks. These are also used for printing a background screen and will show if a document has been attacked with organic solvents.

Erasable inks. These can easily be removed from a paper surface and are used as a printed security screen to show if a document has been attacked with an eraser or scalpel, etc.

Penetration inks. Designed to have a component within the ink which soaks through the substrate, so that even if the surface print is removed there will still be a printed indication in the paper.

Thermochromic inks. Inks which change color at a certain temperature; the change may be reversible or irreversible. Reverse systems are available which can change at low temperatures, e.g. at body heat, which means there is always a means of identification on hand. They will not function at temperatures above body heat.

Irreversible systems change at low temperature to confirm that a product has not been subjected to too much or too little temperature, e.g. deep frozen goods have not thawed or that medical equipment has been sterilized at a sufficiently high temperature.

Photochromic inks. Inks which change color with light. Most photochromic inks change from invisible to a color when subjected to light and can be verified using UV light – which is most useful because the inks do not need low ambient light conditions for viewing.

Optically variable inks. Inks which change color when viewed from different angles. The most effective of these are produced using a very thick film of ink, which can only be achieved using intaglio or screen printing. There are a range of gravure printed optically variable inks which give reasonable effects.

Magnetic inks. These are widely used on documents for machine reading of characters on documents. More use could be made of these inks for covert security features as they are indistinguishable from standard black inks. The existence of a magnetic ink can be verified with a low-cost device.

Luminescent inks. Inks which rely on the property of luminescence whereby certain compounds can absorb energy at one wavelength and emit the energy at another longer wavelength, such as that of UV light. They will activate a fluor to emit light in the visible part of the spectrum. Widely used for both covert and overt security printing applications.

Invisible luminescent inks. These have no color when viewed under daylight, but under UV will show a color, typically blue, green, red, yellow or orange.

SECURITY PAPERS – Papers which include one or more special identification or authentication features, such as metallic strips, threads, colored fibres, reactive dyes, planchettes or watermarks so as to assist in detecting fraud and in preventing counterfeiting. They are papers which are impossible or difficult to reproduce and/or that provide evidence of tampering. Certain special security label papers may contain authentication or identification features.

Many security papers are most commonly used for the protection of currency and high security documents, but can – and are – used in label papers and to protect other products, such as tickets, gift vouchers, lottery tickets, brand labels, etc. A number of derivative label products embodying security papers and often combined with other security elements, are available today. Some of these security paper features are described below:

Anti-copy papers. Papers that cannot be forged by color photocopiers.

Chemical ingredient papers. Papers in which chemicals have been added to the paper or paper surface during the manufacturing process to provide some form of chemical sensitization, i.e. react with other chemicals to change the color of the paper to indicate tampering, or reveal a hidden message when the paper is photocopied.

Papers with fibers. Paper in which colored or colorless fibers are introduced into the paper at the manufacturing stage. These can be made to show up under specific IR, UV or other light sources.

Papers with planchettes. Paper in which minute discs of plastic are embedded in the paper at the manufacturing stage. They can be colored, fluorescent, UV responsive or invisible

Paper with security threads. Paper in which plain or metalized polyester threads – which in turn may contain other security features, microprint, OVD, etc – have been introduced at the papermaking stage.

Watermarked papers. Watermarks can be introduced into paper by either incorporating patterns or designs when the paper is being made, which are used for banknotes, travelers' checks, high security documents – or by a process of chemical watermarking – produced by a printing process applied to the surface of the finished paper. Both types of watermarking have been introduced into some specialist label papers.

A range of proprietary security papers have also been developed and incorporated in label stocks and pressure-sensitive laminates in recent years.

SELF-ADHESIVE – A type of adhesive which, unlike wet-glue adhesives, is pressure-sensitive. When applied at room temperature, self-adhesive products adhere to various surfaces, such as paper, plastic, glass, wood, etc, simply through contact and without requiring more than finger pressure. *See Adhesives, pressure-sensitive and Pressure-sensitive adhesive.*

SELF-ADHESIVE PRODUCTS – Generally refers to any kind of converted pressure-sensitive products that are carried by and dispensed from a release liner, which also serves to protect the adhesive coating until it is needed. This term can also be used to describe self-wound self-adhesive tapes which have adhesive on one side and a release coating on the other side. Self-adhesive products can be found in the medical, building and construction, rubber casting, apparel, tapes and truck and fleet marking sectors. *See also Release liner, non-label applications.*

SELF CHECKING – A barcode or symbol which incorporates a checking algorithm so as to minimize read errors. Barcodes with self checking may have check characters added into the encoded data in addition to the data message.

SELF-WOUND – A roll of pressure-sensitive material with a single ply that has adhesive on one side and a release coating on the other side. When the roll is wound up the face stock functions as the release surface. This describes most tape products, but also so-called 'linerless' or 'backingless' labels.

SEMI-AUTOMATIC – A label application or labeling machine that is part controlled automatically and part by human action.

SEMI-PASSIVE TAG – These are similar to active RFID tags, but in this case the battery is used to run the microchip's circuitry and not to communicate with the reader. Some semi-passive tags may be in a 'sleep' mode until they are 'woken up' by a signal from the reader, so conserving battery life. *See also Active tag.*

SEMI-ROTARY/ROTATIVE – Label presses which

adjust to different repeat lengths by intermittent movement of the label web – found particularly on narrow-web offset presses. May also refer to a 'digital' die-cutting system where the web moves intermittently at the cutting station and the magnetic cutting plate is mounted on a servo driven cylinder, allowing for different repeats without changing the die cylinder.

SEPARATION – A term used in the reprographic industry to describe the films that represent the cyan, magenta, yellow and black (CMYK) content of an image. When the separations are printed one on top of the other, they recreate through the printing process most of the colors of the original image.

SERIAL CODE – A barcode symbol used with a fixed beam scanner. The scanning action is enabled by the movement of the barcoded symbol past the scanning head. The elements of the symbol are evaluated one at a time as the symbol passes the head.

SERVICE DRIVEN SUPPLY – Historically, the label and packaging sectors have been craft or craft/technology based industries. Today, these industries have increasingly had to move towards service driven market supply. Service in which time-to-market is key for the major national and international brand owners and retail groups, with supply chain issues dominating throughout their thinking and strategy.

Design variations too, are geared to market specific differentiation, shorter and shorter design cycles and shorter production runs – it is estimated that some 70 per cent of all label production is now for runs under 50,000 labels.

Collaboration across supply chains, particularly in the food/retail sector is intense and will increasingly have a greater impact on packaging and label suppliers. Planning and approval procedures are also more complex and need to be integrated into label industry management information and workflow systems (where PDF workflows are growing).

Globalization and consolidation are also putting further pressures on brand owner and retail group supplier service, squeezing local and national label converters between global suppliers and global users. Nevertheless, service driven label supply will be one of the main keys to future label industry growth. *See also Globalization, Supply Chain Management, PDF.*

SERVO DRIVE – Servo drive technology for hybrid and single driven label presses may be axle driven with several gear ratios, axle driven by mechanical transmission elements or direct axle driven by assembled motor with no mechanical elements.

Servo drive technology enables manufacturers to produce presses with:

- No main drive shaft
- No gears or gear boxes
- No mechanical main shaft
- No gears or bearer rings.

Such presses offer fast set-up by recalling press data out of the job file, including length and cross cutting, register pre-setting, web tension pre-setting, web thickness setting, as well flying imprinting – that is text or image change without stopping the machine. *See Press design, narrow-web, for further explanation and diagrams showing servo control systems.*

SET-OFF – Set-off occurs when improperly dried or cured inks or coatings enable the still wet ink/coating to transfer to the back of the next sheet in the stack in sheet-fed printing, or to the back of the printed material as it is wound up in roll-fed printing.

SFI – Sustainable Forestry Initiative is a fully independent, charitable organization dedicated to promoting sustainable forest management

SPC – The Sustainable Packaging Coalition, a membership-based, cross-industry group of supply chain stakeholders that work together on a variety of projects to establish a closed-loop packaging infrastructure

SHADOW – The darker areas of an image. May also be called the lowlight.

SHAFTLESS DRIVE – Used to express where a mechanical drive shaft has been replaced by an electronic drive shaft and that single driven motors are in use. *See also Press design and Servo drive.*

SHEAR ADHESION – The time required under specified conditions to slide a standard area of pressure-sensitive adhesive stock from a standard flat surface in a direction parallel to that surface.

SHEAR CUT – Used to describe the cutting of a continuous web of stock using an action similar to the action of scissors.

SHEARING RESISTANCE – The resistance of an adhesive to a static load (weight) acting in the same plane as the label. With the FINAT Test Method (FTM.8 for example), the label is applied to a vertical, standard surface, usually a glass plate and pulled by an attached weight in a downwards direction. Shearing resistance is the measured time taken by the weight to detach the label. *See FINAT Test Method, and Label testing, pressure-sensitive.*

SHEAR TEST – A method used to separate adhesive to adhesive bonded materials by forcing the interfaces to slide over each other. The standard FINAT Test Method (FTM.7) provides a measurement of the shear adhesion of a pressure-sensitive adhesive. It provides an indication of the cohesive strength of the adhesive. *See Label testing, pressure-sensitive.*

SHEET FED – A printing press or other equipment in which individual sheets are held in a stack or pile and fed one at a time into the press rather than being fed from a continuous reel, as with a reel-fed press. Delivery of the printed sheets is back into a stack or pile at the end of the press.

SHEET FORM – Pressure-sensitive label stock that has been packaged in sheets. Designed for use in the graphic arts, point-of-sale, display, etc, markets where the stock will be sheet printed, usually by screen or offset printing.

SHEETING – Converting rolls of printed or unprinted labelstock into individual sheets on a roll-fed press. Sheeting units are located at the unwind end of the press and allow the converting of printed substrates from a reel into individual sheets, as an in-line operation.

The material, in reel form is fed from the unwind, passes through the printing and converting stations into a sheeting station which then sheets-off the printed material into specified lengths.

SHELF LIFE – The time in which a label material, ink, finished roll or applied label can be stored or displayed under specific conditions and yet remain usable and retain its essential properties. Also refers to the length of time a food product can be displayed in a grocery store or on a retail store shelf without deterioration.

SHELL CUP – A device for measuring the viscosity of liquids.

SHOCK INDICATORS – Proprietary Shockwatch™/ Tiltwatch™ self-adhesive label products are used to provide product history details with regard to any mechanical shock that may occur with fragile items- from glass products to microchips or delicate machinery or instruments – in the supply chain.

The 3M Shockwatch Indicator consists of a sealed glass capillary tube containing a red liquid at one end and a dispersive material at the other. The glass tube is housed in an abrasion-resistant plastic mount surrounded by a brightly colored self-adhesive label. When subjected to an impact level sufficient to overcome surface tension forces the liquid disperses, causing the entire tube to turn red, thereby alerting personnel to container mishandling. By varying the internal diameter of the glass capillary tube the impact force required to activate the indicator is changed.

Tiltwatch labels incorporate a small disk that drops when a preset tilt angle is exceeded, revealing a red warning mark. They are used when goods must remain upright to ensure intact delivery, such as with heavy machinery, large computers, electronic panels, cut flowers, batteries, liquids or other products.

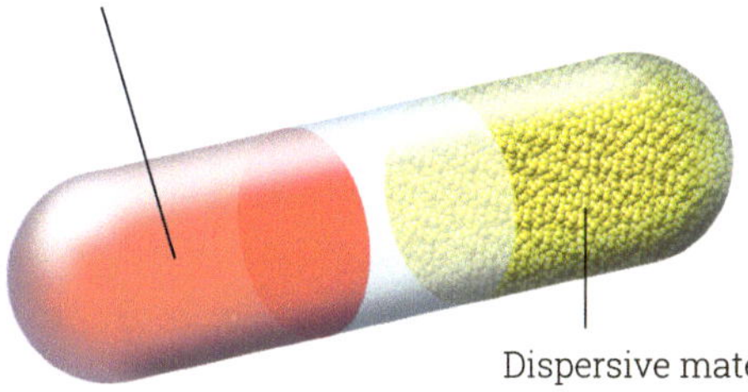

Principle of the 3M Shockwatch™ indicator label.

SHOE SIZING AND MARKING – Specifications for the method of size marking for shoes, applied to all types of shoe, without restriction, are set out in ISO 9407:1991. Labels should conform to this standard or any other relevant national or international regulations or guidelines.

SHORE HARDNESS – A scale (Shore) used to measure and compare the hardness of rubber rollers.

SHRINKAGE AUDIT – An audit examination of store shrinkage; the assessment of the volume and source of actual losses from stores. Shrinkage audits should be carried out prior to introducing electronic article surveillance (EAS) systems, source tagging or other stock loss systems. *See also Shrinkage (retail), EAS and Source tagging.*

SHRINKAGE (MATERIAL) – The reduction in any direction of a material through shrinking.

SHRINKAGE (RETAIL) – A term used to refer to stock losses that occur through wastage, accounting errors, shoplifting by customers at retail level and also theft by staff within the supply chain. Shrinkage is measured in terms of the difference between products actually sold compared to products entering the supply chain. Reduction of shrinkage may be achieved through electronic article surveillance (EAS) or RFID tagging systems.

SHRINK SLEEVE LABELS – Labels produced in sleeve form from specially formulated films which are capable of shrinking biaxially. The application of heat causes them to tighten (shrink) around the labeled product.

The technology of shrink sleeving was originally developed in Japan in the 1960s in order to produce shrinkable cap seals for tamper-evidence on sake bottles. Later, this developed into a method of combining two or more products together for promotional or marketing purposes. Adding single or multi-color printing to the shrink film twin pack or multi-pack promotions soon followed, with the evolution of the technology into a high quality 360 degree method of decoration for unit packaging/ labeling of bottles developing by the 1980s.

Today, shrink sleeve labeling is quite widely used for the decoration of beverages, confectionery, food, home and body-care products, pharmaceuticals, spirits, dairy produce and for a variety of special projects.

Sleeve films are commonly reverse gravure or flexo printed in up to nine colors with photographic images, graphic design, text, color and special finishes and then formed into a tube – which is collapsed for re-winding, handling and shipping.

Origination of the designs and images for shrink sleeving is a special technique as the image for printing needs to be distorted, even differentially distorted, so that the final shrink image on the bottle is viewed correctly anywhere on the bottle.

Shrink sleeve label films were initially made from pre-stretched transparent PVC, which is still used to a varying degree. Now the market also uses orientated polypropylene, polyester and low-density polyethylene in a variety of surface finishes. Films are generally made using either a bubble extrusion or calendering process. The thickness of shrink sleeve films ranges from 35 to 90 microns (1.4 to 3.6 Mils), combining various shrinkage and thickness variations. Additives allow UV blocking or the inclusion of UV fluorescence for security and reconciliation, as well as to provide anti-static or fire retardant features.

Various film specifications are available for a variety of specialist uses, such as low temperature shrink film, high-shrink film (ideal for full sleeving of narrow-necked bottles, and vertical shrink films, which shrink in both directions.

Besides primary decoration, shrink sleeves offer good tamper-evidence features, using tear strips and perforations across and along the sleeve. Integrated holograms can add an additional anti-counterfeiting role.

Shrink sleeve label application. Compared with many other bottle and pack labeling technologies, the shrink sleeve process does tend to have somewhat higher costs. Relatively thick filmic materials are used, the bottle body is completely covered by the sleeve and the sleeve has to be converted into a tube after being printed, dropped over the bottle and then shrunk to fit. As such, raw material costs tend to be high. Furthermore, print costs are also higher; even if only five colors are required on the label it is still necessary to print an opaque white base. Gravure cylinder costs are also quite high. This gives a higher overall print cost for shrink sleeve labels before capital costs and investment are considered. In addition, when used with shaped containers, it is important to pre-distort elements of the design so that a percentage of shrink is taken into account before the sleeve production.

Against this, the technology is currently achieving good success for high quality, high added-value all-round bottle decoration. When used with glass bottles that have been lightweighted, the strength and performance are reinforced by the shrink sleeve. The cost of the sleeve can be offset against the savings in the bottle cost due to the lightweighting.

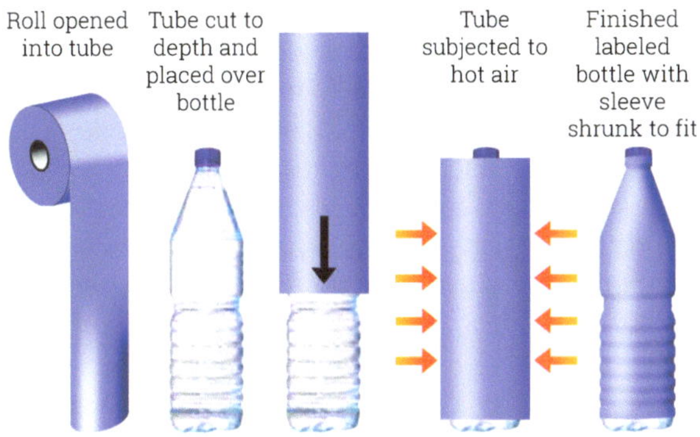

Principle of shrink sleeve labeling.

Shrink sleeves are designed to shrink when exposed to heat and historically the choice of heat tunnel depended on a number of factors, such as: whether the container is being filled or is empty, whether it is glass or plastic, on the shrink film being used and on the line speed. The most commonly used heat tunnels are hot air, infrared, or a combination of the two. There is also an increasing use of steam as the shrink medium.

Benefits of shrink sleeving include:

- 360 degree full length decoration
- 3-Dimensional decoration
- Translucent colors, or transparent windows to view contents
- Total container and print protecting against contamination
- Precise label positioning
- Inventory reduction
- Tamper evidence can be incorporated
- Flexible in application type and size
- Matt or gloss finishes, or a combination of both
- Metalized, pearlized, photochromic and thermochromic finishes
- Tactile sleeves
- Ideal for labeling complex shaped containers
- Ability to incorporate promotional ideas
- An effective alternative to coated containers.

Main end-user markets for shrink sleeves include drinks, toiletries, health and beauty, pharmaceutical, spirits, dairy produce and confectionery.

SHRINK TUNNELS – Shrink tunnels, used in shrink wrapping and shrink sleeving, mostly operate on the hot-air principle. Their successful performance is primarily a function of the speed and effectiveness with which they can transmit sufficient heat to the film to initiate the shrink action. As exposure of the film in the tunnel is only a few seconds, the heat has no effect on the product; the hot-air only affects the film, so causing it to shrink snugly around the product.

Different shrink films have different shrink characteristics, so shrink tunnels need to have a degree of adjustability. Most consist of a conveyor, a heat chamber and a hot-air recirculating system. There are many special types of tunnels available, with some having cooling chambers or pre-heating sections.

SHRINK WRAPPING – A method of packaging in which products are overwrapped with a heat-sensitive polymer shrinkable film and then passed through a heating tunnel to shrink the film tightly against the product. Shrink wrapped products – from rolls or sheets of labelstock, cut stacked labels or packaged goods in food stores – may be subsequently labeled with pressure-sensitive variably-printed labels.

SIGNATURE VERIFICATION – Recent developments in hardware and software can now provide unambiguous authentication of a handwritten signature. Various tablet PCs and pads have the capability to capture the distinct behavioural characteristics of an individual's signature. This includes speed, shape, stroke, pen pressure, timing, etc, all of which characteristics are now able to be securely embedded into a document, card or label to provide secure access. The signatures can be digitized during writing, rather than having to be scanned.

Software developed by the German company Softpro offers a unique approach to the implementation of this technology.

SILICON – An element which has some properties similar to those of carbon. Silicon forms many compounds, including silicates (glass) and silicones.

SILICONE – A polymer material that has adhesive-repelling properties and is used as a release coating in the manufacture of silicone release papers and films. Also used as an ink additive to aid ink flow and improve ink scuff resistance. Silicone can also be found in some pressure-sensitive adhesives capable of withstanding extreme temperatures and is also used as a lubricant within self-adhesive recipes to improve guillotine cutting and can be applied to both sides of a guillotine blade to prevent adhesive build up. *See also Silicone release coatings.*

SILICONE ADHESIVE – Silicone adhesive compounds have a remarkable stability through a very wide range of temperatures. Their main limitations for present use are their high temperature cure and their sensitivity to aromatic and aliphatic fuels. Silicone adhesives can be bonded to difficult surfaces and have outstanding high and low temperature performance. They are also highly resistant to oxidation, ozone and corona radiation.

SILICONE CURE SYSTEMS – Solvent- and water-based silicone coatings are first passed through an oven to remove the solvent or water vehicle. The dry film then enters a second heated zone which activates the crosslinking mechanism.

Solventless silicones can be cured in more compact ovens since the need to evaporate the solvent or water vehicle is avoided.

The more recently developed radiation curable silicones eliminate the need for ovens altogether. The coated web is completely cured simply by passing it through an electron beam or ultra-violet station, similar to those found on many label presses for drying inks.

SILICONE RELEASE COATINGS – First developed by Dow Corning in the mid-1950s, silicones remain the most widely used release agents for pressure-sensitive adhesives applications, dominated by the important label market. A release liner with a surface coating of approximately 1g/m2 of a cured silicone polymer is an essential feature of a typical self-adhesive label construction.

The specific adhesion between the silicone surface and the adhesive can be fine-tuned by the use of controlled release additives to the silicone coating, to ensure the laminate is held together, yet provide appropriate release levels to suit labels that need to be automatically applied at great speed, or that can withstand the turns and flexing in the paper path of a laser printer or copier, without the labels dispensing inside the machine.

The silicone release system will still allow the backing to be removed from the label, regardless of the tack level of the adhesive. Globally, the main silicone release coatings used are solventless systems (65%), solvent-based systems (17%), emulsion systems (13%) and others (5%).

Solventless silicones were first introduced in the mid 1970s and are now the largest and fastest-growing sector of release coatings for pressure-sensitive adhesives. The elimination of solvent recovery or incineration costs and the ability to circulate hot air in curing ovens makes thermal cure solventless silicones an economically attractive alternative to solvent-borne silicones.

Solvent-borne silicones utilize a well-developed and extremely versatile silicone release technology, with a

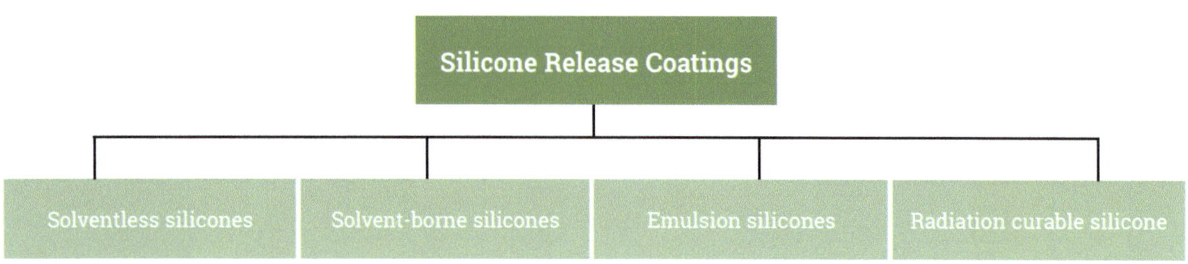

Main types of silicone release coatings

range of additives allowing bath life and cure rate to be controlled accurately within wide limits. The use of solvents enables low coat weights of high molecular weight polymers to be laid down on substrates without difficulty, but solvents are expensive and strict anti-pollution legislation demands that used solvents are either recovered or incinerated. Solvent-based silicone systems are being increasingly displaced by solventless silicones.

Emulsion silicones avoid some of the costs and hazards associated with solvent systems, e.g. no solvent dilution, no fire or explosion risk, and no environmental or operator health problems. There are also some disadvantages. Aqueous systems can swell paper fibres, adversely affecting the dimensional stability of the release liner. Early tin-catalyzed condensation polymer emulsions have draw-backs: limited bath life, shear-induced problems, poor release level control.

Advantages of radiation curing of silicones

Room temperature cure	Processing
No substrate restrictions	Warm application possible
No paper re-moisturizing	Compact - little space needed
No paper shrinkage	High line speed - higher productivity
	Low energy consumption
Absolutely lay-flat	100% product - no solvents
No post-cure	No misting - even at high line speed
Thermal stability	
Long pot life	

Current emulsion systems do not exhibit blocking problems, while the use of controlled release additive emulsions enable the converter to accurately meet specific release requirements.

Radiation curable silicones provide opportunities for new applications which are not easily met by using conventionally thermally curable silicones. Radiation cures at, or only slightly above, room temperature. It allows release liners to be based on thin, thermoplastic film substrates, such as low density polyethylene, polypropylene or PVC which cannot easily be processed through long web path, high temperature curing ovens.

SILICONE RELEASE PAPER – A paper coated with silicone that ensures that the label face material and the matrix (waste face material after die-cutting) can be readily removed from the backing paper of a pressure-sensitive laminate. It also protects the adhesive against any environmental influences which may otherwise adversely affect its efficacy. *See also Silicone, and Silicone release coatings.*

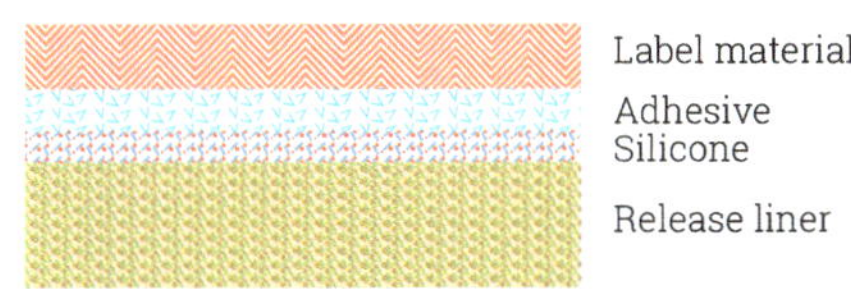

Structure of pressure-sensitive laminate showing the silicone and release liner

SILICONES – Silicones are polymers based on chains featuring alternating silicon and oxygen atoms. Silicone polymers include polydimethyl siloxanes, which are used in formulating silicone release agents. Silicone polymers are very effective release coatings for pressure-sensitive adhesives. *See Silicone release coatings.*

SILICONE STAIN TEST – Laminators perform a stain test on silicone coated paper release liners to assess uniformity of coverage. Where there are pin holes or mis-coating, the aqueous stain will mark the paper at that point.

Label producers can use a similar test on the silicone areas, where a label die-cutting tool may have left an impression in the silicone coating, to determine whether the die has penetrated the silicone coating at any point. The stain highlights uneven die-cutting which may not have cut right through the backing liner, but is sufficient to cause label waste stripping problems, where the label waste matrix tension is uneven, causing it to break.

SINGLE AISLE – A basic electronic article surveillance (EAS) detector system, which is made up of a single receiver and a single transmitter.

SINGLE-FACED TAPES – Pressure-sensitive tapes in which the adhesive is applied to one side of the backing only. Most pressure-sensitive tapes are of this type.

SIX SIGMA – Six Sigma is a management methodology that enables businesses to achieve a goal of increasing profits, improving product quality and enhancing employee morale by eliminating waste, decreasing process variation and by identifying and reducing defects in manufacturing. It uses information and statistical analysis to improve a company's operational performances, practices and systems by identifying and preventing 'defects' in manufacturing and related processes, enabling companies to both anticipate and exceed expectations.

More than just a quality system like ISO or TQM (Total Quality Management), Six Sigma is more about a philosophy or way of doing business in which defects are measured per million opportunities. Developed by Motorola, the Six Sigma methodology can be defined by reference to three levels: metric, methodology and philosophy.

Metric. Commonly referred to as 3.4 Defects Per Million Opportunities (DPMO) provides a guide for considering at least three opportunities for a physical component.

Methodology. The tools and roadmap that provide a structured method of solving problems.

Philosophy. The route to reducing variation in a business by taking customer-focused, data driven decisions.

A wide range of information, support and consultancy services for companies looking at implementing Six Sigma can be found on the Internet. *See also Statistical process control.*

SIZE – Chemical substances such as resins, starch or synthetic polymers, coated on a paper surface to reduce water absorption, scuffing, oil penetration etc.

SKELETON – *See Matrix waste.*

SKIPS OR SKIP COATING – Refers to areas within a pressure-sensitive lamination in which the adhesive or release coating is missing.

SKU – Stock Keeping Unit. Every (even slightly) different product variant manufactured, distributed and retailed will typically represent a unique SKU. It usually has its own designated reference code and storage location and each shipment or sale is accounted for individually.

SLEEVE – Long used on flexo presses to mount plates, sleeves are now making an appearance on narrow-web offset presses, replacing both blanket and plate cylinders with pneumatic shafts and variable diameter sleeves. This allows repeat length to be changed within minutes and without the use of expensive inserts.

SLEEVES, ENDLESS – Instead of pre-mounting a flexo plate onto a sleeve or cylinder, endless (or seamless) sleeves incorporate both photopolymer and an ablation mask. They are imaged, exposed and washed out 'in the round' and used as the print forme. As well as eliminating plate mounting, directly imaged sleeves bring advantages on the press.

No plates means no possibility of plate lift. The uniformity of the imaged sleeve and the ability to run without a gap between plates means that the likelihood of press bounce is much reduced. Combining the reduction in bounce, the elimination of lift and the excellent sleeve uniformity gives the potential for higher press speeds and allows gapless printing of, for example, flexible packaging or foils.

There are different types of sleeve on the market but all share similarities in construction. The basic sleeve is just a thin tube that acts as a carrier. Originally nickel was used as the material but, more recently, composite sleeves have gained in popularity. Whichever type of sleeve is used they are both mounted in the press on an air cylinder (or 'mandrel'). The sleeve must be an interference fit on the cylinder so that it does not move when the press is in operation.

Compressed air connected to the cylinder just acts as a lubricant, so that the sleeve can be slid onto the cylinder or removed.

SLIP – Describes the relative motion, or sliding, between the two members of a braking system. In tensioning, the smoothness of the slip is a critical element in maintaining web tension.

SLIT – The action of cutting rolls of labelstock to specified widths on a slitting machine. Slitting is undertaken using either stationary or rotary knives or blades in a machine with roll unwind and rewind devices, tension control and web tracking. Slitting of master rolls to narrow-web roll widths is normally undertaken by the labelstock manufacturer, whose machinery has lubricated slitting knife cleaning pads on each rotary blade to stop adhesive build up on these slitting tools. These slitting machines also have a special splicing table, before the slitting station, where a diagonal butt splice can be made after removal of pre-marked sub-standard material, plus paper mill and coating machine joins. *See also Slitting.*

SLIT BACK – Slit back or split back refers to sheet pressure-sensitive labelstock in which the release liner has been slit in a pattern to facilitate easy removal of the liner by hand. Not to be confused with scored back patterns – where the backing is not cut through to the adhesive coating – that have to be flexed to break the backing along the scored area.

SLIT OVER PERF – A die profile that creates a finished product with a complete cut through the face material and a perforation on the backing line.

SLITTER REWINDER – The use of slitter/rewinder technology is one of the principal methods of off-line converting rolls of self-adhesive labels, tapes and flexible packaging films into the correct width rolls ready for shipping to customers. The operation may also be combined with web inspection as part of the final stage in a converter's quality control program. Some models will accept a basic flexo printing module.

The slitter/rewinders used in paper mills and label stock manufacturing plants are much larger and tend to have computerized controls that allow fully-automatic cut-off, quick knife settings, unloading and core loading and perhaps a diagnostic facility. Operators' read-outs covering web tension, system status and product data are frequently presented on LCD displays, with touch-screen panels and membrane key pads. *See also Rewinding and inspection.*

SLITTING – Dividing a web of label stock in the lengthwise direction by rotating slitting knives so as to produce two or more narrower webs. Jumbo rolls of label stock are slit into narrower rolls for printing on narrow-web presses; the printed reels coming from the press are then usually slit into even narrower reels for automatic application to a bottle, pack or other surface to be decorated.

Slitting may be carried out with shear or crush cutting. Shear cutting produces a good quality edge at high line speeds with little dust; crush cutting can be economical if speed and edge quality are less critical. Other variables to analyze for each application include the placement of slitting knives, which can be shaft mounted, mounted in individual holders on a bar, or mounted in a combination of the two. Blade materials can be anything from high chrome steel to carbide or ceramic; bevel and grind angles of the slitting blades can be varied, as can cutting angles, while the most important variable is the blade holder. If the holder cannot maintain the proper settings, the entire slitting operation will be problematic.

Decisions to be made before slitting are based on the process material's characteristics, such as weight, thickness, flexibility or abrasiveness. Machine working width, slit width, operational speed, the number of cuts to be made, the quality required, and even the number of set-ups per shift and the number of shifts, are also important. *See also Slit.*

SMART ACTIVE LABELS (SALS) – Consumer demand for freshness, quality and safety is continuing to drive demand for new concepts of diagnostic and active packaging and label technology. Already, smart active and intelligent labels with monitoring capabilities are being used for, say, time and temperature monitoring of cases, boxes and pallets, while the first generations of freshness indicator, anti-microbial and other types of labels and tags are coming to the market. The role and function of smart, smart-active and intelligent labels are shown in the diagram on the previous page.

New generations of SALs already in development will go much further, with additional sensing features incorporated into them for detecting a wider range of gases, bacteria or microbial spoilage, carbon dioxide, humidity changes, for oxygen scavenging, moisture or odor absorbing, vibration detecting, etc.

Intelligent and active functions are also expected to be combined in labels and tags, often invisibly, but able to communicate their status through wireless technology. *See also Intelligent labels, Smart coatings, Smart inks and Nano-codes.*

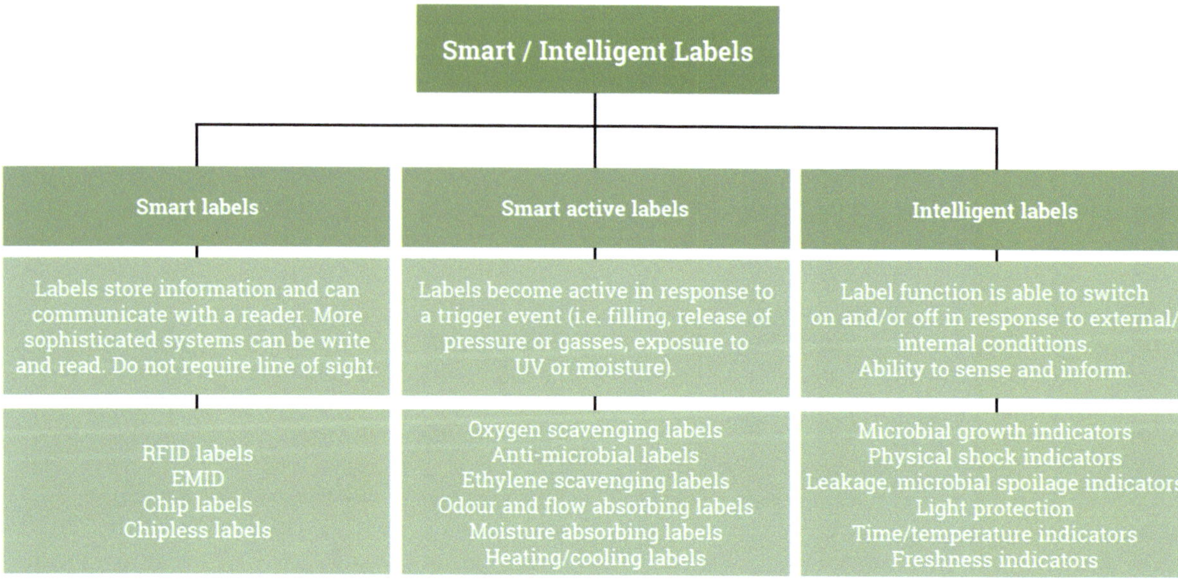

The role and function of smart, smart-active and intelligent labels.

SMART CARDS – Chip cards which contain their own onboard CPU (central processing unit) but which are unable to perform functions on the data stored within the chip. The term is often widened to include all chip cards.

SMART COATINGS – The use of high-performance coatings that can enable surfaces to be more responsive to their environment and add functionality – such as barrier protection, adhesion, release, anti-microbial or other properties. Smart coatings frequently rely on a visible color response to indicate a temperature change or respond to other stimuli, such as a pH change, oxidation, corrosion or electric current.

Applications for smart coatings include tamper-evidence, time/temperature indicators, anti-counterfeiting solutions, anti-microbial properties, barrier performance and freshness indicators. *See also Nanotaggants and Nanocodes.*

SMART INK™ – Ink which contains radio frequency microtaggants, which are responsive to electromagnetic energy, enabling them to be detected when subjected to an energy source. Information contained within the printed ink signature on a label, tag or document can be read remotely when a low-powered energy source is aimed at the printed area.

SMART LABEL – A label that contains an RFID tag. Such tags are referred to as 'smart' because they can store information – such as a unique serial number – and communicate with a reader. *See also RFID tags.*

SMART PHONE ENABLED AUTHENTICATION – The use of solutions and devices that enable consumers to self-authenticate their purchases. Some brand owners are adding serialisation coding to their product labels and packaging so that customers can check the authenticity of the product while in-store using NFC enabled mobile phones. Serialization may be in formats such as barcodes, numerical coding and material biometrics.

In operation the mobile phone reads the NFC tag reference and sends it through GRPS messaging to a database for authentication. Alternatively, on-phone decoding and authentication can take place through descriptions built into Apps within the phone.

SMART SHELVING – Already in trials or being adopted by some store groups, smart shelving – in conjunction with smart packaging/labels – is forecast to be one of the key technologies for future retailing. For a simple shelf replenishment system it is proposed that weight sensing or RFID technology (or both combined) will provide a means of knowing how much stock is on a store shelf at any one time.

Smart shelving is aimed at optimizing inventory management, reducing store shrinkage (stock loss), improving customer relationships and reducing labor costs.

SMEAR TEST – A quick subjective test to determine whether a silicone, varnish or ink has cured or dried sufficiently to be solidified. A finger firmly stroked across the surface will leave no mark if the coating is properly cured, but will leave a streak or smear on an incompletely cured or dried coating.

SMUDGE-PROOFNESS – The degree to which freshly printed ink on paper or film will resist a rubbing or wiping action. *See also Smear test.*

SNAPTAG – Invented by SpyderLynk, a SnapTag is a 2D mobile barcode alternative similar to a QR code but uses an icon or company logo and circular code ring rather than the more traditional square pattern of black dots.

SnapTags are easy to activate and can be used to take consumers to a brand's website and can also enable mobile purchases, coupon downloads, sample requests, video viewing and promotions. *See also QR code*

SNOWFLAKE CODE – A code made up of an array of discrete dots in which more than 100 numeric digits can be encoded within an area of 5mm square (0.0078 sq. inches). The code can still be read even with up to 40% of the code damaged due to user selectable error correction. The code can be applied directly to products or in a label form.

SOFT DOT – Dots are referred to as 'soft' when the halo surrounding them is excessive and where the halo is almost equal in area to the size of the dot itself.

SOFT PROOFING – Used to describe the technique of previewing a page on screen, rather than taking a physical hard copy proof. The challenge is to achieve accurate color representation on the screen proof that will compare with the actual printed job, bearing in mind that one uses additive color and the other uses subtractive color. Professional monitors and systems are now usually sold with calibration capabilities to improve the color match between the two.

SOFT TAG – In electronic article surveillance systems (EAS), an alternative to a hard tag. Soft tags consist of a card insert enclosed in a transparent plastic sheath, which is fastened to merchandise using either a pin or a lanyard.

Soft tags are thin, light, durable and also re-usable and especially effective on lighter weight clothing such as delicate knitwear, lingerie or children's clothing.

SOJUZUPAK – The Union of Manufacturers and Consumers of Packages and Packaging Products in Russia.

SOLID ROTARY DIES – Cutting dies produced from solid bars of steel in which unwanted areas are removed by manual machining, numerically-controlled machining or electro-mechanical discharge methods so as to leave the cutting bevels standing proud of the main body surface.

Solid rotary dies are made of various grades of tool steel, which is determined by the application and the process used to engrave the die. The first step in creating a rotary die is lathing a cut bar of steel. In the next process, the rough milling of the image takes place. Stress relieving is normally done after rough machining, but before final finishing such as cylindrical grinding and/or polishing.

After the heat treating process, where the steel achieves a targeted hardness, the die is ground to a specific diameter. The exact clearance is put into the die to meet the specifications that are required by the end user.

Whatever the system used to produce the die, the actual bevels of the cutting edge should take into consideration the material that is to be converted – standard paper laminates have an even centered bevel; special constructions may have a bevel which is wider at the base; while for non-paper materials or thick materials a side bevel or bevel-and-a-half would provide a keener separation.

These possibilities make it important for the diemaker to be informed of the type of material that is to be converted from a die at the time of ordering.

SOLIDS CONTENT – The percentage by weight of non-volatile components within an ink, adhesive or coating.

SOLVENT – A chemical substance that is capable of reducing or thinning down the viscosity of adhesives, coatings or inks. Solvents are liquids that are able to dissolve other substances.

SOLVENT-BASED ADHESIVE – An adhesive whose components are dissolved in organic solvents for coating purposes, e.g. the adhesive coating on a self-adhesive label.

Solvent-based adhesivesare noted for their fast bond strength, good heat resistance, and versatile adhesion to a wide range of substrates and tolerance to a wide variety of production conditions – including low temperature or high humidity. However, solvent-based pressure-sensitive formulations have been widely displaced by water-based and hot-melt systems for economic as well as ecological reasons. Solvent rubber self-adhesives do not have good heat resistance, which limits their shelf life.

With solvent-based adhesives solvent recovery and/or incineration is now essential to meet clean air legislation requirements. This is expensive and can only be justified by large output operations. *See also Adhesives, pressure-sensitive and Pressure-sensitive adhesives.*

SOLVENT-BASED SILICONIZATION – The process of coating a pressure-sensitive liner material (paper or film) with a solvent-based silicone release coating. The solvent evaporates during curing at high temperature. When compared with solvent-free siliconization, this type of silicone coating is applied relatively thinly and has poorer release properties. *See also Silicone release coatings.*

SOLVENT-BORNE SILICONES – Solvent-borne silicone release technology is well developed and is extremely versatile. A range of additives allows bath life and cure rates to be controlled accurately to within wide limits. The use of solvents enables low coat weights of high molecular weight polymers to be laid down on substrates without difficulty.

However, solvents are expensive and strict anti-pollution legislation demands that used solvents are either recovered or incinerated. For economic as well as ecological reasons they can no longer be vented to the atmosphere. Most solvents used for silicone coating are flammable and present fire and explosion risks.

Solvent-based silicone systems are being increasingly displaced by solventless and emulsion silicones.

SOLVENT-FREE – Refers to the absence of any solvent in a coating, ink or adhesive.

SOLVENT-FREE SILICONIZATION – The process of coating silicone on paper without the use of solvents. Also called solventless siliconization. The silicone is mixed with a hardener and cured at high temperature. When compared with solvent-based siliconization, the coating of silicone is relatively thick, displays better release properties and is mechanically more resistant. *See also Silicone release coatings.*

SOLVENTLESS SILICONES – A silicone release coating that has been applied without the use of any organic solvent or water. Referred to as a 100 per cent solids silicone. Thermal cure solventless silicones are now the largest and fastest growing sector of release coatings for pressure-sensitive adhesives. The elimination of solvent recovery or incineration costs and the ability to recirculate hot air in curing ovens makes thermal cure solventless silicones an economically attractive alternative to solvent-borne silicones.

The chemistry of solventless silicones is similar to that of solvent- and water-based silicones in that all three are based on polydimethyl siloxane. The essential difference between them lies in their molecular weights. Solvent-based and emulsion silicones are high molecular weight polymers which would be far too viscous to coat if they were not dissolved or dispersed in a mobile vehicle. In order to have coatable viscosities, solventless silicones have to have very low molecular weights.

SOLVENT RESISTANCE – The ability of a material, such as a self-adhesive label, to resist the dissolving action of certain organic solvents. The term can relate to the adhesive or the printing.

SORBENT TECHNOLOGY – Sorbent technology is an active packaging innovation that is used to keep food and drug packs fresher for longer. Oxygen absorbent packets, strips, cards or labels are loosely inserted in product packaging, or can be sealed to the underside of a pack using an adhesive layer. The sorbent technology prevents the growth of aerobic pathogens, including molds, and eliminate the need for additives. They also extend shelf life for numerous food and drug products.

SOURCE TAGGING – The application, usually covertly, of EAS tags to products themselves (not the packaging or labels) during or immediately after manufacture.

Ideally, retailers would like to have 100 per cent of the products on their shelves protected by anti-theft technology and especially have electronic article surveillance tags applied at the pre-packing, packaging and labeling or product manufacturing stage. Historically EAS tagging was too expensive to do this because of the unit cost of the tag and also the need to apply the tag to the product, which involved a degree of un-packing and re-packing of at least the transport packaging and perhaps some elements of the primary retail packaging as well.

With retail theft running at a near 2 per cent of turnover the cost benefit of applying or incorporating EAS tags at source has become increasingly interesting to the retailer. The cost of applying an EAS tag at source can be half or even a tenth of the cost of tagging downstream. The need for manual labor in the retail store itself is eliminated, with the tagging cost effectively passed upstream to the product manufacturer's packaging operation or indeed to the label/packaging supplier, depending on the exact point source tagging is carried out.

The cost benefits of source tagging arise primarily from the following three contributing factors, which tend to complement each other:

- Fully automated on-line tag application during a production operation.
- Fractional tag application involving source tagging at a point of manufacture, which facilitates covert tagging. Additional benefits can also be derived from concealing tags, which tends to impede tampering and can also improve appearance as tags can be unsightly.
- Economies of scale in purchasing EAS tags in larger quantities can lead to better prices. The unit cost of tags is closely related to production and order quantity volumes.

In essence, EAS source tagging is a business process. It is much more than a technology issue. Retailer management and departmental teams, perhaps already carrying out piece-meal in store EAS tagging, need to completely change their thinking if source tagging is to be successfully implemented with their suppliers. Implementation of full source tagging solutions requires more than straightforward upgrading or expansion of existing EAS systems.

From a supplier's point of view, key retailer criteria for successful source tagging include the following:

Key retailer criteria for successful source tagging

Tag must be included/added during the manufacturing process (product or packaging).

Tags should preferably be invisible inside the product or pack

Tags must have de-activation capability

Tags must be small

Tags must produce a controllable unique signal to avoid false alarms

Comprehensive specifications and requirements guidelines/checklists for source tagging have been drawn up by several organizations in both Europe and North America.

SPACE – In a barcode. The lighter element of a code formed by the background between the bars. *See Barcode and Barcode symbol.*

SPECIFIC ADHESION – The force needed to remove a pressure-sensitive adhesive from a specific substrate under specified conditions.

Also, a term used to describe particularly high adhesion between a specific adhesive (or class of adhesives) with specific types of substrates.

SPECTRAL RESPONSE – Relates to the sensitivity of a scanner or other device to light of different wavelengths.

SPECTROPHOTOMETER – A measuring instrument used for determining the distribution of light energy reflected by a color or from a printed ink image.

Although most color measurement is done with the human eye, it has its limitations as a device for accurately describing a particular color. Exact measurements mean facts, not opinions on what the printer or customer sees. This is the background to the development of instruments that 'look at color' and describe it in figures or in the form of a spectral curve on a graph.

True spectrophotometers are expensive and are essentially laboratory instruments which provide a precise means of measuring color. Less expensive and more suited to industrial use are photo-electric comparators (sometimes called spectrophotometers) which are not quite as precise as those from true spectrophotometers, but are adequate for many purposes. These comparators may be arranged to reflect light from a printed or transmitted light source. The usefulness of the results depends to a large extent on the quality of the filters and the number used.

The larger, more expensive spectrophotometers may be part of a color computer system made up of a spectrophotometer, computer, visual display, printer and storage units. Such systems provide the computer with a series of values called reflectance values. The values are provided at given wavelengths and can be incorporated into color control, statistical process control or quality management systems.

SPIKES AND SURGES – Fluctuations in voltage that may cause circuits to blow. Spikes may be caused by lightening, thunderstorms, work on cables, supply companies switching lines or intermittent operation of some equipment, e.g. lifts or air conditioning.

SPLICE – A join of the ends of two webs of paper or other material in, or approximately in, the cross-direction of the web using an adhesive, adhesive strip or glue to make one continuous web. Splicing is

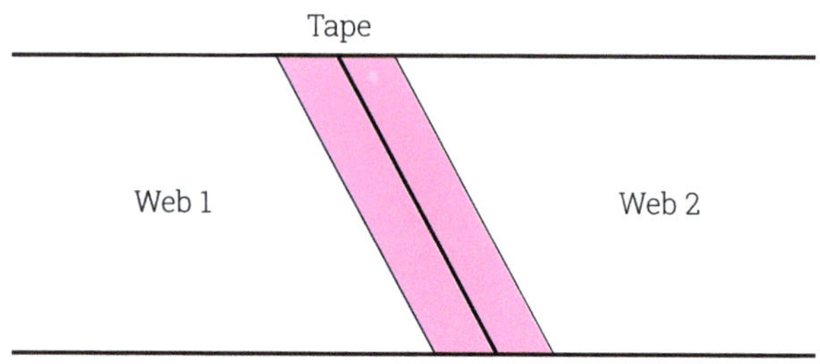

The joining of two webs together using adhesive tape.

usually made to:

- obtain a reel of the desired (longer) length
- permit, in a printing/converting machine, a continuous operation between the end of one reel and the beginning of the next
- no join together the two ends of a web following a web break
- to take a damaged or faulty area out of a printed web before shipping to the customer
- in self-adhesive label material coating and laminating, special butted splices are made during slitting to also remove start up and intermittent faulty or sub-standard material, plus all paper mill and coating machine joins. This permits subsequent continuous label production.

Where splicing is carried out on a converting/printing line automatically, while the machine is still running, this is more commonly referred to as a 'flying splice'. Automatic butt splicing systems make use of tension control, dancer roller and web accumulator devices to provide accurate web control and tension control during the splicing operation

Splicing of a printed web in register is accomplished by using a photoelectric cell (or fiber optics) to sense a printed eye mark. This signal is fed into the splicer logic control which delays the splice sequence until the pattern of the running web is in the correct position to align with the pattern on the new roll waiting to be spliced. *See also Butt splice.*

SPLIT BACK – Slits incorporated into a release liner to make it easier to remove the backing liner. Usually found in pressure-sensitive sheet materials.

SPOT COLOR – The addition of an extra color or any area of color, often in an isolated area within the total label design, that is not a CMYK process color. May be a line or tonal representation. Brand owners often require a separate specific custom made brand or house color – outside of the normal four-color process colors – to distinguish their own products from competitive products on retail shelves. To ensure accurate matching of spot colors, a color measuring instrument should be used.

SPOT VARNISHING – The application of a clear matt or gloss varnish to specific areas of plain web or printed matter so as to improve the quality or performance of the presentation. *See also varnishing.*

SPPES – The Polish Label Association. SPPES, was founded in 1997 to brings together senior people from individual companies connected with the Polish self-adhesive label industry..

SPRAY COATING – Although not a label solution, new developments in the electrostatic spray coating of glass and most recently, PET bottles now look to provide some interesting new possibilities for added-value bottle decoration.

With electrostatic coating technology bottles are automatically sprayed with an electrostatic liquid or powder, giving a range of effects from transparent to opaque, full gloss to matt, silky smooth to stone effects, slippery sleek to rubbery soft, as well as thermochromic or fluorescent images.

Such effects can also provide for a glass etched or frosted look, offering bottles that are colored or textured and which are designed to sell and add value to a product. The various coating options can all be overlabeled or screen printed.

Coatings are extremely thin and can be recycled with the bottle. Being water based, the coatings are compatible with organic inks and reduce the amount of overall packaging used.

By combining spray coated barrier coatings with special effect spray finishes and then labeling (which can make frosted or etched effect areas transparent if required), then electrostatic spray coating could provide bottlers with some very exciting added value effects for, say, speciality niche beer brands.

SPREAD – A pre-press function that compensates for misregister on the press. A spread is a slight increase in size of the inserted image. *See Choke and Trapping.*

SPREADS AND CHOKES – *See Choke and Trapping.*

SPROCKET FEED – Variously called sprocket feed, pin feed or tractor feed. Used to describe a method of transporting continuous stationery or continuous labels through a computer printer using sprocket holes punched in the web that pass over sprocket pins.

SPROCKET HOLES – The continuous series of holes on the outside edges of continuous stationery or continuous labels that are used to drive the paper through the printer via the sprocket pins or tractor feed belts.

SPROCKET PUNCHING – Paper punched longitudinally along both side edges with 5/32-inch (4 mm) diameter engagement holes at a regular 1/2-inch spacing. These holes engage with sprocket wheels or toothed belts, which push or pull the paper through a printer. Better quality continuous paper also includes additional longitudinal slit perforations alongside the engagement holes, allowing the large holes to be torn off the printed page. For punching the holes a shaft with axial movable cutting rings is used.

Normally, the sprocket rings are only cutting the liner and the cut out holes will be removed by pulling up the waste.

STACKER – A device at the take-off end of a press that is used to automatically stack sheeted labels

STACK PRESS – A multi-color printing press in which the color printing units are built vertically or stacked in groups, each with its own impression roller. The total number of units in such a stack may vary, the most common number being six.

The two main advantages of this layout are the possibility of printing both sides of the web through a combination of web threading and the provision of adequate drying facilities in the space between the two stacks. The second advantage rests with the total accessibility thus reducing the time for job change over and running adjustments.

These advantages must be considered against the fact that is not as easy to hold tight color-to-color register on a stack press configuration when running extendable materials, such as film. The layout does however, lend itself to operations such as coating and tinting.

Other press variations are in-line, central impression (CI) and combination process.

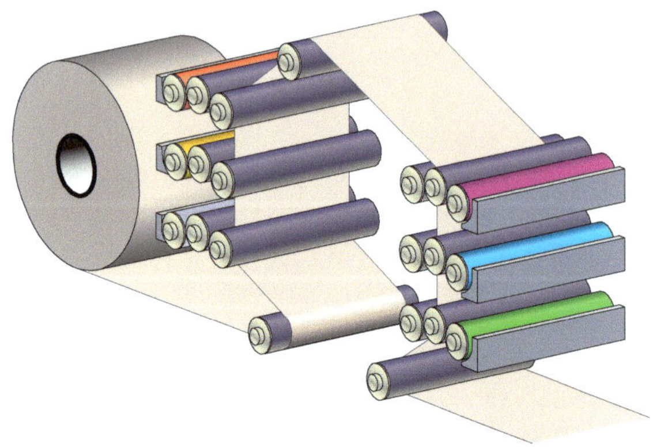

Diagram shows the structure of a stack label press.

STACKER – A device at the take-off end of a press that is used to automatically stack sheeted labels.

STAND-UP POUCHES – Stand-up pouches are a rapidly growing sector of the packaging industry, offering the opportunity to replace a plastic or glass bottle with a pouch and save on warehousing and shipping costs, provide greater flexibility and lower costs. They can be manufactured to provide a range of barrier properties – to moisture, odor, sunlight, etc. – and be used for packaging a wide range of products from liquids to powders, granules and ready food products.

Stand-up pouches keep the stored contents fresh and safe for long periods, and make it easy for products to stand up-right on shelves with falling over.

Depending on products being packaged, pouches may be made from paper-backed materials, clear films, metalized films, foil materials, with different film structures capable of being laminated together to create different barrier properties. Main types of films used for stand-up pouches are polyethylene (around 75 per cent of the films used), polypropylene and polyester (mainly as a backing material). Pouches can also be made to incorporate a range of different re-closure possibilities.

The range of stand-up pouch types includes:

- Stand-up pouches with a window
- Stand-up pouches with a bottom gusset
- Side gusseted pouches
- Flat bottom pouches with a zipper
- Flat bottom pouches with no zipper
- Kraft paper pouches
- Pouches with re-closure caps

Stand-up pouches are available in un-printed or printed formats and for many applications can be customized. Pouches are printed in gravure of flexo in up to 8- or 9-colors. Digital printing may be used for short-run and customized or personalized applications. *See also Pouches.*

Illustration shows examples of stand-up pouches.
Source: Xeikon

START/STOP CHARACTERS – Characters that are included in barcode symbology to indicate the beginning and the end of the code to the reader and also the direction in which it is being read. The start character is normally on the left-hand end of the code; the stop character at the right-hand end. *See also Barcode and Barcode symbology.*

STATIC – An electrical charge that is generated when handling materials and which causes them to cling together. Static charges can jump (or earth) from say, rewind rolls, to humans and so cause an electric shock. Static may also cause a fire if solvents are present.

STATIC ELECTRICITY – An electrical charge that builds up on the surface of a substrate or other surface. Static is most usually created by friction or rubbing and most usually under low atmospheric or dry conditions.

STATIC ELIMINATOR – A device that can be added onto a press or converting line to neutralize static build-up.

STATISTICAL PROCESS CONTROL – Label printing, whether sheet-fed or web-fed, is a manufacturing process that uses variable components such as paper, inks, plates, films, etc., and which is carried out on machinery that will eventually show signs of wear, or can be set up differently by the same or different operators. As such, all these possible variable elements need to be controlled wherever possible.

The system by which the printed label results can be controlled is referred to as Statistical Process Control (more commonly known as Quality Control in the past). Successful SPC depends on the Quality Assurance functions of raw materials specifications and controls, printing process control, on acceptable standards and tolerances and on inspection.

Specification and control of materials. Label printing uses a wide range of raw materials – label stock, ink, adhesives, varnishes, blankets, rollers, cutting dies, etc. – which should all ideally be controlled. Raw materials properties that can be measured, tested and controlled include the moisture content, brightness, gloss, absorption or picking of paper face materials; the tack, gloss, drying, color, yield or flow of inks or varnishes and Shore hardness of blankets, plates or rollers.

Printing process control. The control of printing and processes can begin at the press fingerprinting stage, while color bars, targets, grayscales and other devices or instruments are used to provide objective measurements and numbers to control production. Instruments used for checking at various stages throughout the color origination, proofing, or printing process stages include spectro-photometers and densitometers.

Standards and tolerances. Agreed standards and tolerances should preferably be established between the printer and the customer at the time the job specifications are finalised and accepted. These should aim to agree on tolerances for color variations, density variations, register variations, cutter tolerances, etc. and will vary according to the quality of the labels being produced.

Inspection. The use of inspection procedures is essential in ensuring that SPC systems function properly and that materials control, process control and standards and tolerances are all functioning properly, ensuring that quality is improved and waste reduced.

STEP ACROSS – Measurement from the edge of one label to the same edge of an adjacent label

STEPPED ANVIL – An anvil with the bearer area either higher or lower than the main body.

Stepped up. Bearer is lower than the body, creating a deeper cut.

Stepped down. Body is lower than the bearer, creating a more shallow cut

STEP-AND-REPEAT – The action of reproducing a number of design images onto film and subsequently plate, permitting the label design to be printed as more than one label at a time. This speeds up production by reducing press running time. In digital workflows, the step-and-repeat designs are imaged directly to plate.

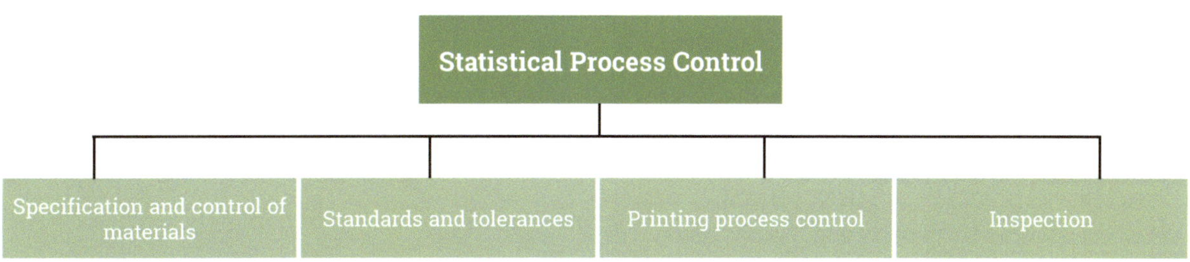

Key elements in the operation of a statistical process control system.

STEP AROUND – Measurement from the edge of one label to the same edge of an adjacent label

STERILIZATION INKS – Inks which change color when sterilization of the labeling/packaging of surgical or medical products using steam or ethylene oxide has been completed satisfactorily.

STICKYBACK – A double-faced adhesive coated material that is used for mounting printing plates onto the printing cylinder on a rotary press. *See also Plate mounting tapes.*

STIFFNESS – Stiffness or rigidity is the property of a material that enables it to resist bending forces.

STOCHASTIC SCREENING – An alternative to conventional halftone screening in which the image is separated into very fine, randomly-placed microdots (measured in microns), rather than the more conventional grid of geometrically aligned halftone cells. Dots are the same shape and size in most versions, but there is varying spacing between the dots.

Sometimes also called frequency modulated or FM screening. Stochastic screening eliminates screen angles and the possibility of moiré patterns and provides greater image detail due to the lack of screen rulings and screen angles. There are no rosette patterns with the process. Results are impressive on both coated and uncoated paper and film.

The stochastic screening process involves imaging dots on film using special randomizing software. The software uses mathematical expressions to statistically evaluate and randomly distribute pixels under a fixed set of parameters.

Stochastic screening produces smoother gradations when vignettes, blends and degrades are involved and enables almost any combination of colors to be used in the creation of subtle or dramatic effects. Interest in the process is being investigated/used in the printing of some labels and packaging using the flexo and gravure processes.

The dot size used in stochastic screening is extremely small when compared to the size of the highlight dot in conventional screening. It is recommended that stochastic screening is used for flexography only after the printer and color separator have undertaken press fingerprinting to determine the ideal dot size and accurate compensation for dot gain.

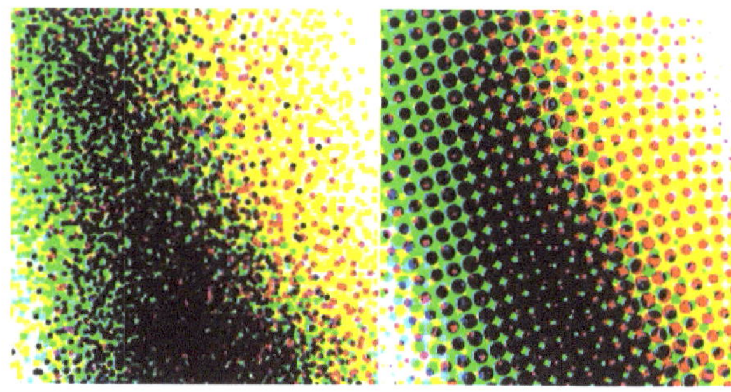

Random placing of dots in stochastic screening compared to AM (or conventional) screening.

STOCK – Paper or other material to be printed or converted. Also referred to as the substrate.

STOCK KEEPING UNIT – *See under SKU.*

STRAIGHT CUTTING – Where labels are printed in sheets on offset presses with, sometimes as many as 100 – 150 labels on each sheet, these sheets of multiple labels must be straight cut, either for use as square or rectangular labels or before cutting out or punching to shape.

The operation of straight cutting requires piles of up to 1,000 printed and properly stacked label

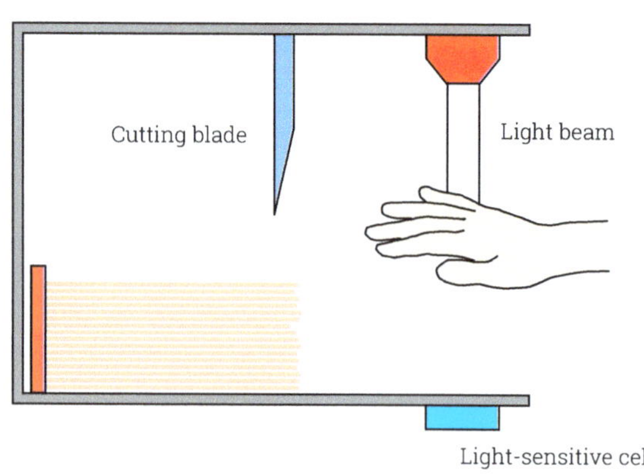

Guillotine cutting of labels sheets. The operator is protected by photoelectric beams

The operation of straightsheets to be fed into the backgage of a guillotine, usually using air glides or an air table which float the paper into position. With the pile against one of the rigid sides of the guillotine the pile is positioned for the first cut.

Once positioned, a clamp descends on the pile to hold it in position while the cut is made with a guillotine knife. The clamp then rises ready for the pile to be moved to the next cutting position. Photoelectric cells are used to protect the guillotine operations.

All subsequent cuts on modern guillotines are carried out using programed gages, with the back gage moving automatically to the correct position for the next cut. Tolerances for straight label cutting are fairly critical to the high-speed application of wet-glue labels.

STRETCH SLEEVE LABELS – The use of stretch sleeve labeling is a rather more recent development of sleeve technology which, instead of shrinking the film tube to fit over the bottle, uses a stretchable film tube which is cut to the depth of the sleeve required and then opened up to slide over the bottle. It then shrinks to fit.

Stretch sleeve technology offers advantages when applied to PET containers in that the flexible sleeve is able to compensate for the expansion of the PET container that occurs after filling. The technique is also well-suited to chilled products, such as fruit juices, where the bottle or container may change shape during filling and the label being subjected to moist conditions in distribution. Returnable bottles can be easily de-sleeved before entering the bottle washer, since the sleeve is not glued to the bottle.

Stretch sleeve label films are Low Density Polyethylene (LDPE) materials. Like the other unsupported film labeling solutions – reel-fed wrap-around, cut and stack wrap-around, shrink sleeving – stretch sleeve films are primarily printed with gravure or flexo technology. However, unlike shrink sleeving, there is no need to use image distortion techniques in the stretch sleeve process. Again, films can be reverse-side printed to minimise ink scuffing or rubbing.

Stretch sleeve label application. Because of the nature of the stretch sleeve labeling process, it can only be used for body decoration of straight-sided containers. It cannot be used with tapering necks or shaped bottles – although it can be combined in an application system with, say, a pressure sensitive neck label. No flexibility in terms of front, back or neck labeling options exists. However, stretch sleeving can be carried out at higher speeds than shrink sleeving – although it is far more limited in application. Costs are lower than for shrink sleeving.

Stretch sleeving is today primarily used for large-size bottles for carbonated drinks. The expansion of CO_2 inside the bottle being accommodated by the stretch label. Generally seen as more relevant where easy removal of body labels on returnable bottles is required. If returnability is a key criterion then the option of stretch sleeving may have higher priority.

Roll-fed stretch sleeves also work well with straight-sided single-serve bottles and are often used with PET bottles in pints or other sizes. Despite its limited applications, an advantage of stretch sleeving is that the elasticity of the material obviates the need for heat shrink tunnels in the production line.

STRIKE-THROUGH – Penetration of an ink or adhesive through the substrate.

STRINGINESS – A condition of a pressure-sensitive adhesive in which it feels very soft and mushy and where the adhesive can be pulled out into 'strings'. The thread-like residues of adhesive that occur when a pressure-sensitive label is removed from a substrate.

STRIPE AND INTERMITTENT ADHESIVE COATING – Stripe or strip coatings can be created by inserting blocking shims at selected points in the die orifice on the coating head. The stripes can be wide or narrow, as desired by the converter or his customer.

This type of coating head can also be adapted to provide intermittent patterns. In this case the distance between the contact lips of the die and the control valves is reduced. The adhesive supply can then be shut off and re-started accurately and cleanly, enabling individual rectangles of adhesive of different length and width to be laid down.

The use of electronic timing devices allows this type of pattern to be controlled and set in 1 mm increments. Such systems are also able to read registration marks allowing a versatile product range to be created.

Screen coating heads are more expensive than slot die heads, but enable more complex adhesive patterns to be coated. Rotary screen coaters are used to selectively apply adhesive to products such as label/form combinations and other multi-layer products with removable elements.

S

STRIPPING – Also called waste stripping. Removal of the face material and adhesive (the matrix waste) from around the die-cut label shape by taking it around a roller, or over a metal bar, prior to being re-wound. *See also Matrix waste.*

SUBSTITUTION ERROR – Refers to a character that has been wrongly decoded when a barcode symbol is being read.

SUBSTRATE – The material, stock, unsupported film or other material upon which a coating of adhesive, silicone or primer is applied, or on which labels are printed. The surface to which a label is applied. Label stock converting/laminating companies also refer to the face stock in the construction as the substrate.

SUBSTRATE COOLING – There is a growing interest in coating films with radiation curable silicones, especially thermoplastic films, for a range of special release liner applications. However, polyethylene and polypropylene films are temperature sensitive and while conditions in a UV-cure chamber are considerably gentler than those found in hot-air curing ovens for thermal cure silicones, they can present problems for thin caliper films. Sensitive substrates can be cooled very effectively by fitting the cure chamber with water-cooled rollers or flat platens. *See Water-cooled.*

SUBTRACTIVE PRIMARIES – The primary colors (hues) used for process color printing. Yellow, magenta and cyan color inks that subtract light and when overlapped, produce other colors and black images.

Subtractive color combinations using printing inks.

SUNLIGHT RESISTANCE – The ability of a material to resist deterioration under the action of sunlight, especially wavelengths that are at the infrared or UV ends of the spectrum.

SUNLIGHT RESISTANT ADHESIVE – An adhesive that contains an inhibitor to resist its destruction by ultra-violet rays.

SUPPLY CHAIN MANAGEMENT – It has been stated that some 30 per cent of the total cost of the manufacturing of consumer products and getting them into retail outlets and on the shelves is in the supply chain, which is frequently beset with problems related to mismatch between materials/products, in materials handling and in information flow. Any lack of accurate and timely information relating to order status, inventory levels or delivery times will both create uncertainty and variability into the supply chain.

The target aim for product, pack and label suppliers and the retail groups is to alleviate these uncertainties and variabilities through the use of new generations of 'intelligent' packaging and labels which can create a direct link between a physical product and all the supporting information needed to manufacture, ship and handle that product right through to the store shelves.

This means using the potential of intelligent technology, such as RFID – together with new standards and network infrastructures enabling tagged products to be connected to real time information – to increase efficiency and effectiveness and reduce costs, in supply chains. Coupling this information into existing business/management information systems can provide immediate benefits in inventory management, from manufacturer, packaging and label producer, through distribution and into the retail environment.

The use of systems based on intelligent (smart) packaging and labels – and the linking of these into a network – will also enable the real-time data and information that is obtained to be used to re-assess the whole supply chain, identify improvements then address inefficiencies.

RFID and other 'smart' systems have the potential to significantly increase supply chain efficiency, improve the speed of distribution, reduce labor costs and minimize 'out-of-stock' situations.

SUPPORT ROLLER – a cylinder positioned below the anvil cylinder in a label press to support the anvil and cutting cylinder, decreasing problems with flexing of the cutting cylinder when increased pressure is applied

SURFACE ACTIVE AGENT – A material used in small quantities which affects the ability of liquids such as water to wet or penetrate. Materials which decrease surface tension are wetting, dispersing, foaming, penetrating or detergent agents. Those which increase surface tension are used as defoaming or demulsifying agents.

SURFACE ENERGY – A measure of the cohesive energy of the surface molecules of a material, usually expressed in dynes/cm. Liquids that have a lower surface energy than a solid will normally wet or spread out on that solid. Liquids or liquid-like materials, such as a pressure-sensitive adhesive, which have a higher surface energy than a solid (e.g. a silicone release coating) do not wet such solids well and so do not adhere strongly to such substrates.

SURFACE PREPARATION – Chemical or physical methods of preparing a surface for further processing.

SURFACE TACKINESS – The stickiness of an adhesive coating. It is determined by rolling a steel cylinder of a certain size along an adhesive strip and measuring the distance it requires to come to a standstill. Not to be confused with 'tack' or 'initial adhesion'.

SURFACE TENSION – The charge generated in plastic surfaces, especially films, by means of a pre-treatment process such as corona treatment or flaming. Surface tension is measured in mN/m (formerly dyne/cm).

The property by which all liquids – through contraction of the surface – tend to bring the contained volume into a form with the least possible area. This is due to molecular forces. For inks to be compatible with a label substrate they must have a surface tension that approximates to that of the substrate.

SURFACE TREATMENT – The use of chemical, flame or corona treatment to increase the surface energy of low surface energy materials (such as plastic polymer films) so as to improve their ability to be 'wetted' by inks, lacquers, adhesives or release coatings.

SURFACTANTS – A generic term to describe all surface active agents.

SUSTAINABLE – To be sustainable, it must be an action/solution that you can keep doing, not just year after year, but decade upon decade, generation upon generation, century upon century. It must not require non-renewable fossil fuels; it must not deplete supplies of fresh water or topsoil; and it must leave stocks of fish and other wildlife at least as rich and as varied as they were when the process started.

SUSTAINABILITY CONSORTIUM (THE) – TSC is an independent group of scientists and engineers from leading academic research institutions who are developing the science to support the indexing of consumer products throughout all phases of a product's life.

SUSTAINABLE FORESTRY INITIATIVE – A fully independent, charitable organization dedicated to promoting sustainable forest management. *See also SFI.*

SUSTAINABLE FORESTS (TREE FARMS) – Plantations containing acutely engineered trees that viably produce high-value pulp for paper, building and furniture construction, taking into account such important issues as the surrounding people and ecosystems, water usage and recovery and harvest yield.

SUSTAINABLE PACKAGING COALITION (SPC) – is a membership-based, cross-industry group of supply chain stakeholders that work together on a variety of projects, utilizing and sharing strengths and proficiencies to establish a closed loop packaging infrastructure. It is a GreenBlue subsidiary.

SWEDISH LABEL ASSOCIATION – *See under GFF Sweden.*

SYMBOL LENGTH – The length of a barcode symbol. The length is measured from the beginning of the quiet area adjacent to the start character, through to the end of the quiet area adjacent to the stop character. *See also Barcode and Barcode symbol.*

SYNTHETIC PAPER – The use of synthetic papers for labeling in niche markets where moisture, most contaminants and harsh environmental conditions would severely damage paper labels, has been undertaken for many years. Synthetic papers generally resist tearing, water, grease and certain chemicals, can withstand extremes of heat and cold, have good UV-resistance and weathering characteristics, are non-toxic and, depending on the specific synthetic material, are FDA approved for food contact.

Synthetic papers are designed to incorporate the best attributes of natural paper and plastic to form a material that delivers strength and durability, yet are printable by most mechanical or electronic printing processes.

Collectively, they can be perforated, sprocket-punched and come in reels or sheets, with or without an adhesive coating, while top coatings on synthetics produce high-quality printing characteristics similar to paper. The most widely used synthetic papers include Polyart, Synteape, Teslin, Tyvek, Papermatch, Valeron and Yupo:

Arjobex's Polyart is a multi-layer biaxially-oriented HDPE, with a clay coating for good print receptivity with an absorbent surface similar to that of matt-coated paper.

Arjo Wiggin's Synteape is a multi-ply film based on polypropylene which contains additives and inorganic fillers. It is extruded and stretched bi-axially, forming micro voids in the web. These, with the fillers give opacity, whiteness and good printability.

Teslin is a mono-layer polyolefin material with a micro-porous structure. It is not coated.

DuPont's Tyvek is a non-woven material made up of spun-bonded HDPE fibres. It has no fillers and offers high strength.

Papermatch from Alpha Packaging Films is a mono-extruded HDPE film containing inorganic fillers to give a micro-roughened surface. It has no coating.

Valeron/PS is a pressure-sensitive version of Van Leer's proprietary high-strength films. Made from extruded HDPE resins and stretched in-line.

Yupo is a multi-layered synthetic paper composed primarily of polypropylene resin along with inorganic fillers and additives. It derives its strength from the base layer, and has an ultra-bright finish and smooth texture from its surface layers.

Applications for synthetic paper labels and tags include the labeling of products which are shipped or stored outdoors; outdoor bar-coding; products that are used on pallets or for garden supplies; for luggage, airline baggage and horticultural tags; as hospital patient identification tags; for chemical drum labels; slaughterhouse meat and carcass tags; tags for outdoor signs or notices; and for in-mold labeling of blow-molded HDPE bottles for under-the-sink products.

Price will always be a deterrent to wider use of synthetic papers, but for the right applications it is a very cost-effective product.

SYNTHETIC RUBBER-BASED ADHESIVE – A pressure-sensitive adhesive whose main component is synthetic rubber. This may be processed as a solvent-based dispersion or hot-melt adhesive. Synthetic rubber-based adhesives display high adhesive strength on rough and apolar surfaces. *See also Adhesives, pressure-sensitive and Pressure-sensitive adhesives.*

SYNTHETICS – Labels materials developed and manufactured through chemical means rather than through the use of natural products. *See also Synthetic paper*

T

TACK – The property, governed by viscosity and adhesion, which renders a film of printing ink, varnish or adhesive sticky to the touch. A high-tack ink has a high separation force and may cause surface picking or splitting of weak papers or paper coatings.

The property of a pressure-sensitive adhesive that causes it to adhere to a substrate with minimum pressure and time.

TACKIFIER – An additive that is used to improve the tack or stickiness of an adhesive.

TAG DEACTIVATION – All electronic article surveillance (EAS) tags must be removed, cancelled or deactivated at the point of sale so that the purchaser does not trigger an alarm in another store or outlet using EAS technology. The method of removing, cancellation or deactivation will depend on the type of tag used. Hard tags and color tags for example, are removed using a special detacher device at the cash point that releases the holding pin from the tag.

Soft tags are removed in the same way if attached with a pin. Where attached with a plastic thread, this is cut with scissors to facilitate removal. Pressure-sensitive EAS labels are not normally reusable and are deactivated at the point-of-sale by passing the label over, or holding it to, a special electronic deactivation pad by the cash desk. In some cases, a deactivation unit is mounted at the mouth of a hand-held barcode scanner.

Automatic hands-free deactivation devices scan the barcode and simultaneously neutralize the EAS tag, giving a high-speed point-of-sale procedure, which provides both pricing and stock control in one operation.

TAG, ELECTRONIC – A unique means of identifying a component. A unit containing some electronics and a unique number that can be read remotely at a reading or charging point. Such tags can be read-only or read-write.

TAGGED IMAGE FILE FORMAT – *See under TIFF.*

TAGGED INKS – Tagged inks provide an effective covert feature to confirm conclusively that a product is genuine. There are various tag materials, from certain chemicals through to micro-biological systems. For a tag to be effective it must be capable of 100 per cent verification at all times and must not be detected by unauthorized personnel.

TAGGERS/TAGGING GUNS – Taggers or tagging guns are hand-held devices that are used with a wide selection of attacher tags, plastic fasteners, locs, tie lines and j-hooks to securely attach a tag to clothing, accessories, fabrics, pairs of shoes, etc, so that the tags will stay in place until cut loose by the customer after purchase.

Taggers come with a choice of needle attaching devices designed especially for anything from delicate

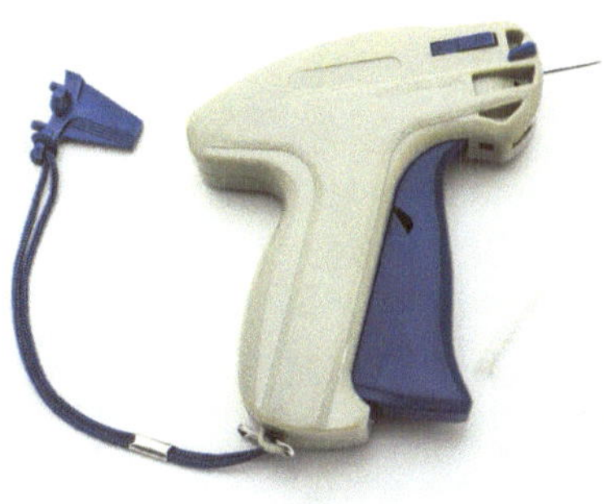

An example of a tagging gun.

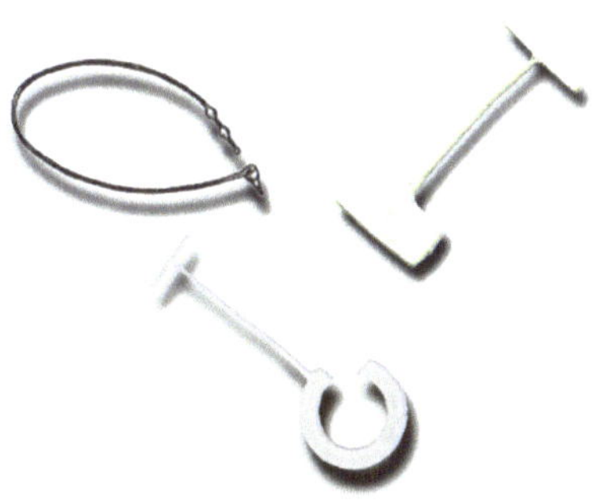

Tagging guns are used with a variety of plastic fastners, tie liners or hooks.

fabrics and merchandise right up to heavy duty materials.

Taggers are easy to use and incorporate a safety cap to help prevent injury.

Attacher tags are available in white or colored, in different lengths, or manufactured to customer's specific requirements. Tags used with taggers come in a variety of shapes and sizes, punched with holes for the attacher tags, and may be supplied blank, mechanically pre-printed and/or electronically overprinted in anything from one to four or more colors. *See also Tag, printed and Hang tag.*

TAG LABEL – Labels produced on heavy paper, card or special tag stock with die-cut holes so that the labels can be folded over a packaging material – as in a header or hanger label. *See also Hang tag.*

TAG POLLUTION – The concept that live electronic article surveillance (EAS) tags could be present within a shopping bag after the consumer has paid for them and left the store. Such un-cancelled tags may then set off alarms in other stores. The risk of this increases with the rise in the tag population and the overall number of detection systems installed by retailers.

TAG, PRINTED – Strong grade of paper board or synthetic material with good bending/folding characteristics, high strength, tearing resistance and selected for printing items where durability is expected, together with a device for attaching the tag to a product. Used to mark or identify items, such as clothing, sacks, agricultural or garden products, etc.

The tag may be rectangular, with square, diagonally clipped or rounded corners, folded, or cut or punched to specific shapes and are typically printed flexo, letterpress or with non-impact technology, or over-printed using thermal transfer, dot matrix, ink jet or laser systems. Many types of tags utilize a die-cut or punched hole, slot, or other means of attaching the tag to the product being marked. This hole or slot may be reinforced with a patch.

The devices used to attach tags to products usually consist of cotton string in any color, elastic string in black or white, a variety of plastic tag sizes and styles, wire, ornamental string in gold, silver or other colors, or nylon ribbon in any color. Taggers are used for attaching certain kinds of tags. *See also Taggers and Hang tag.*

Printed or over-printed tags are used in applications such as agriculture, marine, automotive, airline, garden center, industrial or retail sectors. Newer developments include multiple-layer tags and applications in some of the emerging security markets.

TAGS – Identification solution that is only partially attached to a product or item. Most usually converted through roll-fed production equipment. Tags may be attached by thread, string, plastic, ribbon, metal clip, wire or other device that enables the tag to be removed by the consumer or end user. *See also Hang tag and Tag, printed.*

TAMPER-EVIDENT LABELS – More correctly defined as tamper-resistant, the primary function of labels or seals described as tamper-evident or tamper-resistant is to indicate or prevent alteration by tampering and/or to indicate that an article is genuine and fit for use. If the label is breached or missing, it can be reasonably expected to provide visible evidence to suppliers, retailers or consumers that tampering has occurred. Tamper-evident/resistant labels or seals should not detract from the aesthetic appearance of the product or package and not give a false indication of tampering during normal authorized opening or legitimate use. Tamper-evident/tamper-resistant features are most commonly used to protect documents, packaging or products.

As a general rule however, product tampering cannot be prevented completely. The aim is to make life as difficult as possible for culprits. Tamper protection features can either be visible or concealed; some may be read with the naked eye, while others require technical aids; some have distinctive features which enable anybody (including the consumer) to verify product authenticity. The proper choice of features depends on the purpose to be met by tampering protection. In many cases a combined solution may be recommended.

Numerous types of tamper-evident or tamper-resistant label product configurations have been developed. These include:

- Multi-layered substrate constructions that de-laminate or separate so as to reveal hidden messages
- Fragile or ultra-fragile label stocks that provide

Types of tamper-evident security labels.

tamper-evidence due to their low structural integrity. Usually combined with a high strength adhesive.

- Labels incorporating security cuts which enable the label to break-up when removal is attempted
- Label materials that contain hidden messages.

When peeled back from a surface the adhesive separates from the backing to reveal a visible 'void' or other message pattern on either the film or the surface

- Component laminate adhesive and polyester films which incorporate buried graphics, two-way graphics or customer designs
- Paper or film laminates that de-laminate under certain conditions
- Pilfer indicating seals that incorporate an adhesive activated dye that exposes a hidden message if attempts are made to re-seal a container
- Shrink sleeves that partly or completely envelop the pack closure. The sleeves can also be perforated to enhance tamper-evidence
- Tape seals that contain unique features that make it apparent that seals have been remove

Many of the existing available tamper-evident and tamper-resistant devices are similar in construction and means of tamper indicating, but use different materials and adhesives.

TAMPERPROOF – Although tamperproof is a term used by sections of the label industry to refer to a pressure-sensitive destructible material that cannot be removed intact from a surface or product, thus making re-positioning or re-use of the label impossible, the term is not accepted by the FDA and other industry practitioners. The more correct terminology is to use tamper-resistant or tamper-evident. *See Tamper-evident.*

TAMPER-RESISTANT – *See entry under Tamper-evident*

TANDEM COATING – A process in which different coating operations are carried out sequentially in-line. An example is the process in which a backing is coated with a silicone release composition, the coating is then cured, the release-coated substrate is then coated with adhesive and laminated to the facestock – all in one pass through the coating line.

TAPE, PRESSURE-SENSITIVE – A single one-sided or two-sided adhesive coated substrate wound on a spool for consumer or industrial use, with or without a release liner depending on application. Most tape produced is self-wound packaging tape that does not require a release liner. Tape types also include transfer tapes used for bonding and joining surfaces together, which cover a very wide spectrum of applications. These include the replacement of rivets and screws in white goods assembly, web splicing in the paper and board industry and temporary holding in industrial and medical end uses.

Tapes may be coated paper, cloth, film or metal foil. Release liners are primarily used for double-sided and transfer tapes which in turn require two-sided silicone coating with differential release levels. Double-sided tapes also include the permanent one side/removable other side tapes that are now widely used on resealable garment bags and permanent/resealable mailing envelopes.

TAPE STREAMER – Used for back-up and holding data that is beyond the normal storage capacity of a diskette, a tape streamer is a cassette containing a reel of magnetically-coated narrow plastic tape upon which data can be stored.

TAPPI – The Technical Association of the Paper and Pulp Industry (TAPPI) in the USA.

TARNISHPROOF LABEL – A pressure-sensitive label that is free of any substances that will tarnish discolor or blemish copper or silver.

TEAR STRENGTH – The force required to tear three sheets of paper in either M.D. (machine direction) or X.D. (cross direction). The higher the value the stronger the paper. Usually quoted in milli-Newtons (mN) and described as Tear (3 sheet).

Different self-adhesive label papers will exhibit greater or lesser tear and tensile strengths. Strong paper aids waste stripping during label conversion; a most important factor where free form label shapes are needed. Similarly, if a label is to be specified with removable adhesive, high strength label materials will enable the label to be removed without the label shredding.

Conversely, a weak paper will sometimes give the illusion of greater adhesion, which may be used to advantage in applications where a pilfer-evident seal is required. The choice of a paper or film material with a low edge tear, such as acetate, may then be considered.

TEMPERATURE INDICATORS – Self-expiring indicators and self-destructing labels or similar devices, that record the history of a product/pack temperature profile as it progresses through its life cycle. They are designed to provide a visual indication of the condition of a product/pack and according to the level of sophistication of the particular indicator, will show:

- Whether the product is at the wrong temperature
- Whether the product has, at some stage, been subjected to a wrong temperature during storage or distribution
- Whether the product has, at some stage, been subjected to a wrong temperature during storage or distribution, plus indicate the duration of that wrong temperature exposure.

Low temperatures slow down the growth of micro-organisms and enzyme activity, therefore prolonging shelf-life, quality and safety of food or other products which are temperature sensitive or dependent. Temperature or time/temperature indicators enhance the possibility to control food quality, safety and shelf-life, from producer to consumer.

Temperature indicators are most commonly based on thermochromic ink technology and devices, which thermally react and change color at a specific pre-determined temperature level. More sophisticated time/temperature indicators that can provide a history of a product/pack time/temperature profile incorporate low melting point wax systems, capillary action, colored indicators that react to temperature changes, or to enzymes. *See also Food freshness and Time/temperature tags.*

TEMPERATURE RANGE – Each self adhesive recipe coated by laminators will be designed to adhere to its optimum performance within a specified temperature range. This is often described as the 'Service temperature range' of the adhesive, within which it will essentially remain unchanged for the duration of its shelf life. This time can vary depending on the substrate coated and the labeled surface condition and type.

TENSILE STRENGTH – The lengthways strength of paper, which determines runability and constant throughput during label die-cutting, waste stripping and printing during conversion.

The force required to snap a paper sample of standard width, cut in either the M.D. (machine direction) or X.D. (cross direction). The higher the value the stronger is the paper. Usually quoted in Newtons (N) per 15mm (0.6") paper width.

TENSION CONTROL – During the unwinding of a roll on a converting or web-fed press line the paper, film, foil or laminate material being used is under tension or stress as a mechanical force operates to extend, stretch or pull the web apart. This tension needs to be controlled. Too little tension makes register control more difficult; too much tension may lead to the material stretching or growing in length.

Color to color registration, size of label, position of any perforation or punching, etc. may all be affected by inadequate tension control; the more flexible the material the more sophisticated tension control becomes essential.

The degree of control that should be built into a press may involve a considerable additional investment at the time of purchase, but should be more than repaid through increased productivity throughout the life of the press.

A typical tension control system on a press unwind incorporates sensing rollers mounted in a pivoting floating bracket balanced against the tension setting. The bracket and rollers change the floating position by pivoting in response to changes in factors such as roll diameter, running speed, acceleration of the web and friction changes within the braking system. The pivoting motion of the bracket transmits information that constantly adjusts the brake force to keep the tension in equilibrium.

TESTING, WET-GLUE LABELS – The need to control and test labels destined for use on high speed bottling or canning lines is frequently a requirement of the label specifications, with a wide range of equipment and apparatus now available for carrying out various test procedures to establish individual wet-glue label properties. In day-to-day practice however, most bottling and packaging plants only need to apply a limited number of tests. Some of the more important tests for wet-glue labels and label papers include:

Tear Strength Test Method. Designed as a simple test for use by bottling plants and small printing companies, tear strength testing is normally carried out using a tensile testing machine and a 10 or 15mm (0.4 or 0.6") wide paper strip.

Water Absorption Capacity Test Method. Designed to ensure that labels bond quickly and positively. Water absorption is measured by the 'Cobb' test. Suitable labels should have an absorption capacity of between 7 and 11 g/m2. Labels with a low absorption capacity may have problems due to the edges lifting after application. Labels with excessive water absorption capacity will tend to curl excessively.

Caustic Resistance Test Method. Designed to test the resistance of the label to the effect of caustic, resistance can be tested by placing 120 sq. cm. (18.60 sq. inches) of label paper in a sealed measuring vessel containing 20cl of 1.5% sodium hydroxide solution and shaking vigorously (around 30 times). A sample with adequate caustic resistance will not disintegrate, while there should be no contamination of the solution caused by 'pulping' of the fibers.

Paper Weight Test Method. Designed to test the weight per unit area (basis weight) of a label paper. Approximate paper weight testing can be determined by the use of pocket scales, providing several samples are measured and an average of the value calculated.

Laboratory scales are obviously the most accurate instruments, where a controlled atmosphere ensures the sample is credible. Thickness or caliper can be measured using a standard engineering micrometer, but special paper micrometers control the speed and force of the clamp into which the samples are fed. Again, the average of several measured values should be calculated.

Bending Stiffness Test Method. Designed to measure the flexural stiffness (i.e. bending strength) of paper and labels, bending stiffness is measured using a bending stiffness tester (for example, Thuring-Albert Tester) which provides a comparison of the bending strength of new labels with that of labels known to handle satisfactorily in labeling machines.

TEST PROCEDURES – Almost all label buyers and label end-users today will expect that the labels they purchase and use will meet key criteria in respect of materials, color, print, label line and end-user performance. These criteria will be established with the label printer and set out in the label specifications prior to the commencement of the job and before ordering the label materials, inks, varnishes, etc.

Many packaging and label end-user organizations undertake their own tests during the development of new labeling solutions to ensure that the specifications they draw up for the label printer and converter will:

- meet all the necessary brand identity and image requirements
- meet brand or house color criteria
- meet label line, handling, distribution and end-usage needs in terms of rubbing, scuffing, durability
- meet any product or usage resistance demands
- meet any legislative requirements (three months immersion in sea water for example)
- meet safety, waste or environmental demands

Once established in the specifications, the materials

suppliers (inks, papers or films), label printers, converters and application companies will be expected to check, test and confirm that all criteria and specifications are met throughout the label order.

Test procedures that are accepted by brand owners, label buyers, packaging and printing companies, paper and film suppliers, ink manufacturers and the like have been established over a long period of time by such organizations as PIRA, TAPPI, GATF, or by the relevant trade or industry associations, such as FINAT or TLMI. *See also Statistical process control.*

THERMAL PAPER – A paper coated with a heat-sensitive chemical coating that enables an image to be formed in areas heated by a thermal print head consisting of very many miniature heating elements evenly distributed along its printing width. These elements are selectively heated by pulses of energy which create points or dots of black (other colors are possible with appropriate paper) in the thermal coating and these dots create the image required from the computer's memory.

The image can be variable text, barcodes, or diagrams used for frozen or fresh food labels in supermarkets, industrial barcoded labels and tags, or with other market sectors, such as for fax machines, point-of-sale receipts, ticketing, bank statements, lottery and betting slips. *See Direct thermal printing or Thermal printing.*

THERMAL PRINTING – A non-impact electronic printing process which uses heated elements in a digitally-controlled thermal print head to create a darkened image on thermally coated paper. *See Direct thermal printing.*

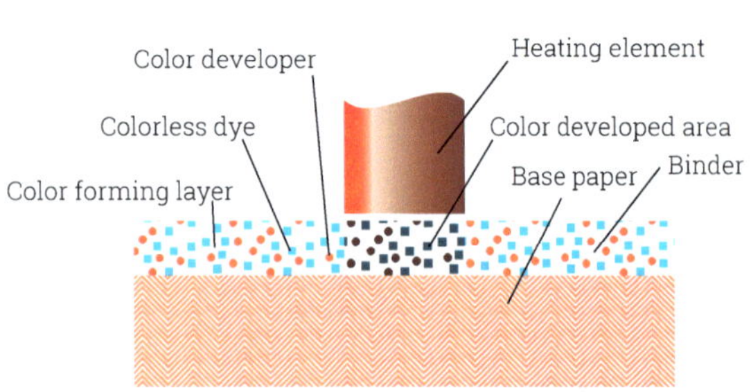

The special heat-sensitive chemical coating on the label substrate darkens under the action of the heating elements.

THERMAL TRANSFER LABELS – Pressure-sensitive labels specifically designed to accept heat activated ink from the ribbon of a thermal transfer printer.

The two essential criteria for satisfactory thermal transfer label printability are, firstly, 'smoothness', so that the label surface, appropriate filmic ribbon and print head can be brought together uniformly to ensure and even distribution of activated ribbon on the label. Secondly, 'receptivity' of the label surface to the activated ink, whether just by absorbing or by a special label top coating to which the transferred ink will have an affinity, which ensures an acceptable scratch or smudge resistant image, depending on the label requirements.

THERMAL TRANSFER PAPER – A face paper or pressure-sensitive labelstock that has been specifically designed to accept heat-activated ink from the ribbon of a thermal transfer printer.

THERMAL TRANSFER PRINTING – A very similar indirect printing process to direct thermal printing, where a heat sensitive ribbon replaces the heat sensitive thermal paper.

Very smooth and receptive conventional paper, or a suitable film, is printed by means of an electronically-controlled printing head which transfers variable data – including barcodes – to the label face material via a heat-sensitive printer ribbon placed between the printing head and the face material.

The thermal transfer printing head makes use of very small resistors or elements which are arranged across the printing width, which are selectively heated and cooled. The heated elements come into contact with a thin film one-pass ribbon, which carries a heat-activatable ink or coating on the underside. By rapidly heating and cooling the resistor dots on the print-head in the required formation, the required character or image is created and so transfers the selected heat-activated ink coating from the film carrier to the substrate according to the pattern or shape of the heated elements.

Thermal transfer printing is used for the variable information printing of batch codes, date codes, sequential numbers, text, diagrams and barcodes onto pallet, carton or box end labels, for warehousing and distribution requirements, for bakery labels and for DIY and industrial labelling. Printers may be incorporated into packaging and/or weighing lines or be stand-alone. Some are also print-and-apply systems. Direct thermal labels can also be printed in thermal transfer printers, by removing the filmic ribbon.

For short term label applications and where there is little chance of label image degradation or contamination, appropriate direct thermal labels are mostly used. Thermal transfer, having indefinite longevity and superior durability, has been widely preferred for industrial bar coding and longer term applications.

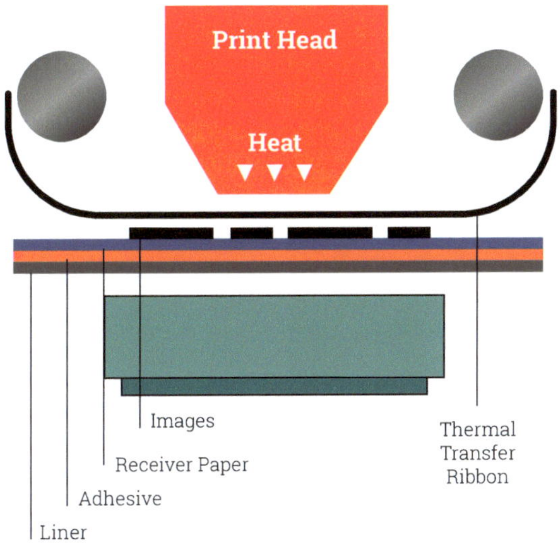

The thermal transfer printing process.

To ensure the optimum thermal transfer printing quality it is vital to choose a ribbon that matches the thermal print head type, with the performance requirements of the label application and label surface. *See also Label printing and production.*

THERMOCHROMIC INKS – Thermochromic inks are inks that change color with changes in temperature. They are available in two main types; reversible and irreversible.

Reversible thermochromic inks are generally used to show when the temperature of a pack or label has reached an optimum level. For example, when the temperature of a beverage (beer or wine) is at the appropriate level and ready to serve the thermochromic ink label will change from white to blue. When the product warms up the message will disappear.

Irreversible thermochromic inks are used as time/temperature indicators in food and other packaging to provide a record of whether the products have been stored at the correct temperature, as well as in anti-counterfeiting or anti-tampering devices.

There are many color-change options, including blue to colorless, orange to yellow and red to colorless. A dual thermochomic ik that rubs through one color and continues to change to a different color when rubbed harder is also available.. Color-to-color change thermochromic ink is used primarily for direct mail, advertising and promotional materials, packaging, labels educational and learning aids, stationery and other novelty pieces,

Thermoforming – A means of shaping or forming a thermoplastic sheet into a tub, cup or pot through the application of heat and pressure. The heated and softened plastic assumes the shape required by being forced against the sides of a mold until it sets and cools. The pressure used for forming may come from a vacuum, positive air pressure, or by mating male and female molds.

In-mold labeling of thermoformed pots and tubs using synthetic label materials has been developed and is used in some applications, e.g. soft spreads. *See also IML.*

THERMOPLASTIC – Materials capable of being melted or deformed by the action of heat.

THERMOSETTING PLASTIC – Plastic materials that once they have become rigid by chemical reaction or crosslinking can no longer be melted or deformed by the application of heat.

THRU CUT/THROUGH CUT/CUT THROUGH – The action of die-cutting through all the layers in a pressure-sensitive laminate. This may take place in just one part of the cutter profile, or it may involve the complete profile. Where the cutting is through the complete cutter profile then some form of collection system for control of the completely cut-out piece – once it is freed from the web or sheet – needs to be incorporated into the converting line.

TICKETS – The use and definition of tickets is extremely diverse, ranging from roll and raffle tickets, through redemption and barcoded tickets, to admission and reservation tickets, garment tickets, price tickets, transport and travel tickets, strung, unstrung and hang tickets, stock control tickets, coat check tickets and valet parking tickets. They may be printed and supplied on roll, in strips, as single cut tickets, as die-cut tickets, punched and strung, perforated and pre-printed, part-printed or part/totally electronically overprinted.

Examples of printed and string tickets.

Many tickets are sequentially numbered and/or barcoded, some carry security features, have tear-off stubs or carry an overall or stripe magnetic coating for automated reading and passage through entry systems. Printing may be anything from one simple black-only, working up to multiple colors or halftone printing.

A wide range of printing processes are used for the production of tickets: flexo, offset, hot stamping, letterpress, screen process, thermal, etc, depending on quality, materials used, whether sheet or roll printed, nature of any overprinting and application. Materials used include paper/paperboard ticket materials, card, thermal stock, laminate structures and some synthetics, and may be over-varnished or laminated.

TIE – The uncut portion of a perforation.

TIE COAT – Coating applied to a face material to improve the anchorage of the adhesive, ink or coating.

TIFF – Tagged Image File Format. A TIFF is a bitmapped graphic file format developed by Adobe and Microsoft with leading scanner vendors, for capturing scanned images and other large bit-maps. Although TIFF is not an official standard, it is a neutral format designed for compatibility with all applications and as such has become the most common interchange format used by DTP software and repro houses.

TIGHT RELEASE – A tight release – caused by a high level of adhesion between the release liner and adhesive – makes the label waste matrix difficult, or impossible, to remove during label conversion and the backing paper equally difficult to remove during label application.

TILT INDICATORS – Used for some packaging and distribution applications where it may be necessary to know whether adverse handling and shipping conditions have been encountered, 3M Tiltwatch™ labels incorporate a small disk that drops when a preset tilt angle is exceeded, revealing a red warning mark. They are used when goods must remain upright to ensure intact delivery, such as with heavy machinery, large computers, electronic panels, cut flowers, batteries, liquids or other products. *See also Shock labels.*

TIME/TEMPERATURE INDICATORS – A proprietary range of integrator tags which provide a visual record of time/temperature history of the tag/product. They are used to monitor the transport and storage of temperature-sensitive items, including chilled prepared foods where shortening of shelf-life is directly related to the length of time that the product has been removed from a correctly chilled environment.

One of the initial manufacturers of time/temperature indicators was 3M. Its MonitorMark™ indicator provides an affordable and widely used solution for the monitoring of products that are sensitive to time/temperature abuse, offering an easy-t-read irreversible signal that estimates the time the threshold temperature has been exceeded.

Developed by 3M, MonitorMark™ Time/Temperature Tags comprise five 'windows'. Once the tag has been activated, a blue color will travel progressively and irreversibly across the windows scale of the tag, from left to right, for as long as the ambient temperature exceeds the pre-set response temperature.

color will pause, only to re-start moving if the temperature should rise again. Activation is achieved by pulling a tab at the end of the tag which reveals a black stripe to show that the tag has been activated.

Other indicators include Freeze Watch, which uses a unique design to provide a clear visual indication that produce has been exposed to freezing temperatures,; TimeStrip, whose core technology is an inexpensive and easy-t-use smart label that monitors elapsed time and is easily adapted for numerous applications n the monitoring of product shelf life and environmental conditions; and iStrip, which identifies if an accidental freezing event has taken place in the cold chain, whereon it irreversibly changes color if exposed to freezing conditions.

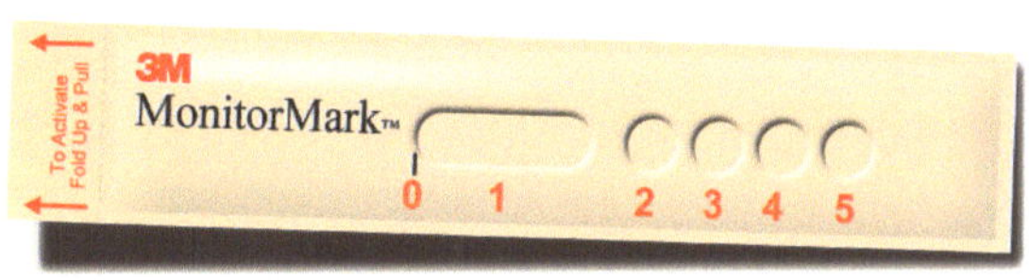

MonitorMark™ Time/Temperature Tag from 3M

TINT (SCREEN) – A solid color which has been reduced in shade by applying a screen during the origination process. The resulting tint is specified as a percentage screen of the original solid color.

TINT (SOLID) – A dense area of color without screening, usually defined as a percentage of a defined solid color, as opposed to a screened tint of a solid color. A solid tint color is obtained by ink mixing, rather than as a pre-press operation.

TINTERS – Tinters enable producers of multiple-part, colored forms and some types of colored labels to cut costs by purchasing only white stock, so avoiding the need to purchase more expensive mill-dyed or colored papers.

Tinters use metering systems to lay down water-based ink coatings of 0.5 to 3.0 grams per square meter at normal press line speeds, compared with flexo print tinting at 5 to 6 grams per square meter. Metering tinters also have the added advantage of

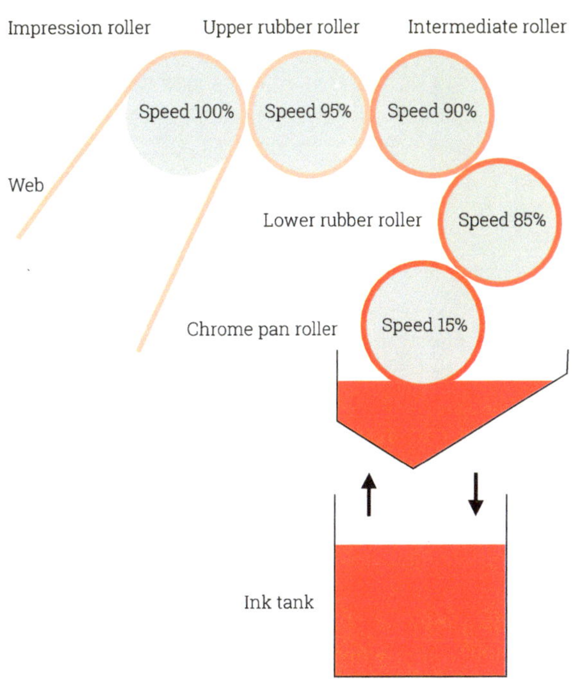

Principle of roll metering tinter.

using less water than flexo, avoiding potential drying, curling and laser printing problems.

Metering tinters are versatile double-sided systems capable of adding value by color tinting both the front and the back of a web, or tinting two colors, or two coatings on one side. The use of engraved rollers allows patterns or logos to be applied to either the front or the back of a web. This can be used to 'print' custom logos on the back side of the label stock while tinting the face with another color.

Metering systems are web driven and can be mounted in any position on a press. This allows coatings such as toner adhesion or UV coatings to be applied after printing before rewinding the labels.

Metering tinters can also apply coatings that are invisible except under UV light, coatings that turn orange in contact with bleach used to attempt eradication of ink, thermochromic inks and other functional coatings.

TLMI – Tag & Label Manufacturers Institute. The TLMI is a non-profit trade association of firms engaged in the manufacture of tags and pressure-sensitive labels, with both international and industry supplier members.

Originally formed in Cleveland in 1933 as the Tag Manufacturers Institute, when representatives of 20 tag manufacturers attended an organizational meeting. The Institute expanded in 1962 to include pressure-sensitive label manufacturers, with the first meeting of the newly formed Tag & Label Manufacturers Institute being held in Indiana in June of that year. Twelve companies attended the meeting. The Institute was further expanded in 1966 with the addition of supplier companies.

The Institute became a self-managed association in 1991 with office headquarters in Iowa City. Its current headquarters are in Glocester, MA.

TOLERANCE – A specified range within which products or printed labels must fall. Tolerances for printed labels may specify color, density, register, barcode readability, etc. within fairly defined limits to which the printer must adhere to avoid possible label rejection by the customer.

TOOLING – Custom-engraved or assembled die cylinders used for butt cutting or to make special perforations, punching or die-cut shapes. More commonly used to refer to all types of tooling required to produce finished labels.

A wide range of standard and custom prepared tooling is available to the label converter, from relatively simple butt cutting blades to complex punching tools and multi-level tools for cutting both face and liner layers in one pass.

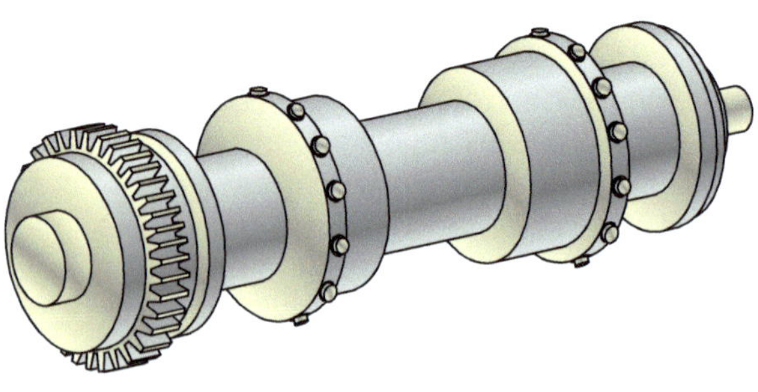

Adjustable E.D.P. punch tooling (can be menufactured to be positioned in cutter system or in place of the anvil for back cutting).

TOOTH COUNT – The tooth count is the number of teeth there are around the circumference of a gear attached to the printing cylinders, die-cutting cylinders or tooling cylinders. Each tooth count refers to separate and actual repeat length.

TOPCOAT – Surface finish applied to label face stock that aids print anchorage and/or durability. May also be used to provide a protective coating.

TORSION – The stress caused when a material is twisted.

TOTALLY CHLORINE-FREE PAPERS – Papers which are bleached without the use of any chlorine compounds

TOTAL PREVENTATIVE MAINTENANCE (TPM) – The introduction of planned levels of maintenance management to improve, over time, the use of equipment and machinery and offer benefits to the bottom line through waste reduction, higher operating speeds, less press or equipment downtime and fewer breakdowns or accidents.

Levels of maintenance management range from only doing maintenance after equipment has broken down up to the level where it should seldom break down due to adoption of a comprehensive program of continuous maintenance improvement. The most common levels of maintenance management are:

Level 1. Emergency repairs or corrective maintenance only. Usually results in high breakdown costs and poor levels of customer service due to unplanned stoppages.

Level 2. Selective maintenance in which the most common, often minor, faults or breakdowns are identified and action taken to keep these problem areas in good condition.

Level 3. Introducing a routine schedule of planned maintenance, as well as reporting procedures to give earlier notice of possible breakdowns.

Level 4. The use of regular monitoring aimed at identifying the condition of parts so as to assess any potential problems well in advance of failure. Level 4 maintenance may also incorporate manufacturer or other inspection tools – perhaps through Internet connection – that can undertake diagnosis without stopping the machine.

Level 5. A program of continuous improvement

aimed at consolidating all the previous levels of maintenance and encouraging local decision making.

TOTAL QUALITY MANAGEMENT – *See TQM.*

TOXICITY – Toxicity is the degree or level of virulence of a substance at which it may be regarded as injurious or poisonous to living tissues.

TOY LABELING – The labeling of toys and children's products must generally comply with the requirements of the relevant national or international safety standards, the most common of which are: International Voluntary Toy Safety Standard; European Standard EN71; ASTM F963.

These regulations require: the identification marking including the full name, trade or identification mark and the address of the manufacturer, the importer or the supplier and any warning or caution with respect to the safe keeping, use, consumption or disposal of a toy or children's product. This information is to be in a conspicuous position on the product, any package, or a label fixed to or on a document enclosed in the package.

Many of the most popular self-adhesive papers, vinyls, polyesters and foils comply with the relevant toy safety regulations, but this should be confirmed with the label stock supplier before use.

Before producing or using toy or children's products labels, the relevant current national, state or international regulations/standards should be assessed and complied with.

TQM – Total Quality Management (TQM) is based on the philosophy that any product or service can be improved upon and that improvements reduce costs, provide better performance and give higher reliability. The approach used by TQM is that any company can continuously improve the quality of its goods and services – manage its quality performance – through the participation of all levels of employees.

Developed in Japan after the Second World War, TQM's management approach is aimed at long-term success through customer satisfaction. It is based on the participation of all employees within a company or group and on improving processes, products and services. Key elements within a TQM program include:

- The customer defines quality. His requirements are confirmed and any problems identified
- Senior management is responsible for taking the 'quality' lead in developing the company's quality culture and strategy. All employees should be involved.
- Quality is dependent on design and execution of all processes and systems to a high standard. It should be designed in from the beginning.
- Error prevention should be built in to work processes and systems
- The 'cost of quality' or the 'price of poor quality or non performance' should be calculated
- Problem solving techniques should be developed and implemented
- Continuous quality improvement is key to achieving higher quality standards
- Suppliers should be involved in the achievement of quality targets
- An ongoing aim should be to shorten response times
- Results should be communicated to employees

Label companies already working to ISO 9000 standards can move to higher levels of quality management through implementation of TQM or Six Sigma. *See also ISO 9000, Statistical Process Control, TQM and Six Sigma.*

TRANSCEIVER – Refers to a terminal that can both transmit and receive data. *See also Terminal.*

TRANSDERMAL PATCH – A pressure-sensitive adhesive patch whose primary objective is to achieve a desired drug dosage transfer to a patient. The patches, supplied with a release liner backing, may be face printed, are die-cut to shape and must meet stability requirements and adhere to the skin for the prescribed length of the therapy. After use, they need to be removed cleanly from the skin.

The choice of adhesive for transdermal patches is therefore critical. Both acrylic and rubber-based pressure-sensitive adhesives are used.

Acrylic copolymer adhesives have a long history of use on skin and good solubility characteristics for a wide range of drugs and can be designed to meet specific solubility and skin adhesion requirements. Rubber-based adhesives are hydrophobic relative to acrylics, so offering a formulating tool at the other end of the solubility spectrum.

TRANSFER METALIZATION – A four-stage production process in which a carrier film is first coated with a transfer varnish. This varnish coating is then vacuum metalized. The extremely smooth surface of the carrier film ensures a high gloss finish to the end product. The metalized layer of varnish is then transferred from the carrier film to the actual label face material.

The transfer metalization process is extremely versatile and can be used for a wide diversity of products. The varnish layer now faces outwards and serves to protect the metalized layer of the laminate.

Transfer vacuum metalizing enables materials to be coated which cannot pass through the direct metalization or hot foil stamping process, such as polyethylene. Paper which is to be metalized does not need a smooth surface because the adhesive layer compensates for any unevenness.

If the substrate is embossed prior to metalization this results in a diffraction (holographic) pattern or special effect substrate. *See also Metalized paper or film.*

TRANSFER ROLLER – A transfer roller is a plain or engraved roller in a flexographic press inking system which rotates in contact with another plain roller or doctor blade so as to transfer variable amounts of ink.

TRANSLUCENT – Ability of material to permit the passage of light through it even though it is not itself transparent. The light is diffused through the material to make the material only partly transparent. Objects viewed through translucent materials are not clearly distinguishable and therefore cannot be described as 'clear' or 'glass clear'.

TRANSMISSION DENSITOMETER – An instrument for measuring the density of transparent materials by projecting light through the materials and measuring the amount of light received by a photoelectric cell. *See also Reflection densitometer.*

TRANSPARENT – The ability of a material to transmit light without any appreciable scattering. Objects viewed through transparent materials are clearly distinguishable and are therefore also described as 'clear' or 'glass clear'.

TRANSPARENT LABEL – Pressure-sensitive label in which the face material, adhesive and any protective coating transmit light and through which objects can be clearly seen. A transparent label on a clear bottle will allow the color, nature and volume of the contents of a bottle to be viewed through the label. *See also 'No-label' look.*

TRANSPONDER – A radio transmitter/receiver that becomes active when a predetermined signal is received. RFID tags are sometimes referred to as transponders.

TRAPPING – A pre-press technique that enables variation in print registration during a press run to be allowed for in the origination by providing an overlap between abutting colors on the desktop. It usually involves expanding the lighter of the two colors to overlap into the darker color.

Trapping avoids unsightly gaps created by registration variations in abutting colors. Designers may build traps into some of their less complex file designs by using simple publication or illustration software; more complex trapping will require the special trapping software used by service bureau or printers. *See also Spread and Choke.*

TRIM – Trim is used to describe the normal edge waste that is removed from a master roll of labelstock. On the printing and converting line, trim describes an action on the press, i.e. edge trim, waste trim.

TURKISH LABEL ASSOCIATION (ESD) – *See under ESD.*

TURNER BAR – A device used to change the direction of a web in a printing press or converting

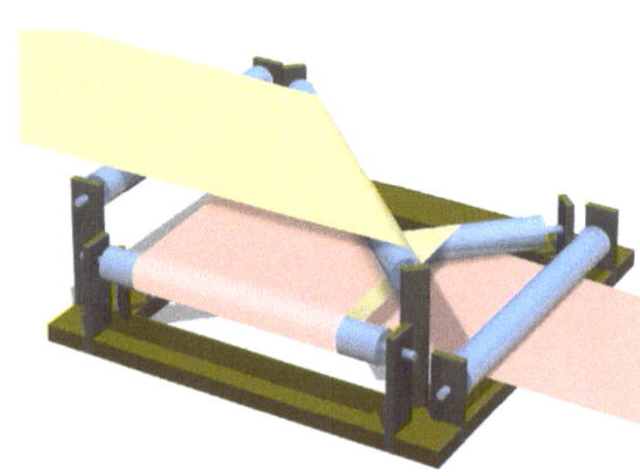

Turner bar changing the direction of a web.

230

line. Turner bars may be adjustable and/or electrically driven and are usually used to change the direction of a web by 45 degrees or 90 degrees. Also used to turn a label web from front to back for printing on the reverse side, such as for sheeted labels where the liner is printed with a trade name or instructions and then turned back again for die-cutting of the face material.

TURRET REWINDERS – Found in-line at the end of a roll-label press, turret rewinders facilitate the creation of successive rolls of self-adhesive labels coming off the press – without any interruption of press speed. Each roll of labels is wound on a core, usually to a specific number of labels per roll or to a specified roll size.

To accomplish this, the press rewinder unit features a turret system from which multiple spindles protrude. Depending on the equipment manufacturer, this could be anything from two to six spindles. As each individual roll is wound to the correct size on a spindle, the turret turns to the next spindle position (onto which a cardboard or plastic core has already been placed), the web is cut and the web end affixed to the new spindle for it to continue with winding the next roll. By this means, continuous rewinding of successive rolls of labels is achieved.

To achieve continuous automatic operation of turret rewinders there additionally needs to be a means of matching the speed of the core to the speed of the web at changeover. This is done by bringing the core speed up to the web speed, cutting away the web from the completed roll and affixing it to the new core – all at the full press running speed. Alternatively, there are festoon systems which accumulate the incoming web, enabling it to be effectively stopped at the point of change; the web can then be cut and a new roll started.

Turret rewinders require a method of affixing the web to each new core. This is commonly achieved with glue, but glue-less rewinding using various other methods to get the web started on the core are also available.

Particularly useful for high-speed, short-cycle winding of multiple small rolls, turret rewinders have become popular with some converters for high volume label jobs, enabling them to increase efficiency and productivity by not stopping the press.

TWIN-WEB CONSTRUCTIONS – Twin-web constructions are comprised of two separate web materials – usually a self-adhesive paper and a bond paper – that are joined together to create a single web that performs two distinct functions, such as providing both a label and a despatch document or receipt.

The press therefore requires a minimum of two web unwinds, each equipped with web guides. The self-adhesive web overlaps the bond paper, so that registration is critically important. Good unwind tension control is required to ensure there are no 'wrinkled' glue joints. The overlap between the self-adhesive and bond webs is typically no more than 3mm wide.

A small bead of glue is applied to the top of the bond/back of the self-adhesive liner, immediately before the lamination nip. This is typically a rapid setting 'cold glue' emulsion adhesive with good adhesion and shear.

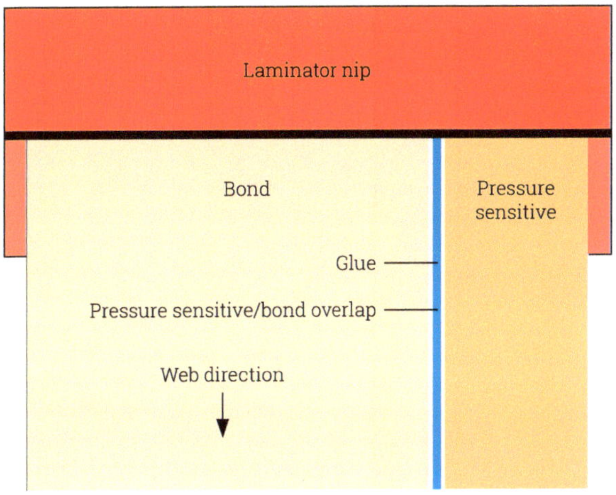

How twin-web form-label combinations are constructed.

The choice of material can be critical, with both the bond paper and pressure-sensitive face paper being printable using the inks on the press as well as by any subsequent variable print method chosen. It should have aesthetically acceptable whiteness and brightness levels, good lay-flat, and sufficient strength to withstand die-cutting, matrix stripping, perforation, hole punching, fan folding, etc.

T

TYVEK – A DuPont registered non-woven, spun-worded polyethylene material, which brings together several of the best properties of paper, textile fibers and film into one synthetic material which is extremely strong and resistant to continuous folding, tearing and weather conditions. Tyvek offers good printability and high resistance to moisture, chemicals and solvents. *See also Synthetic paper.*

U

UHF – Ultra-high frequency. Typically used to define tags that operate between 866 MHz and 930 MHz. UHF tags are able to send information much faster and further than high- and low-frequency tags, but the radio waves cannot pass through items with a high water content, such as fruit. More expensive than low-frequency tags, UHF tags also use more power.

UL – Underwriters Laboratories Inc. (UL) is an independent, not-for-profit product-safety testing and certification organization in North America. It has been testing products for public safety for more than a century and is one of the most recognized and reputable conformity assessment providers in the world.

The services of UL help companies achieve global acceptance, whether this is for electrical equipment, packaging, programmable systems or for a quality process. Since it was founded in 1894 UL has conducted over 100,000 product evaluations and now has customers in 99 countries.

ULTIMATE ADHESION – The ultimate adhesion available from a pressure-sensitive adhesive, determined by the force necessary to remove strips of tape from a surface after an extended period of time.

ULTRA HIGH FREQUENCY – An electronic article surveillance (EAS) system using microwave frequency. *See also UHF.*

ULTRA-VIOLET – That part of the light spectrum that has a shorter wavelength than visible light. Ultra-violet (UV) light is used in the label industry for the curing of inks and varnishes and for the curing of some silicone and adhesive coatings. *See also UV inks.*

UNDERCOLOR REMOVAL – An electronic scanning or retouching method that reduces the amount of cyan, magenta and yellow in a black area and increases the amount of black. The method improves the trapping of colors and reduces C, M, Y ink costs.

UNDER CURE – Where the cure of an ink, varnish or silicone is below the optimum. Under cure may be evidenced by tackiness, softness or migration of the ink or coating or by inferior physical properties.

UNDERCUT – the difference between the diameter of the bearer and diameter of the cylinder body. The difference divided by 2 equals the distance between the cylinder body and the anvil cylinder. Standard undercut is 480 micron (960 microns in diameter).

UNDERWRITERS LABORATORIES INC. – *See under UL.*

UNFEA – Union Nationale de Fabricants D'Etiquettes Adhésive is the French label association. There were originally several earlier French associations which were associated with the label industry, however obliquely, prior to the establishment of UNFEA.

In 1971, all the previous trade associations were dissolved and the Chambre Syndicale Manufacturers of Labels was created. On 1st January 1990, the Chambre Syndicale became UNFEA.

UNIFORM CODE COUNCIL – The Uniform Code Council (UCC) started in 1971 with a domestic North American mandate to administer the UPC barcode system that was created to help grocery stores speed up the checkout process and keep better track of inventory. From there, it grew into an international membership organization administering a global system of commerce used by more than one million companies, between them doing business in over 140 countries in some 23 different industries.

The Uniform Code Council later merged with EAN International to form GS1, a single worldwide organization point for UPC and EAN numbers (now known as GTIN -12 and GTIN – 13. *See GS1.*

UNWIND DEVICES – The mandrel and brake device on a press or converting line from which a roll of labelstock is unwound.

Air shafts have displaced solid steel shafts with locking cones on most label and forms press unwind stands today. These are more operator-friendly and cut out the time taken to unlock and relock cones at every reel change. Modern quick release cones can be fitted to older presses still using solid shafts.

As reel changes take up valuable production time

and cause waste, downtime due to reel changes can be reduced by using roll-lifting gear, as well as unwind units that can handle larger diameter rolls. Automatic web splicing offers even greater time savings for converters with long-run jobs.

UPC – The Universal Product Code (UPC) – A 12-digit code used by manufacturers to identify themselves and their products. The allocation of the manufacturer's identification number and the administration of the code system is done on payment of an annual fee by the manufacturer to, formerly, the Uniform Code Council, now GS1. The allocated code enables manufacturer members of the GS1 to identify trade items, logistics units, assets, locations and service relations so as to bring efficiency into their supply chains.

UPC code

The UPC code consists of:

- A six-digit manufacturer's identification number. This is allocated to the first six digits of the code.
- Followed by a five-digit manufacturer's item number that is allocated to a specific product. The same item number is not used on more than one product. In general, every item that a manufacturer sells, as well as every size of pack and every re-packaging of the item, carries a different item number
- The final digit is a specially calculated check digit, which enables the scanner to determine whether the scanned number is correct. Every time an item is scanned, the scanner checks to see if the check digit matches the digit it has scanned. If the check digit is different to the scanned number the scanner will know that something is incorrect. The item will then need to be re-scanned.

Each manufacturer employs a UPC co-ordinator to assign and keep a record of all item numbers and the products that they relate to. *See also Barcode, Barcode symbol and GS1.*

UPCYCLE – A term that describes adding value to an item by reprocessing one or a few of its materials into something of higher quality

UPS SYSTEMS – Uninterrupted Power Supply (UPS) systems automatically provide back-up power supply to computers using batteries or generators should the main power supply be cut off. UPS systems allow sufficient time to back-up and save work or, in the case of large systems, to continue operating until the power is restored.

UV – UV or ultra-violet is that part of the spectrum in which the wavelengths are shorter than those of visible light. UV light is used in the label industry to cure inks, coatings and adhesives.

UV CURABLE ACRYLIC ADHESIVES – These are a new class of hot-melt pressure-sensitive polyacrylate adhesives which are blended with softeners and tackifiers to obtain a range of peel and shear adhesion properties. They are copolymerized with a photinitiator so that there are no free photoinitiator residues which could cause skin irritation. The adhesive can be coated at temperatures between 110ºC and 130ºC (230°F and 260°F), similar to warm melt adhesives.

UV-CURABLE SBS ADHESIVES – Several adhesive suppliers have developed UV-curable styrene-butadiene-styrene (SBS) block copolymer hot-melt pressure-sensitive adhesive systems. These offer broad adhesion spectrum, providing rather wider formulating possibilities than UV acrylics.

The adhesives can be coated on existing hotmelt coaters, with performance of the uncured product matching that of conventional hotmelts. On curing, the material loses some tack and peel, but its polymer structure allows it to retain a significant level of adhesion.

UV CURING INKS – Inks that are cured (dried) by ultra-violet or electron beam energy. The inks are solventless, give good color consistency and do not evaporate or dry in ink fountains, inking systems or on roller trains, so allowing fast set-ups and job changeovers for short-run work. UV curing inks are used extensively in screen printing, offset printing, narrow-web letterpress and flexographic printing.

UV curable inks use a high-intensity UV light source, most commonly medium-pressure mercury arc lamps, to instantly harden the ink, varnish or coating as the binder components of the ink (primarily monomers, oligomers and photo-initiators) are polymerized, or cross-linked. The photo-initiators act as a catalyst to initiate the photomechanical reaction that instantly turns a liquid UV ink into a dry and durable solvent-free coating.

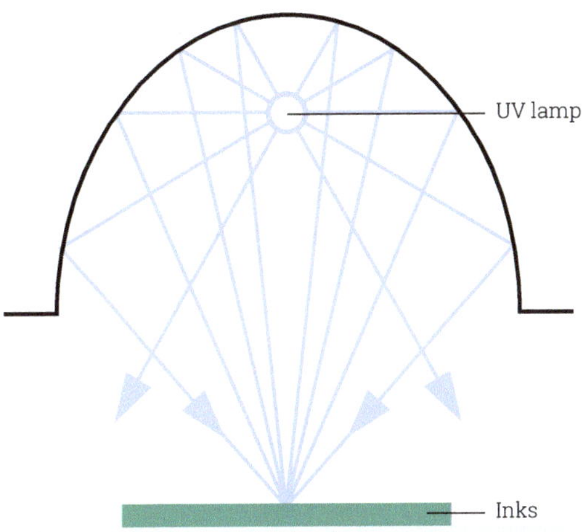

Accelerated ink drying using UV radiation.

All UV curable inks use basically identical ingredients; the rapid polymerization and cross linking of polymer chains are identical. The proportion of ingredients varies to give different ink characteristics. For example, high-opacity rotary screen inks noted for vibrant solids and fine-line detail differ from UV letterpress inks in print performance, while ink film weight and type of pigment are factors that influence curing speed. Research also continues into UV gravure, with prototype units displayed at exhibitions. But difficulties remain with the high viscosity of UV formulations.

The spectral output of the UV lamps used should match that of the ink's photo-initiators. Typical life expectancy of UV lamps is from 1,000 to 2,000 hours before degraded electrodes and internal deposits reduce the cure rate. Some lamps may reach much higher hours of use with good maintenance and careful handling.

UV FLEXO – The development of UV flexo printing in the narrow-web sector began in the early 1990s. Today, it is a process capable of achieving near-offset quality economically, while retaining the characteristics of a relief printing process.

UV flexo inks are thicker and have a higher viscosity than conventional flexo inks. Accordingly, the ink fountain, doctor blade and anilox roll take this into account. Ink makers recommend laser-engraved ceramic rolls combined with reverse-angle doctor blades (hardened steel or composites). Deeper cell volumes deliver less ink to give lower dot gain in screen highlight areas. This gives sharper more consistent results with higher gloss and improved contrast.

There are no VOC (Volatile Organic Compound) emissions and although costing perhaps twice as much as conventional flexo inks, the inks stay open on the press to reduce wastage and wash-up time. Curing starts only on exposure to UV light, so ink never dries on the plate surface and stays open in fountains between shifts and over-night if print units are covered up.

After twenty or so years, much more is understood about the interaction of chambered doctor blades, ceramic anilox rolls, UV flexo inks and plates and the process has become clearly established in both UV flexo and combination process presses.

UV INK SAFETY – Ongoing research and examinations in Europe have concluded that printing using UV inks does not present a higher risk to health than conventional printing – providing workers and the workplace take standard precautions. Advances in photoinitiator technology and UV curing systems mean that monomers currently used in ink formulations pose a much lower risk to health than early formulations.

Although some of the acrylate monomer and oligomer materials used in UV inks may cause skin irritation or sensitization, this can be prevented by wearing gloves whenever UV inks, varnishes or related materials are handled, particularly during the cleaning of printing plates, screens or presses where there is a higher risk than during the running of the press. Gloves should be changed regularly and disposed of immediately if they become damaged. Barrier creams will facilitate the washing off of ink that manages to penetrate gloves or protective clothing, such as disposable overalls, which should be worn when handling uncured ink. Protective clothing should be laundered at a commercial laundry, not at home.

Most monomers and oligomers are also not volatile (they do not evaporate) so the risk of inhalation is low. Similarly, UV ink has no flash-point, so there is a low risk of any fire hazard. However, where solvents are used for cleaning, these may cause some eye, lung or skin irritation, headaches or nausea unless good ventilation is provided on the factory floor. Ideally, fresh air should flow past the worker, then past equipment, before it is extracted from the factory.

Food and drink should not be consumed in the workplace, as this may aid the absorption of chemicals. Smoking too, should not be permitted in the workplace. Gloves, hands and arms should be washed regularly with soap and water – not solvents. All spills and leaks of UV ink should be cleaned up immediately.

Overall, printing with UV inks has a low environmental impact, although ozone emissions can pose a risk to air-quality if they are not dispersed correctly, or ideally extracted outside the factory. However, as occupational health and safety requirements relating to printing plants may vary around the world, label companies are advised to check whether there are any specific UV ink risk assessment protocols or standards issued by their national health and safety authorities.

UV RESISTANCE – The ability of a self-adhesive label or other material to resist UV light (sunlight) without any hardening of the adhesive, strong discoloration or weathering.

UV VARNISH – An ultraviolet light cured coating applied over a printed web and to label surfaces to provide a protective layer or finish.

V

VACUUM DIES – a method of removing small particles of waste by making use of a so-called vacuum die. After cutting, the waste is pulled inside the die and removed through the journals.

VACUUM HARDENING – Vacuum hardening of solid label press tooling is mostly done in a so-called vacuum furnace which is a type of furnace that can heat materials, typically metals, to very high temperatures and can carry out processes such as heat treatment with high consistency and low contamination.

In a vacuum furnace the product in the furnace is surrounded by a vacuum. The absence of air or other gases prevents heat transfer with the product through convection and removes a source of contamination.

A common use of a vacuum furnace is for the heat treatment of steel alloys. Many general heat treating applications involve the hardening and tempering of a steel part to make it strong and tough through service. Hardening involves heating the steel to a pre-determined temperature, then cooling it rapidly.

VARIABLE DATA PRINTING – Many in-plant logistics systems – warehousing, distribution, shipping, storage, tracking, etc – use labels that are printed with variable data. This variable information printing (VIP) may be in the form of variable text, barcodes, sequential numbers, batch codes, date codes, etc.

Variable information or data printing on labels is undertaken with non-impact laser, thermal transfer, or inkjet printers which ensure absolute legibility, or with impact printing systems, such as dot-matrix printers. The diagram shows how these variable data printing systems relate. Brief descriptions of these processes are:

Laser printing. This electrophotographic printing process, which is also widely used in photocopiers, uses fine toner particles to provide the image. The laser beam in the printer creates the image, point by point, controlled by a computer, into a charged pattern on to a pre-sensitized belt or drum. This pattern has opposite polarity to the toner powder and so attracts it, forming the image.

Paper or label material held against the photoreceptor collects the toner image which is then passed through a fusing system to bond the toner to the labelstock.

Most fusing systems use heat in a pressure nip, but variations include radiant heat fusing and flash fusion using halogen and xenon lamps. The latter are also called 'cold' or 'cool' lasers because only the dark toner is heated during the flash fusion process.

Variations of how the image is created from the computer's memory include magnetography and ion deposition (now renamed as Electron Beam Imaging or E.B.I. technology).

Direct thermal printing of labels is the main process used for adding price-weight information, product description and barcodes to supermarket frozen and fresh produce labels – meat, fish, cheese, fruit, vegetables, etc. – which are weighed and priced at food packers, remote from the supermarket, but also in the store for delicatessen, bread and produce labeling.

The print head for direct thermal printing consists of numerous elements in the form of a grid or matrix that are heated and cooled selectively by a microprocessor controller. A special heat-sensitive coated paper is required which, when heated by these elements changes color within the areas of contact to form the required letters, words, numbers or codes.

As the special thermally printable coating is heat-sensitive, direct thermal printing is primarily used for fresh and chill cabinet products that have a short shelf-life in store of several days. It is not normally used for long shelf-life labeled products or in warm or hot conditions.

Direct thermal labels can be unprotected, or have both a top coating and a barrier coating to protect the image from contamination from both sides. They are still only recommended for short term labeling as the thermal coating is sensitive to both heat and light so the coating will darken with age and this can adversely affect the print contrast signal (PCS) of printed barcodes, so they cannot be scanned.

Thermal transfer printing is the most commonly used variable data printing process and again makes use of elements which are heated and cooled selectively. However, this time, rather than using a special thermally-sensitive coated paper; the elements come into contact with a filmic one-pass

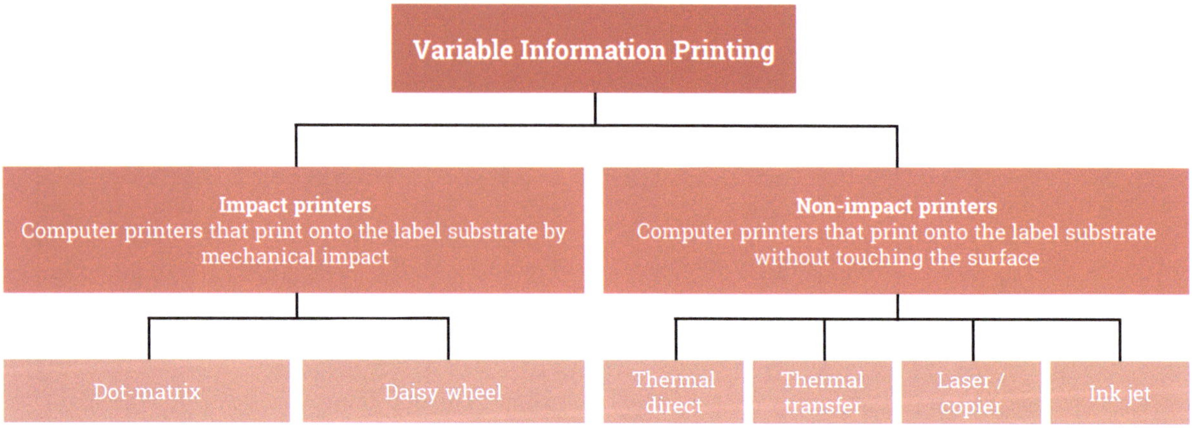

Types of impact and non-impact variable information printing systems.

ribbon (of which there are different types), which carries a heat-activateable ink coating. The required image is therefore created by transferring the heat-activated ink coating from the film carrier to the substrate according to the pattern or shape of the heated elements.

Thermal transfer printing is used for variable information printing of batch codes, date codes, sequential numbers, text, diagrams and barcodes onto pallet, carton or box end labels, for warehousing and distribution requirements, for bakery labels and for DIY and industrial labeling. Printers may be incorporated into packaging and/or weighing lines or be stand-alone. Some are also print-and-apply systems.

Dot matrix printers. The use of dot matrix (impact) printers has been rapidly declining in the labeling industry in recent years. As a printing process, dot matrix printing involves firing pins or hammers against an inked ribbon and thus transferring ink onto a label face material, rather like the early typewriter techniques. Each pin or hammer produces a fine dot and these can be positioned in a grid to produce letters, words, numbers, codes or simple graphics.

Impact printing systems such as dot matrix and line hammer printers have applications for printing onto card or for form/label combinations where an image may need to be created through more than one layer of material, as with carbonless paper systems.

Ink-jet printers. There are two ink-jet printing systems which are called, 'continuous ink-jet' and 'impulse', also called 'drop on demand' ink-jet printing. Continuous ink-jet printing is widely used for printing batch numbers and 'sell by' dates or barcodes directly onto products or in-line, on label printing and continuous stationery machines. This ink-jet printing uses minute droplets of solvent-based inks, usually MEK, which are activated and fired at a label surface by means of electrical charges to form the desired image.

Continuous ink-jet printing that utilizes piezo-electric technology can print in multicolors at press speed.

'Impulse' or 'Drop on demand' ink-jet is used for office A4 printers with both thermal ink-jet and piezo electric methods used to create the ink droplets.

In thermal ink-jet (Canon 'Bubble-jet' and Hewlett Packard 'Desk-jet') each ink drop is generated when it is needed, by the activation of an electric current to a resistor in the wall of each ink chamber, This heats water in the ink causing it to vaporize and expand. A bubble is created, which forces ink out of the chamber nozzle as the pressure increases.

Dye based inks are being replaced by pigmented inks, where improved water resistance and light fastness are needed.

Piezo electric ink-jet systems can utilise either water based or hot melt/solid inks. This technology is also fired by an electrical pulse which is applied to a piece of piezo crystal along the wall of the ink chamber.

The pulse causes the crystal to deform and reduce the area inside the chamber, thereby forcing an ink droplet out through the nozzle.

Film labels must be provided with a special inkjet coating in order to ensure optimum ink absorption and ink keying properties, as well as absolute sharpness of image, this being an essential requirement for the printing of barcodes. *See also Non-impact printer, Direct thermal printing and Thermal transfer printing.*

VARIABLE REPEAT LENGTH – The capability of a printing press to print from various size plate cylinders or printing plates.

VARNISH – A varnish is effectively a thin, clear, transparent ink that contains no coloring pigments or dyes. When printed or coated over the top of a substrate and/or printed matter, the varnish provides a protective finish that enhances appearance and increases durability. Varnishes may be glossy or matt.

VECTOR – Graphics and fonts which are created in outlines and then turned into dots (rasterized) before the printers can utilize them. Vector graphics are defined by points and curves rather than pixels or bitmapped graphics and are stored by a computer as a series of instructions by which the lines may be created. They are scalable and therefore not resolution dependent.

Vector is the general term that has now come to be used for a class or type of graphic drawing systems.

VEHICLE – Used to describe the drying oil, solvent or water part of an ink or coating that allows it to flow and which acts as a carrier for the pigment.

Vehicles enable pigments to be dispersed and give them the ability to 'flow' through the inking system on a printing press, then onto the print image carrier and finally on to the label substrate.

Very simple printing inks consist of little more than pigments dispersed in a vehicle (liquid). More complex inks may incorporate different drying mechanisms, synthetic waxes to improve rub-resistance and/or water repellence, and viscosity modifiers.

VELLUM PAPER – A strong, tough paper with a high quality appearance, originally made to imitate the fine smooth finish of a parchment made from animal skin.

No longer noted for its strength, Vellum has become the generic term for very smooth uncoated wood-free paper utilized by label manufacturers for line and solid printing or thermal transfer overprinting. Individual specifications can vary however and suitability testing is recommended.

VIDEO WEB INSPECTION – In-line video web inspection systems are a key first-line of defence against streaking, plate damage, hickies, color variations, mis-register, varnish faults and similar print defects faced by press operators. They allow the operator to inspect the quality of print in real-time, zoom in or out on images and defects far more effectively than with the naked eye.

Web inspection originally began with stroboscopic xenon flash systems, as well as mirror systems with

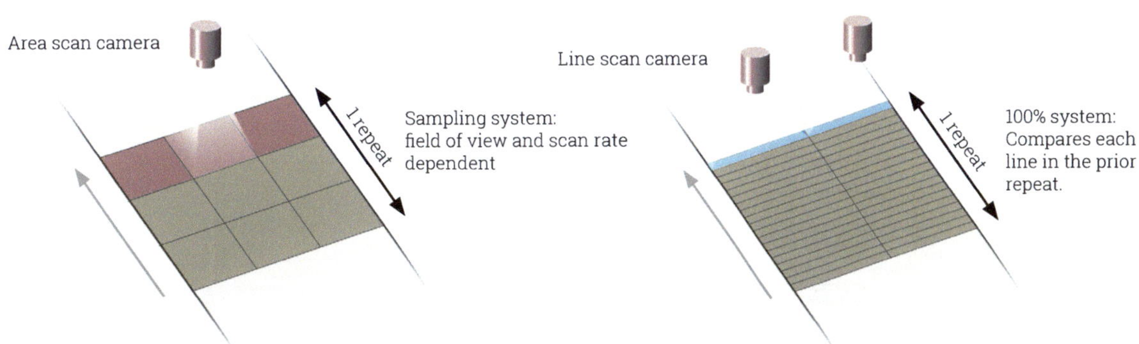

The principles of area and line scan camera systems used in video web inspection. Illustration courtesy of BST Pro Mark.

a rotating drum of front-surfaced mirrors mounted horizontally. Both are still used for certain print inspection applications, but the general trend now favors video web inspection systems, which enable an operator to view a web moving at many hundreds of meters/min (ft/min) on a

Today, video web inspection systems use a CCD (Charge Coupled Device) video camera with a single integrated circuit chip comprising thousands of light sensing elements. Usually they have alternating patterns of red, green and blue (RGB) filters before each element. Three-chip high-resolution cameras have three sets of single color filters. A prism splits the light entering the camera so all three chips see the same subject matter. The cameras can find small defects and display them on large monitors, making them well suited for exacting process color work. By setting alarm thresholds at suitable levels unacceptable color variations can be effectively controlled.

A full video image, or frame, is created by firing a high-intensity xenon strobe light to illuminate the web and act as a shutter. The very short flash intervals are synchronized with the press speed. Electronic motors control the focusing and iris diaphragm that regulates the light entering the camera's optical system. Zoom lenses provide magnification. Sampling methodologies used in video web inspection systems are;

- 100% sampling (inspects 100% of the repeat over time)
- 100% comprehensive (inspects 100% of the material that goes through the machine)

The two main technologies used for defect detection are those using an area scan camera or a line scan camera. Area scan camera systems are effective for trend identification and process management; for detecting random defects; for selective defect sensitivity control. Line scan camera systems are ineffective for trend identification, but effective for random defect identification. They are most effective for identification and sorting, but ineffective as a process management tool.

The electronics and controlling software in the system take digitized images from the camera's electronic chip and save them to a high-volume memory. Split screens can allow operators compare the visual color of two or more images; one as a stored template and the other as a real-time image.

Acceptable color at ± 2.0 Δ E

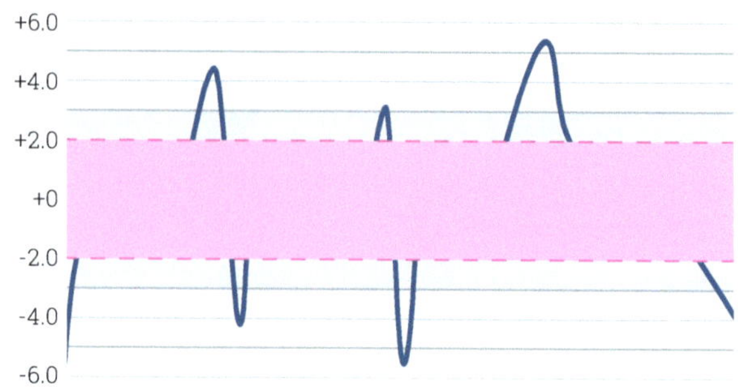

Principle of roll metering tinter.

Acceptable color at ± 2.0 Δ E

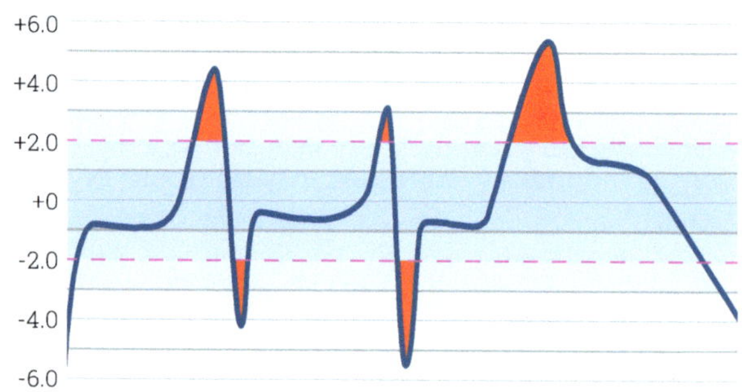

Principle of roll metering tinter.

Acceptable color at ± 2.0 Δ E

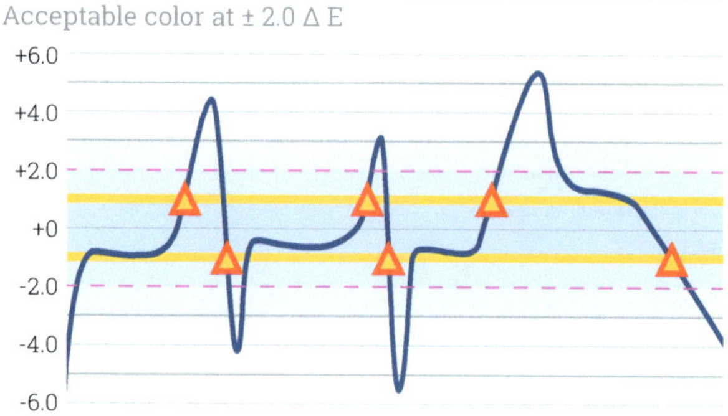

Principle of roll metering tinter.

Acceptable color at ± 2.0 Δ E

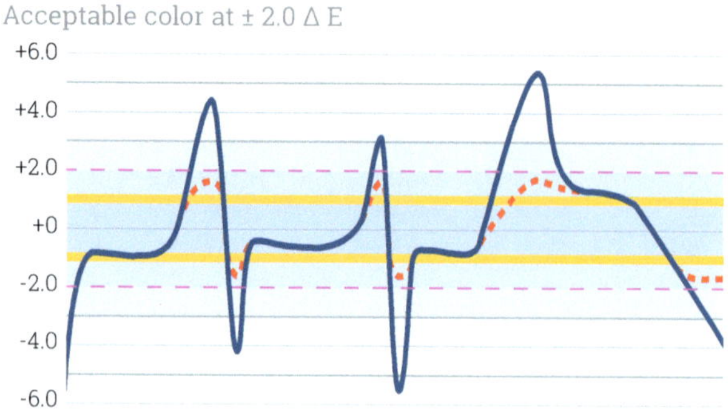

Principle of roll metering tinter.

Software-controlled cameras with a motorized traverse allow movement in both the lateral or linear plane for automated and constant scanning, with the operator programing the interval between each new image and the camera's movement in both directions. Some models can even inspect both sides of the web simultaneously and also handle reflective surfaces like metalized film or paper, as well as metalic inks.

Advantages claimed for video web inspection include:

- Faster press set-up
- Faster press running speed
- Reduced wastage
- Improved quality
- Higher level of customer satisfaction.

On-press process management systems based on video web inspection systems offer converters a substantial competitive advantage and a high return on investment.

VIGNETTE – Describes an image in which the shading of the tonal area fades gradually away until it totally blends into the background or into the unprinted substrate. The transition of a color from one particular value of hue and saturation to another over a specific distance and area.

VIN BARCODE LABEL STANDARD, –Every vehicle has a Vehicle Identification Number (VIN). The VIN number is stamped onto a small metal plate and is attached to the driver's side of the dashboard and in as many as 12 other locations on all new vehicles in the USA.

The VIN number is an automotive industry application standard that defines specifications for printing a Code 39 representation of a Vehicle Identification Number (VIN) on a label to be attached to a motor vehicle. It sets the vehicle apart from the millions of other vehicles on the road.

In recent times it has been reflected in 17 digit characters. It displays a car's uniqueness and heritage and provides a form of 'factory to scrap yard' identification. It can be used to track recalls, registrations, warranty claims, thefts and insurance coverage. Each character or digit has a particular purpose.

VIN numbers are by no means restricted to the USA. They have also become the vehicle identifier across Europe and elsewhere, such as Australia. The VIN regulations were enacted so as to require the marking of 'high theft rate' vehicles with VIN numbers and require a marking system that cannot be removed intact, will leave a residue if removal is attempted and will resist counterfeiting.

The VIN specification is used in conjunction with the Society of Automotive Engineers standards SAE J272 (Vehicle Identification Number Systems) and SAE J1877 (Recommended Practice for Barcoded Vehicle Identification Number Label), which set out the latest information on VINs, their locations, meanings, interpretations and engine numbers, as well as information on motor vehicles manufactured worldwide – including light trucks, vans and utility vehicles.

The 17-character VIN label contains the following information:

- World automotive manufacturer identification
- Type of restraint system
- Manufacturers specifications
- Line, series and body type
- Engine type
- A check digit
- Year of model
- Where the vehicle was assembled
- Production sequence number.

Further information can also be obtained from the Automotive Industry Action Group (AIAG). *See AIAG and Automotive labels.*

VINYL – Vinyl (PVC) films are available as cast or calendered materials in clear, colored, black or white, matt or gloss facestocks. They are supplied in mill reels, or slit or sheeted to customers needs. Most grades have a thickness of 85 to 100 microns, but may be much thicker for long life and/or exterior applications. These include point-of-sale material, outdoor promotional and warning signs, vehicle liveries and fleet markings.

Calendered vinyl is a moderately-priced film with good physical properties and has been used by the label industry for many decades. Flexible PVC is soft and malleable and is highly resistant to scuffing and most chemicals. Semi-rigid vinyls have similar properties, but are more dimensionally stable. Destruct vinyls are brittle and cannot be removed intact after application. Applications include tamper-evident container seals, also computer-imprintable facestocks where security demands non-removable labels.

Vinyl films have a high tear resistance, are easily hot-foil stamped, die-cut and conform readily to squeezable surfaces. They are also noted for rich color and depth of print. Smooth surface grades are ideal for printing with UV letterpress, screen, offset and flexo. Some of the key features and advantages of vinyl materials are shown in the table below.

Advantages and benefits of PVC films

Printing and application performance	End-usage criteria
■ Printability	■ Chemical resistance
■ Die-cutting	■ Weather resistance
■ Registration control	■ Flexibility / squeezability
■ Resistance to tearing	■ UV stability
■ Dimensional stability	■ Scuff resistance
□ Waste disposal	□ Environment friendly
	□ Waste disposal

Key: ■ Good □ Poor or caution

As a chlorine-based product, PVC is perceived as toxic for some applications. This is due to the chloride monomers associated with its production and disposal, especially when burnt. *See also PVC face materials.*

VIP – Variable Information Printing/Variable Image Printing. *See also Variable data printing.*

VIRTUAL MEMORY – Memory space in a computer that is separated from the main memory (physical random access memory), such as the hard disk space. Virtual memory enables the operator to work on large documents without having a large amount of RAM (Random Access Memory).

VISCOMETER – An instrument used for measuring or comparing the viscosity or resistance to the flow of an ink, varnish, oil or other viscose liquid. Viscosity is measured with instruments called viscometers.

The first group of viscometers measure the time it takes for a fixed volume of liquid to flow through a narrow tube or orifice. Typical of these are the Ostwald viscometer and the Ford cup. The Ostwald viscometer is used for comparing the time taken for samples of liquids drawn into a bulb to empty under the force of gravity through a capillary tube immediately below it.

Ford cups are also made with different sizes of aperture. They are exactly filled to the top, when the cup is full the time taken for it to empty through the aperture in the bottom is taken and can be compared with the time taken by another liquid. The cup is regarded as being empty when the thread of liquid flowing out first breaks.

For workshop use, comparison between the flow rates of low-viscosity inks, such as those used in high-speed gravure, measurements are made by means of a Zahn cup. Held by a long handle, the cup is immersed in the ink and the time taken for it to empty compared with standard times already established.

Viscosity measurement can also be taken with rotational type viscometers. Here the outer cylinder is rotated by a motor and exerts a drag on an inner cylinder. The deflection of the scale depends on the viscosity of the sample between the two cylinders.

VISCOSITY OF INK – Viscosity is a measure of the thickness of an ink, particularly of a thin, fluid flexo or gravure ink. Viscosity is measured by recording the time it takes for a given amount of ink to pass through a regulating aperture in a special measuring cup. *See Zahn cup.*

VOC – Volatile Organic Compound. Volatile Organic Compounds are potentially harmful gases released from solid and liquid materials such as building materials, inks and solvents.

VOC EMISSION CONTROL – In recent years international, national and federal regulations on solvent emissions have been looking to improve capture of volatile organic compounds (VOCs) and achieve higher levels of VOC destruction. Regulations that have been produced apply to those areas of the printing industry and printing processes where solvents are used. Solvent consumption includes the solvent content of inks, solvents used for dilution and solvents used for cleaning. These solvents are neither halegonated nor very toxic.

Levels of solvent consumption set out in regulations usually relate to the amount of solvent that is purchased and used. Solvents which are recovered and reused for the same activity are not counted as consumption. Compliance with regulations can involve a special environmental permit and reduction of emissions according to 'Best available techniques'.

Printing processes considered to be in scope of emission control regulations include those where liquid inks are used in gravure, web-fed screen printing, flexography, varnishing or laminating processes and which dry through evaporation. Letterpress, sheetfed offset and UV-curing printing processes do not come within emission control regulations unless fountain solutions, cleaning solvents and other materials within emission control calculations are used.

It is advisable to check with the relevant regulations for the country or state if there is any doubt about whether a printing operation or plant needs to comply with emission control procedures.

VOID – A bare uncoated area. Also refers to a void label used to display tamper evidence which is indicated by the word 'Void' left in the residue adhesive after a security label has been removed.

VOID MATERIALS – These are anti-tamper label products, usually polyester-based, that are applied to a labeling surface and when attempts are made to remove the material from the labeling surface, the pigmented or metalized layer on the back of the label – between the polyester and the adhesive – fractures to leave a message on the labeling surface and the same message in relief on the back of the polyester.

The effect of 'void' material

There are a wide range of 'void' materials and they are becoming more sophisticated to meet the growing requirements for security/authentication in the automotive, electronic or industrial components sectors. Top coating with a printable thermal transfer, laser, or other receptive coating is frequently required. Many materials must have Underwriters Laboratory (UL) or other approvals, meet approvals for indoor and outdoor printed products and have compliance on specified surfaces, including small diameters and temperatures up to 80 degrees Centigrade (176ºF).

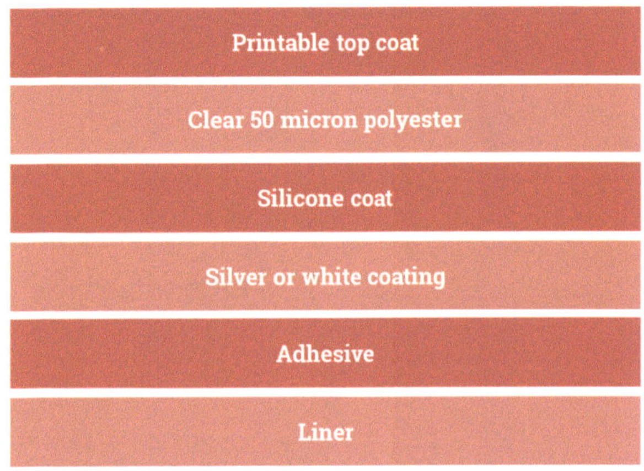

Structure of a 'void' material

Void label materials can also be customized, with an end-user selecting combinations of materials in the layers to specifically protect his own products. Customization options include UV print behind the polyester, colored voiding polyethylene, stretchable face materials, etc.

VOLATILE ORGANIC COMPOUND – *See under VOC.*

VSKE – Verband Der Hersteller Selbstklebender Etiketten Und Schmalbahnconverter E.V., the German label association, was established and registered in 1980 as a specialized federation within the HPV (Federation of Paper and Plastic Converters). It was mainly due to the efforts of Gerd Mayntz that this new association came into being. Fourteen label converters made up the founding companies.

In 2010 the formal interaction between FINAT and the national federations was revived and is still going strong.

W

WALMART SUSTAINABILITY INDEX – Introduced to determine then reduce the retailer's environmental impact in July 2009. Initially introduced in the form of a 15-question survey, it has been extended to more than 100,000 of the retailer's global suppliers.

WAND – A wand or light pen is a hand-held scanning (reading) device that is passed across a barcode symbol to read and/or capture the data from the code. Wands may be contact or near contact devices.

WARM COLOR – Warm colors are those that are in the orange to red area of the color range or spectrum.

WASTE – *See Matrix waste.*

WASTE AND ENVIRONMENTAL ISSUES – Increasing concerns about the environmental impact of packaging have been driven by three groups: governments, the global retail supply chain, and consumer pressure.

Governments around the world have introduced measures ranging from container deposits, packaging levies, control of solvent emissions, bans on certain types of packaging and mandatory recycling rates. Some retail chains have banned what they consider to be environmentally unacceptable types of packaging. They are also responding to pressures from customers and governments.

Consumers and consumer groups are showing ever-more interest in the environmental credentials of the products that people buy and the companies that they buy from.

Environmental legislation, bringing these various issues together, has been a key issue for the packaging and labeling industry from the late 1980s, through the 1990s and into the current decade. *See also VOC emission control and Waste control systems.*

WASTE & RESOURCES ACTION PROGRAMME – *See WRAP.*

WASTE SKELETON – *See Matrix waste.*

WASTE-TO-ENERGY – The recovery of materials deemed as 'waste' to be reprocessed into a new form that can be used as an alternative to fossil fuels such as biomass.

WATER-BASED ADHESIVE(S) – Water-based adhesives have had a dominant place in the labeling industry for many years and are used in wet-glue, self-adhesive and gummed paper formulations. They are made up of materials or compounds that can be dissolved or dispersed in water to become tacky, then form a bond and dry by losing water through evaporation or by penetration into the label substrate. At least one surface must be absorbent or porous to form a strong bond.

Water-based adhesives are available in a variety of chemistries and compositions and are categorized as either natural or synthetic polymers.

WATER-COOLED UV CHAMBERS – In coating applications where temperature-sensitive substrates such as thermoplastic films are used with UV-curing systems, it may be necessary to fit the cure chamber with water-cooled rollers, reflectors, cylinders (chill drums) or flat platens.

Films such as polyethylene and polypropylene are temperature-sensitive and while conditions in a

Water-cooled reflector and cylinder UV-curing system.

UV-cure chamber are considerably gentler than, say, those found in hot-air curing ovens used for thermal cure silicones, they can present problems for thin caliper films.

WATERLESS PLATE – An offset printing plate in which the non-image areas have a silicone rubber coating and can be printed without the use of a dampening solution. *See also Offset lithography.*

WATERMARK – A design, symbol or pattern imparted in paper by a raised wire design on a roll (a dandy roll) as the wet paper is being formed on a mesh as it passes through the wet-end drainage process on a paper-making machine. The watermark can be seen in the finished dry paper by holding the paper up to the light. May be used as a security feature or to denote a high quality writing or business paper. Some very large organizations may have their own watermark design.

A similar effect can be created by impressing the formed paper web with a raised image using a stereo, while the paper is still quite wet. These are called rubber marks and enable greater machine speed.

WATER SOLUBLE ADHESIVE – a pressure-sensitive adhesive in which all the components of the adhesive are soluble in water.

Early recipes were extremely hydroscopic and their adhesion performance varied according to the changing relative humidity. Sealing in moisture proof packaging was essential.

Where a wash-off, but permanent self adhesive label product is needed, most recipes are now described as 'water-removable' and they are not water soluble. These adhesives have additives to aid removability when immersed in water, wash-off times being shortened in hot water, or pressured steam.

WATER-VAPOR PERMEABILITY – The degree to which water vapor can pass through a substrate which is intended either as a supporting surface for a self-adhesive label or is itself a self-adhesive material.

WEAR RESISTANCE (TOOLING) – A tool is in constant movement, either against another tool or work material. This constant movement can wear away the surface of the tool. The tool will need therefore to have some amount of wear resistance. Wear resistance is the steels ability to resist erosion.

WEATHEROMETER – A machine designed for testing or evaluating the resistance of a material to various simulated weather conditions.

WEATHER RESISTANCE – The ability of a self-adhesive label designed for outdoor application to resist the influences of water, sunlight, variations in temperature and humidity without any significant deterioration of its essential properties. Also referred to as weatherability.

WEB – A continuous roll of flexible material, usually paper, plastic film, foil or laminate in a roll form. Continuously wound rolls or reels are used on web-fed printing presses, coating and laminating lines and converting machines. The webs may be re-wound for subsequent operations, slit into narrower webs or rolls, despatched to customers, or cut into separate sheets.

Web-fed presses may be referred to as narrow-web or wide-web according to the width of the web being printed and converted.

WEB CLEANER – A device or unit on a web-fed press that is designed to remove paper edge dust, surface debris, or airborne dust from the web prior to it entering the printing unit(s).

When selecting a web cleaner, whether contact or non-contact, one of the crucial factors is to make sure that the cleaner breaks the layer of air that is held by the moving web and which can hold contamination on the substrate surface. To break this layer, different systems are used including powerful airflows, air turbulence or actual contact with the web.

Some web cleaning systems use a tacky roller, which is in contact with the web. Any debris that is picked up by this roller is transferred to a second, more adhesive roller.

The benefits of web cleaning technology is that it is relatively easy to install and gives good results.

WEB GUIDE – A device on a web-fed printing press or converting line that keeps the web travelling straight and true as it passes through the machine. Web guides play an important part in the trouble-free progress of a web through a press and may consist of photo-electric, UV light or ultrasonic sensors that are focussed on the edge or the center of a web.

The web guide reaction speed should match the line

speed of the press to avoid 'hunting' in which excessive movement causes over-correction so that the guide has to be continually readjusted. Hunting causes the web to move from side to side, which creates problems during printing and finishing operations.

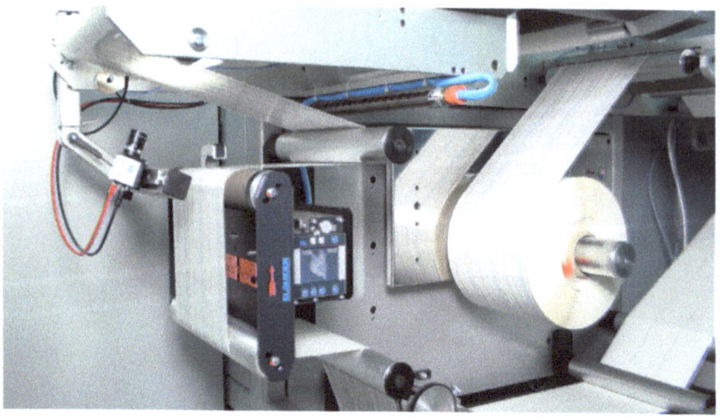

Web edge guidance. *Source: Erhardt+Leimer*

WEB PRESS – A printing press that prints from a continuous roll or web of material. May also be referred to as a roll-fed press, a roll-label press or a roll-to-roll press.

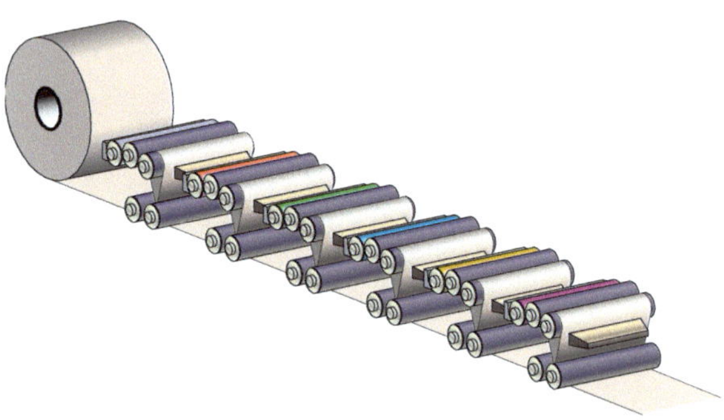

Web-fed press printing from a continuous roll.

WEB TENSION – Amount of pull applied to a substrate in the web direction by the action of a web-fed press. Poor tension control may result in register problems, web stretch, loose rolls and problems with sheeting and/or fan folding. *See also Tension.*

Web-to-print: A software system allowing a file to be taken directly from a customer to a digital printing engine, and which may include a number of components such as a digital storefront, pre-flighting and automated invoicing modules. *See Preflighting.*

WEB TREATMENT – The treating of webs of filmic materials is undertaken so as to achieve good adhesion of inks through the creation of a high level of surface tension on the film. In the label industry this is commonly achieved by corona treatment.

In order to achieve good adhesion when printing, the surface energy of the polymer film should be at least 10 dynes higher than the surface tension of the ink. Chemical changes caused by corona treatment include polarity and therefore increased wettability on the material surface. Treatment occurs as a very thin and ordered surface layer to the film. If the surface tension of a plastic substrate is raised by treatment to a higher level, then adhesion properties can be improved.

Corona treatment is created by applying a high frequency, high voltage signal to an electrode separated from an earthed plane by an air gap, the substrate and a layer of dielectric material. When a voltage is applied to the electrode, virtually no current will pass through the air gap until electrical breakdown of the air takes place. When this occurs, free electrons are accelerated towards the positive electrode with such high energy that they are capable of displacing electrons from molecules in the air gap, resulting in current flows across the air gap – so producing large ionizing currents which bombard the surface of the film.

Corona treatment systems produce ozone, which should be extracted out of the production area to minimize toxic or corrosive effects.

WEB WIDTH – The width of material being fed through a (label) press

WET CONDITIONS – Hot-melt and solution pressure-sensitive adhesives can be used for labels that are applied to moist surfaces or have to withstand wet conditions once applied. Acrylic and synthetic emulsion adhesives, being water-based self-adhesives, can absorb some light condensation, but excessive moisture will adversely affect the adhesion and rubber-based self adhesives are preferred for these demanding applications, where lower temperatures are often found.

WET-ON-WET – The printing of several different colored inks, one wet ink on the other, on a substrate in one pass through a printing press. *See wet trapping.*

WET TRAPPING – The printing of multiple colors, in which each overlapping ink film is printed on to the previously printed open (not yet dry) ink film. The printing of a wet ink on top of a wet ink.

WIDE WEB – There is no actual agreed web width at which narrow-web is differentiated from wide web. Common practice however, is beginning to increasingly define presses with web widths between 600 mm (24 inches) and 800 mm (32 inches) as mid-web, and above that size as wide-web.

WORKFLOW – Initially the term used to describe pre-press workflow, but in turn now being extended to cover the entire process from the moment a customer asks for a quote or places an order, right through to its delivery and to job invoicing – all of which are now considered as one entity, rather than separate stand-alone stages – and to provide a means of communicating right across the factory floor, into the office, and throughout the complete supply chain.

Many software systems can already provide trapping, ripping and imposition tools, but not all are capable of linking such tools with MIS, digital asset management systems, accounting systems and desktop applications into one comprehensive toolkit. Future workflow systems will increasingly create a synergy between all partners in the label industry supply chain – materials suppliers, converters, distributors, brand owners, etc – and have connectivity to other systems. PDF, XMP and JDF will be used as the standards for workflow parameter exchange.

All elements of integrated workflow, from design and specification, through all production stages to final distribution and invoicing working together, will significantly increase throughput and eliminate communication and operator errors in the total label industry supply chain. *See also JDF, MIS and PDF.*

WORK STATION – A powerful desktop computer (more powerful and expensive than a basic PC) which is usually networked with other computers or work stations. Work stations are used for processing more intensive tasks, such as digital design and artwork, CAD, 3D modelling.

WORLD BUSINESS COUNCIL FOR SUSTAINABLE DEVELOPMENT (WBCSD) – The WBCSD is a CEO-led organization of forward-thinking companies established to galvanize the global business community to create a sustainable future for business, society and the environment. The WBCSD is made up of some of the globe's leading corporations.

In conjunction with the World Resources Institute (WRI) the WBCSD has developed the GHG Protocol, a widely used accounting tool for government and business to track supply chain and product life-cycle impacts.

WORLD LABEL ASSOCIATION AWARDS – The World Label Association Awards (WLAA) is a competition which has been running for over 20 years. It began as a means of comparing the technical aspects of the best labels produced by the major world label associations, but now encompasses more end-use applications. Entrants to the WLAA must have won an award in their own domestic competition in the past year.

WRAP – Waste & Resources Action Programme, a not-for-profit company founded in 2000. to help recycling take off in the UK and to create a market for recycled materials.

WRAP-AROUND FILM LABELS – Wrap-around film labeling is a relatively new variant of the wrap-around glue-applied paper label used for can and some bottle labeling. Wrap-around film labeling was developed to allow the labeling of PET bottles and beverage cans with reel-fed, tear-proof plastic film labels, rather than paper, but required new developments for film handling, glues and gluing. The labels, which can be reverse-side printed to provide a non-scuff, scratch-resistant print quality, can also provide a clear, 'no-label' look solution on plastic bottles

Wrap-around label films. These are usually coated or uncoated bi-axially orientated polypropylene films. Uncoated films come in thicknesses from 19-50 microns, while coated films range from 30-70 microns. Super white BOPP films are made with a cavitated core and specially developed outer layers, which provide a bright white high gloss finish. Metalized BOPP films have a vacuum deposited aluminum layer on one side and an acrylic coating on the other.

Wrap-around film label application. Wrap-around film-labeling machines incorporate a label feed system with computer cut control, rotary cutting, gluing station and a mechanically assisted vacuum gripper for transfer of labels to the container. Large capacity reels minimize the operator workload.

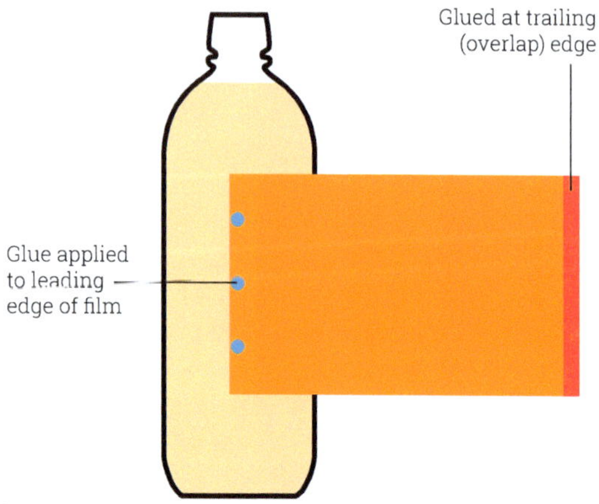

Glue pattern used with wrap-around labels on returnable PET bottles.

In operation, glue is applied onto the leading edge of the film via mechanically-driven rollers. The film is cut to length and wrapped around the bottle or can and then glued on the trailing (overlap) edge to form a complete wrap. Glass, plastic and metal containers – square, cylindrical or oval can be handled. Label application speeds up to 800 bottles a minute are achievable when using two applicator stations.

Label removal on returnable PET bottles. A newly-developed variant allows wrap-around labeling of returnable PET bottles with very thin plastic film labels. A small amount of cold glue is applied to the bottle for label pick-up. This cold glue crystallizes after curing, which facilitates subsequent label removal from the returned PET bottle. The lap overlap is sealed with hot-melt adhesive. Label removal is carried out using a de-labeling machine, which cuts the label vertically with a knife. The label can then be flushed from the bottle using a high-pressure water jet, with the label being discharged into a label removal system.

Body labeling. The wrap-around reel-fed labeling system is primarily designed for wrap-around body labeling of straight-sided bottles or cans, possibly with

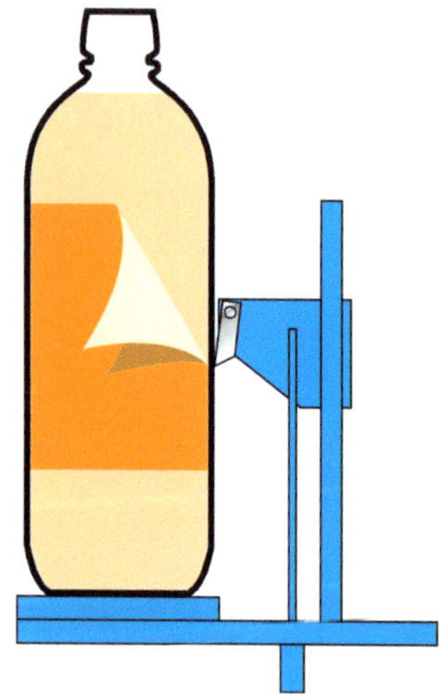

Labels are removed from returnable PET bottles using a vertical knife.

a light contor but, in their standard format, will not apply a neck label onto a tapered bottle neck. However, this can be achieved with a combination applicator line that can apply a wrap-around body label and a separate wet-glue or pressure-sensitive neck label on the same line.

As far as cans are concerned, wrap-around film labeling provides a number of advantages over the use of pre-printed cans, including reduced can inventory, reduced lead times, greater flexibility for sales/marketing activities (special offers, proof of purchase promotions, etc.).

With the addition of a second carousel in the labeling system it is also possible to positively shrink the label film to fit contoured cans or bottles. Label technology for wrapping-around or rolling-on film from reels, followed by a light shrink, may also be referred to as Roll-On Shrink-On (ROSO) labeling. *See also ROSO.*

WRINKLES – Wrinkles are small creases or folds in a smooth surface. Paper affected by humidity changes may become creased when passing through press or nip rollers.

WRITE RATE – The rate at which information can be transferred to an RFID tag, written into the tag's memory and then verified as being correct.

X

X-HEIGHT – The height of the main body of a piece of type – not counting ascenders and descenders – so that letters such 'a', 'o' or 'e' are the x-height measurement between the base line and mean line.

XEIKON – The brand name for dry toner-based digital color printing systems manufactured by Belgian company Xeikon International. One of the early digital label press pioneers (1993) and the early market leader, Xeikon offers total solutions involving: workflow package X-800, digital press and in-line or near line converting units.

In operation Xeikon presses print from five print stations using a dry toner. Each print station consists of a central electrophotographic drum. During each print cycle, the drum is cleaned, charged, imaged with a 1200 dpi LED array and developed with the toner particles. The developed toner image is then transferred to the substrate using heat and some pressure, after which the whole cleaning, charging imaging, developing and printing cycle is repeated.

Xeikon 3000 label press series architecture is based upon a full rotary press with a variable repeat length which guarantees that a converter's productivity is not dependent on the number of colors or on the dimensions of the labels. The QA dry toner, specially developed for label and package printing, allows the use of standardized substrates, provides the highest lightfastness and is compliant with FDA standards for indirect and direct dry food applications. There is also a dedicated self-adhesive label toner called ICE that expands the substrate range by enabling PE and direct thermal labels to be printed. Both toner versions offer a one-pass opaque white.

The toner and toner fusing mechanism in Xeikon color printing systems provide good gloss results which closely resemble the offset printing process.

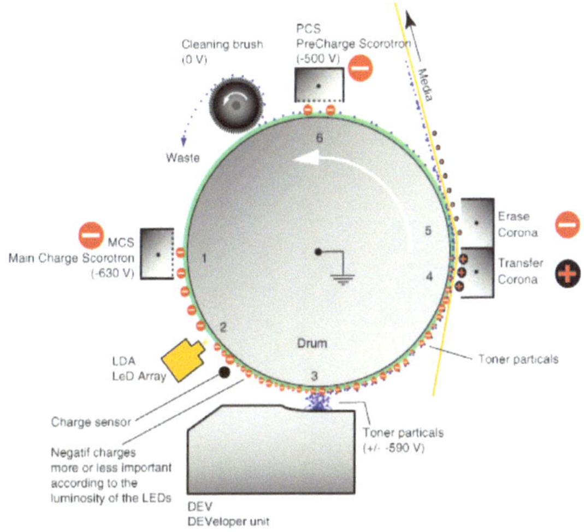

The diagram shows the imaging process that takes place on each of the five print stations in a Xeikon 3000 series press.

XEROGRAPHY – An electrophotographic imaging process used in copying and printing. The process of xerographic printing. The process involves electrically charging a belt or drum, removing the charge by laser or mirrors from the areas not to be imaged, applying toner to the charged areas, transferring the toner to paper and then fusing the toner to the paper surface.

Y

YELLOW – Yellow is a hue of a subtractive primary and a four-color process ink, which reflects red and green light and absorbs blue light.

YELLOWING – A gradual color change in the appearance of a pressure-sensitive face paper, characterized by a white paper developing a yellowish or brownish hue.

YIELD– The area of product provided by unit weight, expressed internationally in terms of square meters per kilogram.

Z

ZERO (CORRECTED) ANVIL – Means the body and bearer position of the anvil roller have been ground to the same diameter

ZAHN CUP – A device in the form of a cup shaped container with a hole in the base that is used for measuring the viscosity of flexo or gravure inks. Viscosity is measured by the time taken for a controlled amount of ink to flow through a standard-sized hole.

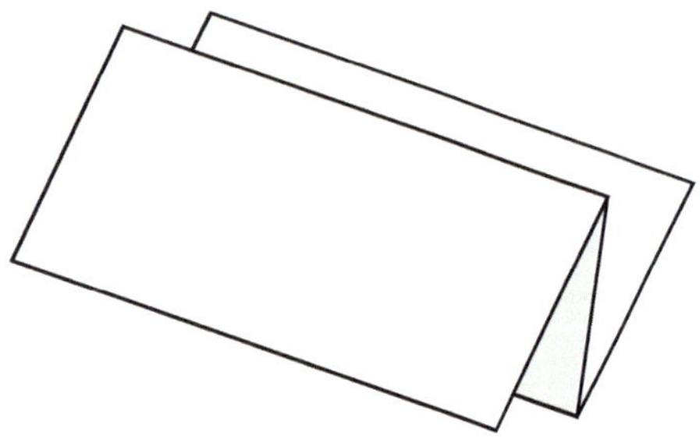

Diagram shows a typical Z-fold.

Z FOLD – A description for a type of fold where the folds in a web produce a Z shape. *See also Zig zag and Fan fold.*

ZIG ZAG FOLDING– A method of folding continuous labels or continuous forms in a zig zag style. *See also Fan fold.*

ZONE COATING – *See Pattern coating.*

ZOOM – The ability to examine image detail on a display screen or on a printed web without mathematical interpolation. On screen, each pixel can be examined and modified as required. Zoom capabilities are used with web-based video inspection systems. *See Video inspection systems.*

www.ingramcontent.com/pod-product-compliance
Lightning Source LLC
LaVergne TN
LVHW070949210926
848482LV00002B/149

9781910507001